SCIENCE AND MEDICINE IN IMPERIAL RUSSIA

Second Edition

Anatoly Bezkorovainy, J. D., Ph. D.

Professor Emeritus

Department of Biochemistry
Rush University
Chicago, Illinois 60612

Lulu

Lulu Enterprises, Inc.
3101 Hillsborough Street
Raleigh NC 27607
2018

ISBN: 978-1-64255-801-2

I dedicate this book to

Marilyn Grib Bezkorovainy

my faithful wife for fifty-three years so far

1964–2017

and many more, God willing

TABLE OF CONTENTS

PREFACE
for the first edition

An author was once asked on a promotional television program appearance why he wrote whatever book he had written. His reply was that one generally writes a book to either make money or to champion a cause, and he, the author, had done it for the latter purpose. I often thought about this reply when I was in the process of gathering data for the present treatise. I certainly wasn't going to make any money by these labors, not in terms of dollars or cents, or academic promotions. Was it then to push a cause or ideology? I have finally come down to the conclusion that it was indeed the latter case. What then was this cause, in the pursuit of which I was obliged to spend all of my free time over a period of three years? It began with mere curiosity or interest in the history of the land of my forefathers, the land of Russia. And since I happen to be a scientist by profession, science in Russia naturally became the principal object of my interest. Having discovered over a period of years many Russian names that are now immortalized in the annals of science and medicine, I became painfully aware of the habit of some ignorant journalists to point to Tsarist Russia as an example of backwardness, ignorance, and tyranny whenever they need an analogy in these fields of interest. We thus have seen Tsarist Russia being compared to African countries on the day of their independence (1), the declaration that Russia was "a land of serfs, feudalism, tyranny…" (2), and the discovery that "the scientific revolution was in Russia only a feeble reverberation of the explosion in the West" (3). The latter statement was made in spite of the two Nobel prizes in medicine that went to the Russians during the first ten years of the prize's existence. All of this journalistic drivel prompted me to embark upon a systematic collection of data from sources available right here in the United States on the various health-related sciences of prerevolutionary Russia with the intent of publishing these data in a book form for the benefit of my colleagues who have little if any awareness of the stature and contribution of science and medicine in Russia in those years.

After the manuscript was finished, various university presses were contacted and refusal letters received. Some claimed that my narrative was too-pro tsarist (and this we must never allow), others lamented my failure to use Soviet sources more extensively, still others called the work too encyclopedic, and one felt that there was too much science and not enough sociology in the manuscript. It is too bad that the historians were the ones to review the manuscript, for it wasn't written for them, and there is nothing in it that would help them. Their minds have been made up. Instead, the manuscript was designed for my fellow bench scientists—that is, for those who like to get their hands wet and enjoy reading about how others got theirs wet many years ago. So, my hope is that my colleagues in the health sciences will enjoy reading this work, and will learn to appreciate the great contributions that Russian scientists have made toward the advancement of our profession.

And lastly, I would like to assure my readers that private publishing has its advantages. Had a university press accepted this manuscript for publication, I would have never got away with writing the above preface, nor with many of the views expressed below. Long live the freedom of expression, even if it does cost a lot of money to express oneself.

Among the persons whom I would like to thank for their help and encouragement are Mr. Paul Ross, who gathered much of the data presented in chapter 1; Mr. William Kona, formerly chief librarian of the Rush Presbyterian St. Luke's Medical Center for his help with obtaining the necessary references herein; to my former graduate student, Dr. Dietmar Grohlich, who called my attention to many sources that I otherwise would have missed; to the editors of the Rush Presbyterian St. Luke's Medical Bulletin, where many of my articles on Russian science and medicine made their appearance in the years past; and last but not least, to my wife Marilyn, who patiently tolerated my long nights at the typewriter, the uncut lawn, and unpainted woodwork.

REFERENCES

1. New York Herald Tribune, Section II, July 2, 1963.
2. Chicago Daily News, June 13–14, 1970.
3. Robert Wallace. THE RISE OF RUSSIA. Time-Life Books, New York, 1971, p. 12.

A. Bezkorovainy, J. D., Ph. D.
1978.

PREFACE
for the second edition

It is now the third day of February 2017. Some thirty-five-plus years have passed since my opus, *Science and Medicine in Imperial Russia,* saw the light of day. I attempted to have it published by official science- or history-publishing establishments, but had no luck. For various reasons, they did not wish to touch it. And thus, I decided to "self-publish" my work. Due to the extraordinary costs of such endeavors, I put together a typical home-made entity similar to my MS (1958) and PhD (1961) dissertations, obtained an ISBN (0-9607600-0-8, new ISBN 978-0-9607600-0-8), and had one hundred copies printed. The book was ready for distribution. About forty or so were given to friends, relatives, and colleagues, and the other sixty were sold for $20 each. The money thus generated paid for all costs (except for my time spent in preparation thereof). Readers expressed mostly pretty good opinions; one customer called me and said that my book gave the best story about Russian ophthalmology that he had ever seen. This was nice to hear. As time elapsed, self-publishing became less and less expensive, and the quality of such literature was difficult to distinguish from that done by official publishers. And as I neared my date of retirement (in the late 1990s), I decided to prepare a second edition of the book thereafter using the available modern publishing methodologies. My plans also included writing and publication of two additional books: an autobiography and a history of the Russian Empire. And thus, after I had retired in AD 2005 and had moved with my wife Marilyn to Galena, Illinois, I wrote and published *All was not Lost* (2008) (1) and *History of Imperial Russia. A Layman's Perspective* (2014) (2). My intention was now to work on the second edition of *Science and Medicine in Imperial Russia.* However, an unforeseen intervention had delayed my plans.

Around 2010–2012, my wife and I were visiting Russia and the Baltic states including my place of birth—Riga, Latvia. There, we met Mrs. Tatiana Peters who formerly lived in the Chicago area and had attended the Holy Trinity Orthodox Cathedral where we were also parishioners. Mrs. Peters introduced us to Professor Alexander Gavrilin of the Latvian University, who was writing a biography of Archbishop John (Garklavs) (1898–1982) of blessed memory. He had married my wife and me in 1964, and thus we had been very fond of him. Prof. Gavrilin was writing his book in the Russian language and was looking for a suitable person to translate his work into English. And to make a long story short, I agreed to do the translation. This was done, and Prof. Gavrilin's book, *I have Received Much More than I have Deserved,* was published in the English language by Lulu Press in 2015 (3). I could now think of doing the second edition of my book.

The cosmetic (*samizdat'*) aspect of the book was not the only reason I decided to do its second edition. Much has changed in Russia since the 1990s, the time when its communist system was abolished. Many new sources about Russia's science and medicine became available, and a number of newly published books and articles on the subject were written since then, e. g., the work of Prof. Daniel Steinberg (4), who reported how Ignatovski in 1909 and later Nicholas Anitschkow showed how cholesterol in the diet resulted in the development of atherosclerosis in animals. Not much attention was paid to their work by their Western colleagues at that time.

In the book, I had written almost nothing about Russia's accomplishments in the areas of physics and related sciences, geology, astronomy, and agriculture, though the accomplishments of Russia's scientists in those areas were enormous. I've had little contact with these sciences and, to avoid a fiasco, I have avoided mentioning anything about them (except for some aspects of agriculture). For this, I am humbly begging my readers' forgiveness and hope that someday, an able science history author will pick up my omissions. And so, the story begins *(scribere est agere)* (1, p. xiii).

REFERENCES

1. Anatoly Bezkorovainy. ALL WAS NOT LOST. Authorhouse Press, Bloomington, Indiana, 2008.
2. Anatoly Bezkorovainy. HISTORY OF IMPERIAL RUSSIA. Lulu Press, Raleigh, North Carolina, 2014.
3. Alexander Gavrilin (transl. by Anatoly Bezkorovainy). I HAVE RECEIVED MUCH MORE THAN I HAVE DESERVED. Lulu Press, Raleigh, North Carolina, 2015.
4. Daniel Steinberg. THE CHOLESTEROL WARS. Elsevier-AP Press, New York, N. Y., 2007.

A. Bezkorovainy, J. D., Ph. D.
February, 2017

Illustrations

This volume relates the stories of medicine and some basic medical sciences in Imperial Russia. As indicated in the Preface, the author was not sufficiently competent to include the stories of other sciences and engineering technologies in this book, which were not less successfully pursued in Russia than the ones he managed to describe. As an example, Bezkorovainy chose to mention the stories of Igor Sikorsky, K. Tsiolkovsly, and N. Lobachevsky as examples of scientists whose works could not be included in this book.

Constantine Tsiolkovsky (1857-1936) "deserves to be called *father of Russian space travel.*" He was a mathematics teacher in Russian schools and a promoter of space travel. He developed the basic equation for space travel: $dV=I_0\ln(M_1/M_0)$, where V is rocket velocity, I_0 is a function of impulse fuel, and M_1 and M_0 are weights of the rocket at beginning and end of the trip.

Nicholas Lobachevsky (1793-1856), discoverer of the non-Euclidian geometry, was a professor of mathematics at the Kazan University from 1816. In 1827, he was elected "rector" of the university by the faculty and served in that position for 19 years.

Igor I. Sikorsky (1889-1972) was in Russia an airplane designer and builder. He first designed and built a passenger 4-engine airplane he called "Le Grand." It could carry 16 passengers. During the First World War, the Le Grand was modified to serve as a bomber, and was named "Ilya Mourometz." A total of 73 of these bombers were built, and they served well against the German and Austrian enemy. Only 1 (some say 3) were lost to enemy action. During the war, they were located at airports in Tourerkahlen in today's Lithuania and in Riga in today's Latvia. The bombers were used largely to destroy the enemy's communications and transportation systems, especially railroad stations. After the Bolshevik coup in Russia, Sikorsky emigrated to the United States, where he established an airplane factory that invented and manufactured helicopters. Sikorsky had previously designed such flying machines while in Russia, but it wasn't until he came to the U. S. that he was successful in building helicopters that actually flew.

Igor Sikorsky

Sikorsky's "Le Grand"

Most illustrations in this book are portraits of Russian physicians and scientists. They appeared on Soviet postage stamps over the years, a commendable method for honoring a country's successful professionals. Several other illustrations were taken from E. Nilov's book *Botkin* (Molodaya Gvardiya, Publ., Moscow, 1966), and from M. Kolesnikov's book *Lobachevskii* (same publisher as above, 1965), both in the Russian language.

Chapter I

EDUCATION AND SCIENTIFIC INSTITUTIONS IN IMPERIAL RUSSIA

Introduction

It is generally agreed that higher education and scientific/medical research cannot flourish in a society with a weak economic base. Some would even say that optimal development of scientific inquiry and accomplishment cannot take place unless scientists can work in a climate of freedom and security. All this, of course, brings up multiple questions in regard to the social, political, and economic state of affairs as they existed in Tsarist Russia and the influence of these factors on Russian science and medicine. This subject matter is largely beyond the scope of this book and the author's expertise, and the reader can be referred to a number of publications that address these aspect(s) of life under the Tsars (1–13). For orientation purposes, however, it is possible to give the reader some statistical information on Russian scientific and educational institutions, learned societies, degrees of literacy among Russia's populations, and Russia's economic base. These data have been gathered from publications of the appropriate periods (14–26) that are available in most American libraries. Some economic indicators in Tsarist Russia are given in Tables 1 and 2, which indicate that Russia was tops in agricultural production, while lagging significantly in industrial and technical capabilities, especially if Russia's huge area is taken into account.

Literacy in Imperial Russia

Russia ranked rather low in the civilized world with respect to the literacy of its general population. Table 3 summarizes its distribution according to age, sex, and place of residence, as shown by the last Imperial census of 1897. The overall literacy rate was thus 24 percent; however, its distribution among the various population groups was quite uneven. The westernmost possessions of the empire were populated by reasonably educated folks: the Poles, Germans, Latvians, Estonians, Jews, while the populace of Siberia and Caucasus, with their nomadic inhabitants, fared much worse. Similarly, a large literacy gap existed between the males and females of the empire, especially in rural areas. The educational level of urban populations was immeasurably greater than that of the peasantry. Yet, this rather poor situation was improving rapidly, as shown by the literacy rates among army recruits between year 1875 and 1913: while only 21 percent of recruits were literate in 1875, forty percent were so in 1895, and 73 percent in 1913 (Table 4). Even these numbers were perhaps on the low side, because many educated young men such as teachers were exempted from military draft.

Education in Imperial Russia

Primary and Secondary Schools. The fortunes of Russian educational systems often fluctuated with the will of the tsar. A formal school system was begun in Russia by Peter the Great, who believed that schools were necessary for the preparation of its military and civil service leaders. By the force of circumstances and exigencies of war, such schools carried a utilitarian and technical character. In 1786, Catherine II introduced a concept of education for everyone and decreed that every district town (county seat) was to have a two-year primary school, and every provincial capital was to have a five-year secondary school (gimnaziia). The school system was to be free for everyone, including the serfs. At the end of her reign, some 250 towns and cities had established such schools. Alexander I supported this idea, and at the end of his reign in 1825 there were 60 secondary schools, 370 primary schools, and 600 church-supported schools in Russia. In addition there were 18 girls' secondary schools and 300 private schools for privileged classes.

During the relatively reactionary reign of Nicholas I, universal education was considered to be harmful; consequently, it was decreed that in every school the subjects of instruction and the very methods of teaching were to be in accordance with the future destination of the pupils. No one was to aim to rise above the position in which it was the person's lot to remain. By the statute of 1827, the serfs were forbidden to enter the gimnaziia and the university. Church schools were intended for the peasants, the district schools for the merchants and townspeople, and the secondary schools—for the gentry and civil servants.

The ascendancy of Alexander II to the throne brought a rebirth of the idea of universal education, and the entire educational world of Russia experienced a tremendous growth during that time. Education of women was especially encouraged with the establishment of several women's colleges. This growth was temporarily arrested following the assassination of Alexander II and subsequent reign of Alexander III. It was resumed some ten years before the start of World War I during the reign of Nicholas II. In 1910, there were some 90,000 primary schools (including Orthodox Church schools) and 16,000 non-Orthodox

Church schools in Russia with 6.5 million students. In 1897, this number was only 3.5 million.

The backbone of secondary education in Russia was the gimnaziia, or the classical high school. These were separate for girls and boys. In addition to classical high schools, there were the technical, commercial, pedagogical, theological, and military secondary schools. Children of the gentry and other privileged classes (by birth or money) could attend private lyceums, university laboratory schools, and institutes for girls. In 1894 and 1914, there were 224,000 and 733,000 students respectively in Russian secondary schools. Table 5 gives the numbers of the different secondary schools in Russia in 1910.

With the advent of the parliamentary system in Russia's government, plans were made in 1906–1908 to completely eradicate illiteracy in Russia by 1920. To accomplish this, it was proposed to build 10,000 new primary schools and sixty secondary schools per year. The curricula of all schools were to be standardized, and Russian was to be the language of instruction in all schools. A fund of 500 million rubles was allocated to get the work started. But the start of World War I and the subsequent revolution in Russia laid to rest all such plans, yet to be implemented much later by the Bolsheviks.

Education and the Russian Military. Since the degree of illiteracy in Russia was high, the Russian military maintained a vast network of elementary and technical schools, so that a recruit, if he happened to be illiterate or undereducated, learned not only to read and write, but could learn a trade as well. Perhaps this explains, at least in part, why male literacy levels in the general population were higher than those of women. Men were drafted and if illiterate, were taught how to read and write. But girls were not drafted, and if they had not attended a civilian school, they did not have such an opportunity. The same was true with learning a trade. For instance, selected recruits were taught the trade of a *rotnyii feldsher* (paramedical specialist). Upon discharge from the military, the military feldsher could open a civilian practice, since there was a serious shortage of physicians in Russia, especially in the countryside where the bulk of the Russian population resided. The feldsher thus became an important component of the medical care machinery in prerevolutionary Russia.

The length of military service in the later years of Imperial Russia was three and a half years, though only 30 percent of the eligible men were ever called up. Men with two years of secondary education served only one year and those with college education served nine months. Teachers and ministers or priests of all faiths were exempted altogether.

In order to provide officers for the armed forces, the military maintained secondary school-level cadet corps and university-level military academies. There were thirty-seven cadet corps-type schools and seven military academies in Russia, with a total enrollment of 20,000 in 1913.

Institutions of Higher Learning in Russia. The first university in Russia was established in 1725 as an appendage to the Imperial Academy of Sciences. Lack of students soon forced the university to close its doors. Later, universities were founded primarily for the purpose of training civil servants; however, with time, these institutions grew into real nurseries of the Russian national culture. One peculiar characteristic of the Russian university was that pure knowledge and theoretical subject matter were emphasized to a greater degree than anywhere in Europe. This attitude reflected the general characteristic of pre-World War I Russian culture, namely, a search for the human ideal.

Russian universities represented that liberal cultural thought that lay between revolution and reaction. The universities thus seemed frequently to be at odds with the more conservative autocracy; however, with few exceptions, the universities always seemed to gain the upper hand until the Bolshevik coup. The statutes of 1863 recognized full autonomy of Russian universities, yet this situation of complete academic freedom lasted for only twenty years. Upon the assassination of Alexander II, control of universities was tightened by the government, whereby the appointments of university presidents and even the deans had to be approved by the Ministry of Education. This edict resulted in frequent student disorders and chronic feelings of discontent among the university faculties. Yet regardless of any statutes, the Russian universities did in fact develop independently of anyone to meet the spiritual needs of the people and in accordance with the views of their students and faculties.

Organization of the Russian universities was probably most similar to their German counterparts. Typically, there were four "faculties": medical, law, historic-philological, and physical-mathematical. St. Petersburg University did not have a medical faculty; instead, it had a faculty of oriental languages. Tuition in Russian universities was comparatively low, some 50 rubles ($25) per year. Additionally, many students were granted full or partial tuition scholarships. Thus, of the 4017 students at Moscow University in 1899–1900, 1957 students were excused from paying their tuitions, and an additional 874 received other forms of financial help. Generally, some 50 percent of the student body in any given Russian institution of higher learning received such help. Regardless of the liberal monetary policy in Russian educational system, Russian students were considered to be most impecunious ones among their European colleagues. According to Prof. M. M. Novikov, the last elected rector of Moscow University, poverty was quite normal in the lives of Russian students. This poverty was due, in part, to the fact that the Russian university was open to all and could be attended by anyone, prince and peasant alike. Because of this policy, the Russian university was known for its democratic character in addition to the poverty of its students.

In 1897, there were 10 universities with some 15,000 students in Russia. In 1904, this number had grown to 20,000. Germany, the scientific Mecca of the pre-World War I era, had 46,000 students in 21 universities in 1907–08. By 1914, the number of students in Russian universities was 40,000.

In addition to the universities, Russia maintained a large network of specialized institutes, such as the various engineering schools, agricultural academies, lyceums, and women's colleges. Education of women was especially well developed in Russia. The peculiar characteristic of women's colleges was that, with few exceptions, they were indebted to public initiative, mostly to women themselves, for their very existence. The idea of higher education for women originated during the reform period of Alexander II, subsequent to which many women's colleges were founded. The growth of the female population is aptly illustrated by the growth of attendance at the Women's College of St. Petersburg (Bestuzhev School): in 1905, there were 1,600 students, whereas 5,177 were registered in 1909. Table 6 gives a list of institutions of higher learning in Imperial Russia.

Growth of the Russian educational system in the late nineteenth and beginning of the twentieth centuries was quite rapid. Table 7 lists the growth of some Russian institutions during that period. In 1914, two Russian universities were among the ten largest ones in the world, three Russian engineering schools were among the ten largest in the world, and four Russian schools were among the world's ten largest specialized institutions (Table 8). These facts notwithstanding, Russia still lagged behind the more advanced countries in Europe with respect to the numbers of their populations (Table 9), though this situation was by no means catastrophic. Recognizing this problem, Russia's minister of education urged in 1916 that Russia open ten additional universities in various parts of the country. Another issue that needed to be solved was the unequal distribution of institutions of higher learning in Russia. For example, the vast Siberian territory had only one university and one engineering school, both in Tomsk. It was decided to establish new universities in Perm (Ural Mountain region), Tiflis (Caucasus), Simferopol (Crimea), Rostov-on Don (Little Russia), Irkutsk (Eastern Siberia), and to enlarge the Saratov University by addition of historical-philological and physical-mathematical faculties. In addition, it was planned to establish an agricultural institute in Omsk (Western Siberia), a polytechnic institute in Tiflis (Caucasus), two commercial institutes, one in Kharkov, the other in St. Petersburg, and a statistical institute, also in St. Petersburg. The Ministry of Education and the Duma (Russia's parliament) also agreed to establish two experimental multi-faculty universities: one in Tashkent (Turkestan) with a medical, social science, philological, military science, agricultural and engineering faculties. The other would be located in Ekaterinburg (Ural Mountain region) with mining, pedagogical, medical, and social science faculties. Such plans of the tsar's ministers were eventually implemented by the Soviets: in 1925, we find new universities in Ekaterinburg (renamed Sverdlovsk), Rostov-on-Don, Tashkent, Irkutsk, and Tiflis, an agricultural academy in Omsk, and a pedagogical institute in Simferopol.

What Russia lacked in quantity with respect to education was certainly made up with quality. The standards to which both the student and professor were to confirm were high indeed. The first degree of *Candidat rerum naturalium* was awarded a student after four years of intensive study, writing a thesis covering an original investigation, and passing oral and written examinations. A student wishing to pursue an academic career had to take an examination in his chosen subject no sooner than two years after receiving the *Candidat* degree. Passing this examination permitted the student to pursue research toward a thesis, whose successful defense resulted in awarding the *Magister* degree. Such degree holder had a right to be appointed to the faculty of a university with the title of *Privat-Docent*. Further promotion depended on whether or not the scholar would obtain the highest and final degree of *Doctor of Science*. This degree could be won by writing an extensive dissertation and defending it no sooner than two years and no later than ten years following the acquisition of the *Magister* degree. Some time between the two degrees, the candidate usually spent two to four years working abroad, usually in Germany or France, with a prominent scientist or scholar. This period of foreign exposure was almost mandatory if the candidate wanted to land a decent academic position in a Russian university. The expenses of such foreign training were usually paid by the government.

And even after establishing themselves in Russian universities, the professors made frequent trips abroad for further study or research, usually at government expense and encouragement.

Educational reforms of the 1900s included a complete revision of Russia's university statutes. The Ministry of Education and the Educational Committee of the Duma collaborated in drafting the new document, which was completed in March of 1915. Under the new statutes, the universities were granted full autonomy, including a complete independence of the university elective organs (the selection of faculty and administration), the guardianship of the district superintendents over the universities in their districts was abolished, and the students were given the right to organize student corporations and unions. These changes were adopted by the Russian congress of academic workers (scientists and scholars) in June 1917 without any alterations.

It may be argued that a recitation of the uncompleted plans of the tsarist government and the Duma may be irrelevant in a historical narrative that is supposed to concern itself with facts only. On the other hand, the imperial Russian government has at times been accused by Western historians, and especially by the Soviets, of being interested only in its own well beings and preservation, and of being totally oblivious to the needs of the people. The above recitation of the growth of Russian educational system and the government's plans for the future expansion thereof may serve to present the other side of such controversy. The ways of the imperial government were at times slow and cumbersome to be sure, yet progress was being made consistent with the high cost of such programs, and without the benefit of slave labor camps and mass executions. It can be safely assumed that had the plans of the Duma and the Ministry of Education been realized without the intervention of social upheavals, Russia would have emerged by 1925–1930 as a power with an educational system second to none in both quality and quantity.

Scientific organizations in Imperial Russia

The Imperial Academy of Sciences. The best known scientific establishment of both Imperial Russia and the Soviet Union is the Academy of Sciences. It was conceived by Peter the Great and established soon after his death in 1725 with the name of St. Petersburg Imperial Academy of Sciences. Similar institutions in England (Royal Academy) and Germany (Koenigl. Akad. Der Wissenschaften) were established in 1660 and 1700, respectively.

The purpose of the academy was to advise the government in scientific matters, to publish scientific periodicals and books, to organize scientific expeditions and research projects, and to award prizes for scientific excellence. The first academicians were, of course, foreigners. Among these were Leonhard Euler, the mathematician; Joseph Delisle, the astronomer; Daniel Bernoulli, the mathematician-physicist; and Johann Gmelin, the botanist-chemist. In 1913, the

Academy had 50 full and 238 corresponding members. Its budget was 300,000 rubles per year.

The scientific periodical of the academy, the *Commentarii Academiae Scientarium Petropolitanae,* was established in 1728. It later split into several sections according to subject matter. Among such journal published were *Melanges Biologiques* (since 1849), *Melanges Mathematiques et Astronomiques* (1853), *Melanges Chimiques et Physiques* (1854), and *Melanges Geologiques et Paleontologiques* (1894). The academy had an extensive scientific library with some 500,000 volumes in 1913, and a collection of all books ever published in Russia. The Royal Society had a library with 75,000 volumes, the German Academy had 30,000 volumes, and the Smithsonian (US) had 300,000 volumes at the same time.

The academy maintained several scientific research laboratories and astronomical and meteorological observatories. In 1913, the former included a physics laboratory, a chemistry laboratory (Paul Walden, director), a zoology laboratory, a physiology laboratory (Ivan Pavlov, director), and a botanical laboratory. The observatories administered by the academy included the Ekaterinburg Observatory, the Irkutsk Observatory (est. 1886), Central Physical Observatory in St. Petersburg (est. 1849), the Nicholas Observatory in Pulkovo (est. 1839), and the observatory complex in Pavlovsk, which consisted of magnetism, meteorological, and higher atmosphere sections. In addition, the academy administered a biological experimental station in Sevastopol (est. 1871).

Research in the Universities. Most scientific research in Imperial Russia was done in laboratories and institutes connected with universities, technological institutes, and agricultural academies. These research laboratories had independent budgets and directors who were university professors but not necessarily heads of faculties. For example, in 1913, the chemistry research laboratory of St. Petersburg

University had an annual budget of 17,400 rubles with A. E. Favorsky as its director. The organic chemistry research laboratory of the Kiev University had S. N. Reformatsky as its director. Other examples include the Bacteriological Institute of the Kazan University and the Physiological Institute of the Military-Medical Academy in St. Petersburg. Most universities and technological institutes also had astronomical observatories and botanical gardens where the pertinent research work was conducted.

Medical research was carried out in numerous clinics, research hospitals, and institutes associated with medical faculties of universities. Thus, the Military-Medical Academy of St. Petersburg administered the Military Clinical Hospital with 500 beds. Its director in 1911 was Nicholas A. Velyaminov. The privately-endowed Wylie Hospital and the Psychiatric and Neurological Clinic were also administered by the academy. Moscow University administered the surgical, therapeutic medicine, psychiatric, dermatological-venereal, obstetric, pediatric, and gynecological clinics. It also had a cancer research institute (est. 1903), hygiene, pharmacologic, and pathological anatomy institutes. Similar arrangements were found in all Russian medical schools. In addition, much medical research was done at municipal and private hospitals, such as the famous Obukhov Hospital in St. Petersburg (est. 1784).

Under the Soviets, basic research activities at the universities were diminished, and were transferred to institutes established under the auspices of the Academy of Sciences. The main function of the universities became teaching.

Independent Research Institutions. There were several research organizations in Russia that were not formally connected with either the academy or the universities. The Imperial Botanical Garden in St. Petersburg was established in 1713 for the purpose of cultivating medicinal plants. By 1913 it had some 2 million specimens, and it published its own botanical research journal. Another important botanical garden was in Tiflis (Caucasus), which was established in 1890. The Imperial Institute of Experimental Medicine was established in 1890, and it rapidly grew into a first-rate research institution with an annual budget of 190,000 rubles and 100 spaces for graduate students. Among its directors were such luminaries as Pavlov and Vinogradsky. Another medical research establishment was the Grand Duchess Helena Pavlovna Clinical Institute (est. 1885). Its mission was apparently to provide postgraduate training for practicing physicians. The Obstetric-Gynecological Institute, established in 1791, was a center with a worldwide reputation, and its activities are described more thoroughly in Chapter II. During World War I, the Institute of Physiotherapy in Moscow (1914) and in St. Petersburg (1916), and the Institute of Nutritional Physiology in Moscow were established. Other specialized research institutes included the Bacteriological Institute in Kiev (est. 1896) and the Institute of Milk Technology in Fominskoye (Vologda Province, est. 1911).

An Arctic experimental station was established in 1881 in Alexandrovsk on the White Sea. Its proprietor was the St. Petersburg Society of Naturalists. There were also several biological experimental stations: the Volga Biological Station in Saratov (est. 1900), which was administered by the Saratov Society of Naturalists; the Glubokoye Ozero Station (est. 1891); and the Sevastopol Biological Station administered by the Academy of Sciences and the Odessa University. There were three independent observatories: the Astronomic and Seismographic Observatory in Tashkent (est. 1878); the Kronstadt Naval Observatory (est. 1857); and the Naval Astronomical Observatory in Nikolayev (est. 1821).

Some applied science establishments included the Imperial Bureau of Weights and Measures in St. Petersburg, whose founder and first director was Dimitry Mendeleyev, the inventor of the periodical table of elements. The Geological Committee, administered by the Ministry of Commerce (est. 1882), was assigned to coordinate and provide expertise for the search of useful minerals in Russia and to publish geological maps and periodicals. The Central Statistical Committee of St. Petersburg (est. 1806) was concerned with gathering statistical data in Russia. The Imperial Archeological Committee in St. Petersburg (est. 1859) was administered by the Ministry of the Court, and was responsible for the coordination of archeological excavations and the distribution of historic objects among the museums of Russia.

The Scientific Records Commissions. A peculiar scientific institution in Russia was the local *Uchennaya Arkhivnaya Kommissiya*, or the Scientific Records Commission. These organizations usually existed in provincial cities that had no other scientific establishment in town. Their budgets were made up of local public funds and membership dues. Their function was to support the exploration of their local regions, to maintain scientific libraries and museums, and to serve in an advisory capacity to local governments. Such commissions existed in the cities of Yaroslavl, Nizhnii Novgorod, Orel, Ryazan, Simferopol, Tambov, Tver, Vitebsk, and Saratov.

Scientific Societies in Imperial Russia. There were numerous scientific and medical societies in Imperial Russia; however, few of them were national in scope. Most were associated directly or indirectly with the universities or other institutions of higher learning. Most of them were organized after the first all-Russian congress of scientists in St. Petersburg in 1867. Thirteen such conventions took place between that year and 1913. Although the membership of such societies was relatively small, several of them still managed to publish excellent journals in their fields. Thus, the Russian Physical-Chemical Society in St. Petersburg University had 600 members in 1913; it published the oft-quoted *Journal of the Russian Physical-Chemical Society.* The

American Chemical Society had 6400 members at that time. Russian scientists were also members of foreign scientific societies. Thus, the German Anatomical Society with a total of 236 members in 1889 had 14 members from Russia and only 9 from the US. A partial list of Russian scientific societies is given in Table 10.

Scientific Publications in Imperial Russia. Many accounts of original research carried out in Russian laboratories were published in foreign journals, especially the German ones. Such journals of that time were considered to be truly international in scope, and many world scientists used the German media to communicate the results of their work. The physiologists preferred *Pflueger's Archiv*, the immunologists and pathologists published in *Virchow's Archiv*, and the chemists used the *Liebig's Annalen* and *Berichte*. Thus, if one selects at random volume 29 (Jan.–April 1896) of the *Berichte*, one finds therein ten articles originating from Russian laboratories.

Though the Russians published heavily in foreign journals, they published even more heavily in domestic periodicals, many being obscure publications of local scientific societies. As a result, much original Russian work went unnoticed by the outside world, only to be rediscovered years later by the Western world. Every university and most technological institutes had their own journals to report on the work of their faculties. In addition, the scientific societies associated with the universities published journals in their respective specialties. Every Russian province (*gubernya*) had a medical society, which was often also publisher of a medical journal. And almost every province had a journal dealing with sanitation and public health. Thus, Russian scientific publications were both numerous and diverse, and no attempt will be made to provide a comprehensive list thereof in this volume. Table 11 gives the names of publications that are most representative of specific fields of scientific research, and which were national in scope and distribution.

Anatoly Bezkorovainy, J. D., Ph. D.

REFERENCES

General
1. S. W. Baron. The Russian Jew under the Tsars and Soviets. McMillan, New York, 1964.
2. Editorial. *Censorship in Russia.* Brit. Med. J., 1909 (1), p. 1027.
3. Editorial. *Doctors in Russia.* Brit. Med. J., 1907 (2), p. 1675.
4. Editorial. The Lancet, 1915 (1), p. 1054.
5. J. Lawrence. *A history of Russia.* New American Library Press (Monitor), New. York, 1965.
6. A. Petrunkevich. *Russia's contributions to science.* Trans. Conn. Acad. Sci., 23: 211, 1920.
7. V. Robinson. *Pathfinders in medicine.* Med. Life Press, New York, 1929, p. 784.
8. L. Shapiro. Review of Salo Baron's Book (ref. 1), Jewish Social Studies, 30:109, 1968.
9. J. Turkevich. *Soviet chemistry and physics..* In *Soviet Science* (C. Zirkle, ed.), AAAS, Washington, 1952, p. 70.
10. M. Vishniak. *Antisemitism in Russia.* In *Essays on Anti-Semitism.* (Koppel and S. Pinslon, eds.), New York, 1942.
11. A. Vucinich. *Science in Russian culture.* Vols. I and II. Stanford University Press, Stanford, 1963.
12. S. A. Waksman. *W. M. W. Haffkine.* Rutgers Univ. Press, New Brunswick, 1964.
13. C. Zirkle (ed.). *Soviet science.* AAAS, Washington, 1952.

Sources of Statistical Information
14. H. C. Bolton. *A Catalogue of Scientific and Technical Periodicals 1665–1895.* Smithsonian Inst., Washington, 1897.
15. A. deGoulevitch. *Czarism and Revolution.* Omni Press, Hawthorne, 1962.
16. *Efron-Brockhaus Bol'shaya Entsiklopediya.* Leipzig-St. Petersburg, various years until 1914.
17. P. N. Ignatiev, D. M. Odinets, and P. J. Novgorodsev. *Russian Schools and Universities in World War I.* Yale University Press, Hartford, 1929.
18. S. N. Yuzhakov. *Bol'shaya Entsiklopediya.* Prosveshcheniye Press, St. Petersburg, 1904.
19. E. Lyons. *Workers' Paradise Lost.* Funk & Wagnalls, New York, 1967.
20. *Minerva, Jahrbuch der Gelehrten Welt.* Volumes 1-23 (1891–1914), Verlag von Karl J. Trubner, Strassburg, Germany.
21. F. Pavlenkov. *Entsiklopedicheskii Slovar'.* Trud. Publ. Co., St. Petersburg, 1904.
22. S. H. Scudder. *Catalogue of Scientific Serials.* Harvard University Press, Cambridge, 1879.

23. W. Smith, A. Kent, F. Lawrence, and G. Stratton. *World List of Scientific Periodicals Published in the Years 1900–1950.* 3rd ed., Academic Press, New York, 1952.

24. A. Suvorin. *Russkii Kalendar'* (years 1885 and on). Suvorin Press, St. Petersburg.

25. *Witaker's Almanac 1900–1915.* London.

26. *World Almanac and Book of Facts 1900–1915.* Newspaper Enterprise Assn., New York.

Tables and Illustrations

Table 1. Agricultural production of Imperial Russia and other nations
of the world (1912).

Product	Units	Russia*	U. S.	Germany	France
Wheat	Mill. Brit. tons	21.6 (1)**	20.9 (2)	-	9.6
Barley	"	11.1 (1)	5.6	3.6	-
Oats	"	17.0 (2)	21.4 (1)	9.0	6.1
Rye	"	29.0 (1)	1.0	12.2	1.4
Beet sugar	Mill. met. tons	2.1 (1)	0.5	1.5 (2)	0.5
Wool	"	380 (3)	322	-	-
Wine	Mill. of gallons	100 (5)	-	-	1000 (1)
Tobacco	Mill. of lbs.	200 (3)	1100 (1)	-	-
Cotton	Mill. of bales	2 (3)	14.1 (1)	-	-
Cattle	Mill. of heads	48.9	69.3	-	
Sheep	"	74.1	61.7	-	-
Pigs	"	13.5	64.7	-	-
Horses	"	33.2	21.0	-	-

*Russia's exports ͘   ͘912 were, in mill. Brit. tons, 2.5 of wheat, 3.0
of barley, 0.5 o͘   ͘͘͘s, and 0.5 of rye.
**Number in parentheses refers to rank in world

Table 2. Industrial capability of Imperial Russia and other nations of
the world (1910).

Product	Units	Russia	U. S.	Germany	France
Pig iron	Mill. met. tons	2.7 (5)*	28.0 (1)	15.0 (3)	4.0 (4)
Coal	"	24 (7)	411 (1)	217 (3)	38 (5)
Oil	Mill. barrels	70.3 (2)	210 (1)	-	-
Railroads	Miles x 1000	46 (2)	255 (1)	39	31
Telegraph	Miles x 1000	127 (3)	260 (1)	142 (2)	114
Telephones	Units x 1000	173	7700 (1)	1000	230

* Number in parentheses denotes rank in the world

Table 3. Literacy in Imperial Russia (1897).

Population group	% literacy	Population group	% literacy
Entire population		Urban population	54.3
(age 9 & older)	24.0	Male	63.3
Male	35.8	Female	39.3
Female	12.4	Age 9–49	55.6
Age 9–49	26.3	Male	65.5
Male	39.1	Female	43.1
Female	13.7	Age 50 & over	34.9
Age 50 and over	13.3	Male	48.7
Male	20.5	Female	23.1
Female	6.5	St. Petersburg	64.0
Central Russia	22.7	Male	74.0
Little Russia (Ukraine)	18.8	Female	54.0
Belorussia	21.7	Moscow	56.0
Poland & Lithuania	31.8	Kiev	44.0
Latvia & Estonia	76.3	Baku	32.0
Siberia (Asian Russia)	6.0		
Northern Caucasus	9.0		
Rural population	19.6		
Male	31.1		
Female	8.6		
Ages 9–49	21.7		
Male	34.2		
Female	9.6		
Age 50 & over	10.5		
Male	17.0		
Female	4.1		

Table 4. Literacy among the recruits of the Russian Imperial Army

Year	% literacy
1875	21
1880	22
1885	26
1890	31
1895	40
1900	49
1905	58
1913	73

Table 5. Secondary schools in Imperial Russia (1910).

Type of school	Numbers
Classical high school	794
Technical high school	142
Theological (prepared priests)	441
Agricultural	206
Trade	677
Commercial	113
Pedagogical (normal)	82
Specialized professional	135
Military (cadet corps)	37
Other	226

MOSCOW UNIVERSITY, 19th CENTURY

Table 6. Institutions of higher learning in Imperial Russia.

Institution	Location	Yr. founded	Students	Year
Universities				
Imp. Univ.	Moscow	1755	10,399	1912
Imper. Univ.	Dorpat	1802 *	3039	1909
Imp. Univ.	Kazan	1804	3000	1909
Imp. Univ.	Kharkov	1804	4537	1909
Imp. Univ.	St. Petersburg	1819	9886	1912
St. Alexander Univ.	Helsingfors	1827	3532	1913
St. Vladimir Univ.	Kiev	1832 *	3000	1909
New Russia Univ.	Odessa	1864	3257	1909
Imp. Univ.	Warsaw	1869 *	2257	1913
Imp. Univ.	Tomsk	1888	1300	1909
Nicholas II Univ.	Saratov	1909	420	1913
Imp. Univ.	Perm	1916	?	

* The Dorpat University was established in 1632 during the Swedish
occupation of Livonia. It was closed during succeeding wars and was
reopened by the Russians in 1802. The Kiev University was the successor to the Vilna University (est. 1588), which was moved to Kiev following the Polish rebellion of 1831. The Warsaw University (est. 1816)
was closed at the same time, and reopened in 1869.

University-level military academies				
Naval Engineering	Kronstadt	-	-	-
Naval	St. Petersburg	1827	36	-
Artillery	St. Petersburg	1855	62	1897
Engineering	St. Petersburg	1855	110	1897
Army (General Staff)	St. Petersburg	1832	323	1897
Law	St. Petersburg	1866	76	1913
Commissary	St. Petersburg	-	90	1897

Theological academies				
Armenian Orthodox	Etschmiadzin	1874	-	-
Russian Orthodox	Kiev	1701 (1589)	200	1909
Russian Orthodox	St. Petersburg	1797	280	1913
Russian Orthodox	Kazan	1797	222	1913
Russian Orthodox	Moscow	1814 (1685)	250	1913
Roman Catholic	St. Petersburg	1842	66	1914

Agricultural academies				
Forestry Institute	St. Petersburg	1803	560	1909
Agricultural Inst.	Novo-Alexandria	1894	450	1909
Agricultural Inst.	Moscow	1894	1000	1909
Agricultural Inst.	Voronezh	1913	?	
Agricultural Inst.	Saratov	1913	?	

Table 6 (cont'd)

Agricultural academies (cont'd)

Agricultural Inst.	Saratov	1913	?	
Veterinary Institute	Dorpat	1848	350	1909
Veterinary Institute	Kharkov	1873	500	1909
Veterinary Institute	Kazan	1873	402	1913
Veterinary Institute	Warsaw	1889	?	

Engineering schools

Institute of Mines	St. Petersburg	1773	640	1909
Institute of Mines	Ekaterinoslav	1899	480	1913
Institute of Mines	Ekaterinburg	1916	?	
Inst. for Engineers of Ways & Communications	St. Petersburg	1810	1384	1913
Same	Moscow	1896	580	1909
Civil Engineering Institute	St. Petersburg	1842	810	1913
Same	Moscow	1904	?	
Electrotechnical Institute	St. Petersburg	1886	750	1909
Surveying Institute	Moscow	1835	200	1900
Nicholas I Technological Institute	St. Petersburg	1828	2525	1913
Peter I Polytechnic Institute	St. Petersburg	1902	5215	1915
Technological Inst.	Kharkov	1885	1400	1909
Technological Inst.	Tomsk	1896	1607	1909
Technological Inst.	Riga	1862	2088	1913
Polytechnic Institute	Kiev	1898	2500	1909
Polytechnic Institute	Warsaw	1898	600	1909
Don Polytech. Inst.	Novocherkassk	1907	702	1913
Polytechnic Institute	Tiflis	1916	?	
Imp. Technol. Inst.	Moscow	1832	3000	1909
Finnish Technol. Inst.	Helsingfors	?	458	1913

Miscellaneous institutions of higher learning

Demidov Lyceum	Yaroslavl	1803	980	1909
Nicholas Lyceum	Moscow	1868	277	1913
Alexander Lyceum	St. Petersburg	1811	290	1913
Imp. Law School	St. Petersburg	1835	330	1913
Music Conservatory	St. Petersburg	1861	?	
Music Conservatory	Moscow	1866	?	
Archeological Inst.	St. Petersburg	1877	542	1913
Oriental Institute	Vladivostok	1899	165	1909
Lazarev's Inst. of Oriental Languages	Moscow	1815	150	1909
Historic-Philological Institute	St. Petersburg	1867	134	1913

Table 6 (cont'd)

Miscellaneous institutions of higher learning (cont'd)

Bezborodko's Historic-Philological Inst.	Nezhyn	1875(1815)	131	1913
Imp. Institute of Fine Arts	St. Petersburg	1757	314	1898
Geographical Institute	St. Petersburg	1916	?	
Commercial Institute	Moscow	1906	4261	1913
Commercial Institute	Warsaw	1906	?	
Nekrasov Pedagogical Institute	St. Petersburg	1872	?	
Pedagogical Institute	St. Petersburg	Before 1884	?	
Pedagogical Institute	Moscow	Before 1884	?	
Military-Medical Academy	St. Petersburg	1783	900	1909
Medical Institute	Ekaterinoslav	1916	?	

Colleges for Women ("Higher Courses" for women)

Women's Medical College	St. Petersburg	1872 (1897)	1600	1909
Women's University	St. Petersburg	1878	5897	1913
Natural Sciences and Literature School	Kiev	1878	2500	1909
Women's University	Moscow	1900	6477	1913
Women's College	Kazan	1906	1010	1913
Women's Agricultural College	St. Petersburg	1904	700	1913
Women's Medical College	Ekaterinoslav	1915	?	
Pedagogical Institute	St. Petersburg	1910 (?)	548	1913
Polytechnic Institute	St. Petersburg	1906	225	1909
Pharmaceutical Institute	St. Petersburg	1891	?	
Historico-Literature & Law Institute	St. Petersburg	1906	?	
Chemical-Pharmaceutical Institute	Odessa	1915	?	
Women's College	Dorpat	?	?	
Women's College	Kharkov	?	?	
Women's College	Odessa	?	?	

Abbreviations: Inst.--Institute; Univ.--University; Imp.--Imperial;

Table 7. Growth of Russian institutions of higher learning.

Institution	Number of students				
	1890	1902	1909	1912	1914
Moscow University	3473	4344	9516	10399	9760
St. Petersburg University	2200	3775	9000	9889	7455
Kazan University	--	873	3000	--	--
Nicholas I Technol. Inst.	--	1207	2127	2300	2525
Tomsk University	270	540	1300	--	--
Odessa University	450	1116	3257	--	--

Table 8. World's largest institutions of higher learning in 1913.
Russian institutions are marked with an asterisk (*)

Universities	Technological Institutes	Specialized Schools
Paris (17556)	Belfast (6550)	Moscow Women's (6477)*
Berlin (14178)	St.. Petersburg Poly.(5212)*	St. Petersburg Women's
Moscow (9760)*	Manchester (5000)	(5897)*
Cairo (9540)	Vienna (3143)	Moscow Commercial
Vienna (8784)	Moscow (3000)*	(4251)*
Budapest (7814)	Berlin (2884)	Drexel Inst. (3000)
Munich (7718)	Munich (2804)	Chicago Art Institute
St. Petersburg (7455)*	Prague (2759)	(2886)
Univ. of Minnesota (6955)	Zurich (2549)	Kings College, London
Univ. of Chicago (6802)	St. Petersburg	(2745)
Naples (6600)	T ech. (2525)*	Univ. of Pennsylvania
Columbia (6090)	Kiev (2500)*	(2534)
	Riga (2088)*	Kiev Women's (2500)*
		Vienna Export Academy
		(2211)
		London School of Econo-
		mics (2137)

Table 9. Number of students in institutions of higher learning (1913).

	Austria-Hungary	Germany	Russia
Total population	45 mill. (1900)	68 mill.	182 mill.
Universities	40,000	67,000	45,000
Technological Schools	14,000	17,000	25,000
Agricultural Institutes	3300	3500	3300
Commercial schools	2900	6200	4300
Misc. liberal arts	2500	3900	4000
Specialized medical	--	700	2500
Women's colleges	--	350	18,000
Totals	62,700	98,650	102,100

Table 10. Scientific and other learned societies of Imperial Russia

Name of society	Location	Year established	Membership (1909)
General interest			
Society of Naturalists	Moscow[1]	1805	620
Imperial Society of Naturalists	Moscow[1]	1863	-
Society of Naturalists	Kharkov[2]	1869	127
Society of Naturalists	Dorpat[3]	1853	187
Ural Society of Naturalists	Ekaterinburg	1870	381
Society of Naturalists	Kazan[4]	1869	173
Society of Naturalists	Kiev[5]	1870	250
New Russia Soc. of Naturalists	Odessa[6]	1869	300
Imperial Society of Naturalists	St. Petersburg[7]	1869	300
Society of Naturalists	Riga	1845	420
Society of Naturalists	Saratov	1895	-
Warsaw Society of Naturalists	Warsaw[8]	1889	160
Finnish Society of Naturalists	Helsingfors[9]	1838	80
Warsaw Scientific Society	Warsaw	1908	102
Crimean Society of Naturalists	Simferopol	1910	-
Soc. of Naturalists & Physicians	Tomsk[10]	1889	104
Geography and archeology			
Imp. Russian Geographic Society	St. Petersburg	1845	1200
West Siberian Section	Omsk	-	-
East Siberian Section	Irkutsk	1851	190
Soc. for the Exploration of the Amur Region	Vladivostok	1884	-
Imp. Moscow Archeological Soc.	Moscow	1864	-
Imp. Russian Archeological Soc.	St. Petersburg	1846	400
Physical sciences and mathematics			
Physical-chemical Society	Kharkov[2]	1873	65
Mathematical Society	Kharkov[2]	1869	160
Physical-mathematical Society	Kazan[4]	1880	172
Physical-mathematical Society	Kiev[5]	1890	304
Mathematical Society	Moscow[1]	1867	110
Russian Physical-Chemical Soc.	St. Petersburg[7]	1864	
Physic Section			213
Chemistry Section			363
Russian Astronomical Society	St. Petersburg[7]	1890	313
Imp. Russian Mineralogical Soc.	St. Petersburg[7]	1816	-
Medical and biological sciences			
Russian Botanical Society	Moscow	1805	-
Russian Entomological Society	St. Petersburg[7]	1860	300
Imp. Russian Acclimatization Society	Moscow[11]	1864	-

27

Table 10 (cont'd)

Medical and biological sciences (cont'd)

Name of Society	Location	Year established	Membership (1909)
Imp. Vilna Medical Society	Vilna[12]	1805	249
Finnish Medical Society	Helsingfors	1835	504
Physical-Medical Society	Kiev[5]	1899	-
Dermatologic Society	Kiev[5]	-	-
Obstetric-Gynecologic Society	Kiev[5]	1886	110
Psychiatric Society	Kiev[5]	1898	40
Pediatric Society	Moscow[1]	1892	125
Physical-Medical Society	Moscow[1]	1804	90
Moscow Therapeutic Society	Moscow[1]	1876	250
Moscow Venereal Disease and Dermatologic Society	Moscow[1]	1891	62
Surgical Society	Moscow[1]	1880 (?)	-
Riga Society of Physicians	Riga	1822	-
Moscow Dental Society	Moscow	1899	-
Anatomical & Anthropologic Soc.	St. Petersburg[13]	1908	80
Soc. of Normal & Pathologic Psychology	St. Petersburg[13]	-	-
Psychological Society	Moscow[1]	1885	209
Russian Anthropological Society	St. Petersburg[13]	1888	150
Warsaw Medical Society	Warsaw	1821	400
Pirogov Russian Surgical Society	St. Petersburg	1881	-
Society of Russian Physicians	Moscow	1861	-

Technological societies

Imp. Russian Technological Soc.	St. Petersburg	1884 (?)	-
Polytechnic Society	Moscow[1]	1884 (?)	-
St. Petersburg Soc. of Architects	St. Petersburg	1884 (?)	

Affiliations: 1--Moscow University; 2--Kharkov University; 3--Dorpat University; 4--Kazan University; 5--Kiev University; 6--Odessa University; 7--St. Petersburg University; 8--Warsaw University; 9--St.Alexander University; 10--Tomsk University; 13--Military Medical Academy. 11--Administered the Moscow Zoological Garden; 12-Said to be the oldest learned society in Russia .

Table 11. Representative scientific periodicals of Imperial Russia

Periodical	Location	Year established
Publications of general interest		
Proc. of Imp. Academy of Sciences	St. Petersburg	1728
Bulletin of the Society of Naturalists	Moscow	1829
Proc. of the Imp. Society of Naturalists	St. Petersburg	1870
Notes of the Kiev Society of Naturalists	Kiev	1870
Notes of the Ural Society of Naturalists	Ekaterinburg	1870
Notes of the New Russia Soc. of Naturalists	Odessa	1872
Notes of the Crimean Soc. of Naturalists	Simferopol	1911
Proceedings of the Academy of Sciences	Ekaterinoslav	-
University publications		
Scientific Reports of Kazan University	Kazan	1834
Scientific Reports of Moscow University	Moscow	1880
Proc. of the Society of Naturalists	Warsaw	1889
Medical publications		
Military-Medical Journal	St. Petersburg	b.1829
Moscow Medical Journal	Moscow	b. 1848
St. Petersburg Medical Journal (weekly)	St. Petersburg	1857
Medical News	St. Petersburg	b.1862
Modern Medicine	St. Petersburg	b.1861
Annals of the Surgical Society	Moscow	1873
Medical Reviews	Moscow	1874
Proc. of the Soc. of Russian Physicians	Moscow	1878
Russian Physician	St. Petersburg	1880
Medical Journal	St. Petersburg	1880
Practical Medicine	St. Petersburg	1885
Proc. of the Ural Medical Society	Ekaterinburg	1890
Surgery	Moscow	1897
Kharkov Medical Journal	Kharkov	1906
Velyaminov's Surgical Archives	St. Petersburg	1885
Contributions of the Psychiatric Society	St. Petersburg	1880
Modern Psychiatry	Moscow	1907
Journal of Neuropathology & Psychiatry	Moscow	1901
Problems of Neurology & Psychiatry	St. Petersburg	1896
Herald of Clinical & Legal Psychiatry	St. Petersburg	1833
Neurological Herald	Kazan	1893
Russian Journal of Venereal & Dermatologic Diseases	Kharkov	1901
Proc. of the Moscow Venereologic and Dermatologic Society	Moscow	1891
Journal of Obstetrics & Gynecology	St. Petersburg	1887
Proceedings of the Obstetrical & Gynecologic Society	Kiev	1885
Pediatrics	St. Petersburg	1910

Table 11 (cont'd)

Medical publications (cont'd)

Herald of Ear, Nose and Throat Diseases	St. Petersburg	1909
Herald of Ophthalmology	Kiev	1884
Russian Archives of Pathology, Clinical Medicine and Bacteriology	St. Petersburg	1886
Journal of the Russian Public Health Society	St. Petersburg	1891
Contributions of the Society for Scientific Medicine and Hygiene	Kharkov	1885
Dental Herald	St. Petersburg	1885

Biological sciences

Contributions of the Imperial St. Petersburg Botanical Garden	St. Petersburg	1871
Acta Horti Botan. Univ. Imp. Yuriev	Dorpat[1]	1900
Herald of Russian Flora	St. Petersburg	1914
Scripta Botanica Petropolitanae	St. Petersburg	1886
Russian Botanical Journal	St. Petersburg	1909
Materials of Mycology and Phytopathology	St. Petersburg	1915
Arbeiten d. Pharmakolog. Inst. Dorpat	Dorpat	1888
Zoological Herald	St. Petersburg	1916
Russian Zoological Journal	Moscow	1916
Ornithological Herald	Moscow	1910
Russian Fauna	St. Petersburg	1911
Entomological Herald	Kiev	1912
Herald of Russian Applied Entomology	Kiev	1914
Russian Entomological Review	St. Petersburg	1901
Contributions of the Russian Entomological Society	St. Petersburg	1861
Reports of the Moscow Entomological Society	Moscow	1915
Archives of Biological Sciences	St. Petersburg	1892
Contributions of the Physiology Institute of the Imperial Moscow University	Moscow	1888
Journal of Microbiology	St. Petersburg	1914
Contributions of the Volga Biological Station	Saratov	1906
Russian Archives of Anatomy, Histology and Embryology	St. Petersburg	1916

Agriculture and forestry

Journal of Experimental Agriculture	St. Petersburg	1913
Herald of Agriculture	Moscow	1910
Pedology	St. Petersburg	1899
Reports of the Agricultural Institute of Novo-Alexandria	Warsaw	1877
Veterinary Reviews	Moscow	1899
Veterinary Life	Moscow	1907
Herald of the Veterinary Society	St. Petersburg	1889
Journal of Scientific and Practical Veterinary Medicine	Dorpat	1906

Table 11 (cont'd)

Agriculture and forestry (cont'd)

Agriculture and Forestry	St. Petersburg	1842
Reports of the St. Petersburg Forestry Institute	St. Petersburg	1913
Practice and Economics of Forestry	Tambov	1912
Forestry Life	St. Petersburg	1871

Physical sciences

Annals of the Kiev Observatory	Kiev	1884
Annals of the Nicholas Physical Observatory	St. Petersburg	1835
Reports of the Pulkovo Observatory	St. Petersburg	-
Reports of the Russian Astronomical Society	St. Petersburg	1892
Materials on the Geology of Caucasus	Tiflis	1874
Materials on the Geology of Russia	St. Petersburg	1869
Reports of the Geological Committee	St. Petersburg	1882
Mining Journal	St. Petersburg	1825
Notes of the Mineralogical Society	St. Petersburg	1842
Journal of Medicinal Chemistry and Pharmacy	St. Petersburg	1895
Journal of the Russian Physical-Chemical Soc.	St. Petersburg	1869
Reports of the Kharkov Mathematical Society	Kharkov	1884
Proceedings of the Physical-Mathematical Soc.	Kiev	1885
Proceedings of the Physical-Chemical Section of the Society of Naturalists	Warsaw	1889
Journal of Pharmacy	St. Petersburg	1879
Mathematical Reviews	Moscow	1866

Engineering and technology

Reports of the Moscow Institute of Technology	Moscow	1907
Reports of the St. Petersburg Polytechnic Inst.	St. Petersburg	1904
Reports of the Kiev Polytechnic Institute	Kiev	1910
Reports of the Don Polytechnic Institute	Novocherkassk	1912
Electricity	St. Petersburg	1880
Notes of the Russian Technological Society	St. Petersburg	1867
Journal of the Russian Metallurgical Society	St. Petersburg	1912
Technological Hersld	St. Petersburg	1906
Reviews of World Technology	St. Petersburg	1908
Engineering Life	St. Petersburg	1860
Notes of the Kiev Section of the Imperial Technological Society	Kiev	1871
Notes of the Moscow Section of the Imperial Technological Society	Moscow	1878

[1] Russian name for Dorpat was Yuriev. Some translations: Reports from Izvestiya or Zapiski; Annals from Letopis'; Notes from Zapiski; Reviews from Obozreniye or Zbornik; Proceedings from Protokoly; Contributions from Trudy; Herald from Vestnik; Problems from Voprosy; Materials from Materialy; Journal from Zhurnal.

Chapter II

MEDICAL CARE IN IMPERIAL RUSSIA

The Middle Ages

Many authors concerned with the history of medicine in the Middle Ages and the early modern period feel compelled to include a paragraph or two on early Russian medicine, or, to be more precise, a lack thereof. With the little information that is available on that aspect of Russian history, the following general picture may be constructed: through the centuries of Russian pagan past, its masses developed several classes of healers including the *kolduny* (warlocks), the *kudesniki* (soothsayers), and the *znakharki* (wise women). These "professionals" combined superstition, ritual, and herbs to treat the maladies of their patients. Some princes, on the other hand, imported Greek (Byzantine), Jewish, and sometimes, Arab "physicians" to provide medical care for themselves. Most likely, the results of treatments by the Russian "professionals" were not much worse than those of foreign physicians and their herbs.

Introduction of Western European Medicine

With the advent of Christianity in AD 988, the Russian folk healers were gradually replaced by monastic clergy who were often of Greek origin and were reasonably well versed in modern medical techniques of that time. They treated their patients with herbs and prayers. Not much is known about Russian medicine during the 200 years of Tatar domination, but it may be assumed that folk healers were once again in a dominant position. With the emergence of Moscow as the dominant Russian principality, medical expertise began to be imported from the West rather than the East, and by the middle of the sixteenth century, Western-style medical profession was already highly respected in Russia. By the turn of the eighteenth century, Russian physicians were known even to Goethe, who in his *Dichtung und Wahrheit* (part two, tenth book), mentions a certain Dr. Pegeloff who was, at that time, attempting to heal Herder's (the philosopher's) tear duct disorder by pushing a horse's hair through it. Goethe apparently valued highly both the professional skills and the intellectual powers of Herder's physician.

Though Western medicine was known in Russia since the reign of Ivan III, most historians agree that it did not affect the lives of both the peasant and nobleman until the beginning of the eighteenth century. This fact *per se* was probably not much of a tragedy for the Russians and their folk healers, since Western medicine before the eighteenth century had very little to offer to the ill and the infirm. And indeed, the Russian peasant with his faith in the icons, the herbs, and prayers of the monks may have done better than his Western cousin whose evil spirits were being released from his organism by bleeding or help by barber physicians. Yet when new knowledge and techniques based on solid scientific principles did become available through men like Jenner and others, there came in the West a foundation upon which modern medicine could be built reasonably fast. No such foundation existed in Russia, and the introduction of modern medicine and public health measures on a large scale thereto was an extremely slow and painful process. An objective account of early Russian medicine is given by Garrison (1), and another review on the subject was written by Gantt (2). An excellent article on Russian medicine and its delivery under the Tsars was published in *Lancet* (3) in connection with the Twelfth International Medical Congress that took place in Moscow in 1897. Public health efforts by the Tsarist government are very ably reviewed by Winslow (4).

Western medicine in Russia was at first represented by foreigners, who came to serve either as court physicians or in the military. Among such men one may mention the English doctor brothers Jacobi, who went into the service of Ivan IV (the Terrible) in 1581, and Mark Ridley, who served Tsar Feodor from 1594 to 1598. Others were Hartmann G. Gramann, a German, who went into Russian service in 1639; Jeody du Gour (1766–1840), a Frenchman; Abram E. Ens (d. 1770), a Dutch physician; and Elias Ensholm, a Swedish physician in Russian service from 1786 to 1820. Thus, physicians of practically all European nationalities were represented. The first physicians of native birth were also either in the military or were associated with the court. One of the more prominent doctors of this group was Karl v. Espenberg (1761–1822), who was born in the Estonian province of Russia. He joined the Russian navy and made the world cruise with Admiral Krusenstern (1802–1806). A peninsula near the Cape of Good Hope is named after him (Cape Espenberg). A physician who took part in many events of historical importance was Leonid Nagumovich (1792–1853). A military physician, Nagumovich was with the troops in their battles against Napoleon at Borodino, Maloyaroslavets, Lutzen, Kulm, and Leipzig. Later, he participated in the storming of Warsaw that ended the Polish rebellion of 1830–1831. Nagumovich published numerous works in the field of military medicine. Among the early court physicians, the best known is probably Pajanota Condoidi, a Greek by birth, who was brought to Russia in his childhood. He studied medicine in Leyden (Doctor of Medicine, 1733), and in Russia he occupied the position of head of the Medical

Chancellery in addition to his duties as court physician. Another such doctor of some accomplishment was Ivan Enokhin (1791–1863), who accompanied Tsar Nicholas I on many of his military expeditions.

Early Russian Medical Organizations

The first official medical establishment in Russia was apparently the Aptekarsky Prikaz (Pharmaceutical Committee), which was established in 1620. It was an outgrowth of a pharmacy established in Moscow in 1581 by an Englishman named James Frenchman. Its function was at first to manufacture and dispense medicinal preparations to the court and to supervise court physicians. However, this organization soon added another branch whose functions were to provide medicines and treatment to the common people. And it was even involved in attempts to control epidemics. The Aptekarsky Prikaz eventually became a large organization employing physicians, pharmacists, chemists, and botanists. It manufactured all kinds of materials including drugs, varnishes, and paints. It was eventually renamed the Medical Chancellery, and in 1811 it was incorporated into the Ministry of the Interior. With the then created Medical Council, the Medical Chancellery continued to regulate medical affairs in Russia until the Bolshevik coup.

Medicine received its greatest impetus during the reign of Peter the Great, who recognized the importance of good medical services in his army and navy. Peter encouraged Russians to study medicine abroad, and brought numerous foreign doctors to Russia. His purchase of the Ruysh anatomical collection has been noted by many historians. Many of the foreign doctors were admitted into the membership of the newly created St. Petersburg Academy of Sciences (est. 1725), and through this organization, medicine acquired the dignity of a scientific discipline. Its growth and development in Russia was thus assured.

Prior to the reign of Peter the Great, few hospitals worthy of their name existed in Russia. The first "modern" hospitals established were for the benefit of Russia's military. These were the Court Hospital in Moscow (est. 1706), the Admiralty Hospital (est. 1716), and the Infantry Hospital (est. 1717), the latter two both in St. Petersburg. Catherine II was probably the most active pioneer hospital builder. She established the Catherine, Pavlovsky, and Galitsin general hospitals in Moscow and the Obukhovsky General Hospital in St. Petersburg (1784). In addition, insane asylums were built in each city. A venereal disease hospital was opened in St. Petersburg in 1763, and foundling hospitals were established both in Moscow (1764) and St. Petersburg (1770).

Russian Medical Professionals

Before the advent of university-type medical education, Russian doctors were trained either through apprenticeships under practicing physicians, or they learned their trade in schools operated by hospitals. Johann Dorndorf (1761–1803) was an example of an apprentice physician. He became a very accomplished surgeon in the city of Riga. The first medical school in Russia was associated with the Court Hospital in Moscow. It was established in 1707 and was administered up until 1735 by Nicholas Bidloo, a Dutch physician. The school operated until 1799. Its first class was graduated in 1713. By 1733 there were three other medical schools in Russia, all associated with St. Petersburg hospitals. As more universities became established in Russia, the hospital-based medical schools ceased to exist and the responsibility for admission to practice medicine exams was gradually transferred from the Medical Chancellery to the universities. The oldest Russian medical school still in existence is the Moscow University (est. 1755), which graduated its first medical class in 1794.

The Medical-Chirurgical Academy was started soon thereafter (1798), and was later reorganized into the famous Military-Medical Academy. By 1805, there were five university-level medical schools in Russia: the two mentioned above, the Dorpat University (1802), Kharkov University (1804), and Kazan University (1804). The medical schools in existence in Russia at the time of its fall are listed in Chapter I.

Medical education in Imperial Russia spanned a period of five years in a university, after which the student was admitted to take an examination. Upon its successful completion, the student was awarded the title of *Lekar* (Physician) with the right to practice medicine. The degree of *Doctor of Medicine* was awarded only to those who endeavored to write and defend a thesis based on original scientific research. Such additional work usually required two to three years, and was thus much easier to attain than the degree of *Doctor of Science.* However, starting with the early 1900s, the physician aspiring to become a *Doctor of Medicine* was required to undergo training in one of the medical specialties (surgery, internal medicine, etc.) in addition to the thesis requirement. The degree of *Doctor of Medicine* thus became much harder to attain. Medicine was probably the most popular field among university students. In 1913, medical students accounted for about 50, 60, 45, and 25 percent of the student bodies in the Dorpat and Kazan universities, Tomsk, Odessa, and the Moscow and Warsaw universities, respectively.

The fully trained Russian physician was probably second to none in quality among his or her European peers. Yet their numbers were inadequate to satisfy the needs of their country. Physician shortage was especially acute in the countryside. The Russians chose to meet such needs by establishing a semiprofessional called the *feldsher*. The first feldshers were discharged army medical corpsmen who, in civil practice, were called the *rotnye feldshers* (*rota* in Russian is the army company). The somewhat better trained civilian feldshers were termed the *shkol'nye feldshers* (5). In 1894, the feldshers outnumbered the doctors in Russia. These semiprofessionals practiced mainly in the villages, theoretically under the leadership of doctors, but more often than not, independently. They knew how to set fractures, how to handle most agricultural accidents, and how to treat infections. They had the right to prescribe drugs for their patients. In addition to the feldshers, there were the fully trained and licensed midwives who did most of the obstetric work in the villages. The civilian-trained feldsher (shkol'ny) had to undergo a four-year course of training. A four-year secondary school education was required for admission to the feldsher school. In 1897 there were twenty-one civilian and four military feldsher schools in Russia. Table 1 shows the number of civilian medical professionals. This information was gathered from official Russian statistics (6) or as reported in Western medical journals (7, 8).

It should be pointed out that most doctors in Imperial Russia were concentrated in the cities, and that most physicians in the countryside were employees of the Zemstvo service system (see below). In 1912, there was one physician for every 1,450 persons in towns and cities, and only one doctor per 21,600 persons in the rural areas. The average came to one doctor for every 7,200 persons of Russia. The areawise distribution of doctors in Russia was also unequal. In central Russia, there was one doctor per 7,300 persons; in Belorussia, there was one doctor per 13,000 inhabitants; in Asian Russia, there was one doctor for every 9,800 persons; in Poland and Lithuania, there was one physician for every 6,000 persons; and in Latvia and Estonia, one doctor per 2,900 persons. Dismal as these figures were by modern standards, one must not forget the feldsher who, as stated above, had his or her practice mainly in the villages. If these semiprofessionals are included into the above statistics, one then gets the figure of one medically-trained practitioner (excluding the dentists) for 3,200 persons in rural Russia. Some 60 percent of Russian hospitals had fewer than fifteen beds, indicating large numbers of rural hospitals and dispensaries. The mortality rate of Russian hospitals was at some 4.5 percent. In addition to the general and mental hospitals in Russia, there were twenty-eight Pasteur stations that in 1912 treated 38,000 patients, of which 136 died of rabies poisoning.

The Zemstvo Organization

Medical care, when available, was provided free of charge to most of Russia's population. The peasants were under the auspices of the Zemstvo organization (4), whereas factory workers were covered by a statewide insurance plan (9). The Zemstvo system of local governments was an institution peculiar to Russia. It was a locally elected organization, established at the time of emancipation of the serfs in 1861 by Tsar Alexander II, and its function was to bring the benefits of modern technology and medicine to the enormous numbers of Russian population in the countryside (some 136 million souls in 1912, or 85 percent of the entire Russian populace), which was scattered from the German borders in the west to the Pacific Ocean in the east. The Zemstvo organization had the power of taxation, as well as the power of drafting workers for limited periods of time to do local public projects. It concerned itself mainly with the maintenance of rural schools, provision of insurance against natural disasters, providing agricultural expertise to the peasants, and delivery of medical care. Each Russian province with the Zemstvo form of government was divided into medical districts, each with a hospital or medical dispensary manned by trained medical personnel. In 1910, each district had a radius of about 15 kilometers. In 1870, there were only 756 Zemstvo physicians; by 1890, this number had grown to 1,805; and in 1910, there were 3,100 doctors in Zemstvo service. It had been stated that the Soviet system of health care delivery in Russia is based upon that of the Zemstvo concept (10). The growth of Zemstvo medicine in Russia is summarized in Table 2.

Winslow's article (4) presents some data on two model Zemstvo medical operations, one on the rural level, exemplified by the province (gubernia) of Saratov, and the other on a city-provincial level, as in the city and gubernia of Moscow. In the Saratov province in 1911 (population 3 million), there were a general referring hospital with 200 beds, a psychiatric hospital with 460 beds, and a Pasteur station, all maintained by the provincial Zemstvo. There were 123 medical districts maintained by local Zemstvos with at least one doctor per district. There were 78 district hospitals with a total of 1,100 beds, and there were 9 inpatient and 526 outpatient cases per 1,000 inhabitants per year in these districts. The Moscow province, exclusive of the city of Moscow, maintained 100 public hospitals with 20–60 beds each, and each having an average of 2 physicians, 4 feldshers, and 4 nurses in attendance. Each hospital had an outpatient clinic with an average of 100 visits per day. The city of Moscow had 24 general public hospitals with a total of 7,000 beds, 2 tuberculosis hospitals, one venereal disease hospital, and 3 infant welfare stations in 1915, all in addition to several private hospitals and clinics. Some 15 percent of the city's budget was earmarked for medical services. There were a total of 3,500 physicians in the city of Moscow in 1914, many of them in private practice.

Medical Care and Death Rates in Russia

The admirable quality of medical care delivery in the Saratov and Moscow provinces was, of course, not universal. In fact, such facilities were rather poor in many places, especially in those provinces where the tax base was inadequate. This may be deduced from the death rates in Russia (2, 4, 6, 7, 11), which do not compare favorably with those of other European countries (Table 3). Nevertheless, as can be seen in this table, a tremendous progress toward lowering the death rates in Russia was being made between the years 1885 and 1914. Death rates were quite unequal among the various nationality groups of Russia, ranging from 33/1000/year in central and Little Russia to only 12/1000/year among the Jews in 1909. The relatively high death rates among the Russians was attributable to some extent by a habit of Russian peasants to bathe in their steam baths that were kept at 45–50°C, then emerging naked into cold weather to cool off. It was estimated that as much as 1/3 of the infectious diseases contracted in Russia was the result of such activity (12).

Infant Mortality Rates

The largest single contributing cause of the high mortality rates among the Russians was their enormous birth rates (Table 4) and the accompanying high infant mortality rates (Table 5). In fact, some 60 percent of all deaths in Russia were attributable to deaths among children before the age of five. Infant death rates were not uniform among all Russian provinces: the central and Little Russian provinces had the highest infant death rates, 300/1000 births, whereas in the western provinces and among some minority groups these were low: 128 deaths/1000 births among the Jews, and 134 deaths/1000 births in the Estonian province. Even in the city of St. Petersburg, infant mortality rate ranged from 122 deaths/1000 births to 279 deaths/1000 births depending on the neighborhood.

The high infant mortality rate in Russia has been attributed to the ignorance and uncleanliness of the peasant mother. For instance, it was the practice among the peasantry to wean infants at the age of one month. And one can see that in the absence of proper food sterilization techniques, only the hardiest of the children could survive that. Thus, some 32% of the infant death rate was due to digestive tract disorders, and only 5% were due to infectious disease. Russian society was not very successful in decreasing the high infant mortality rate in the period between the time when proper pre- and post-natal techniques became available and the time when the Bolshevik revolution took place. This failure was probably due to the enormity of the effort that was required to educate the peasant masses in the vast territory that was Russia. As can be seen from Table 5, France and Italy were able to accomplish this task in relatively short amount of time. A beginning to tackle the problem was made in the cities of Moscow and St. Petersburg through the establishment of infant welfare stations, children's hospitals, and maternity hospitals, but this program was expanded into the rural areas only after the Bolshevik revolution.

Infectious Diseases

Though less important, another cause of the mortality rates among the Russians was the prevalence of infectious disease. Leading causes of mortality in this instance were tuberculosis, measles, scarlet fever, typhoid fever, diphtheria, smallpox, and whooping cough, in that order. Progress was however being made toward control of infectious diseases as shown by the fact that there were 645, 467, 397, and 230 deaths per 100,000 persons from infectious diseases in 1910, 1911, 1912, and 1914 respectively (the latter figure for cities only).

Of special concern were the gastrointestinal disorders such as dysentery, typhoid fever, and cholera. The latter was especially troublesome in Russia for over 100 years, not because of its prevalence in comparison to other infectious diseases, but because of its rapid spread, high mortality rates, and association with the supernatural. Cholera made its first appearance in Russia in 1823. It apparently originated in India and invaded Russia via Astrakhan, the traditional point of contact between Russia and the Middle East and Asia. Russia was apparently the first European country to be struck by cholera on a massive scale. It is said that for the next 100 years, there were a total of 5.6 million cholera cases with some 2.1 million deaths. The Russian physicians during the first widespread cholera epidemic (1829–1834) were apparently quite dedicated, and for their time very competent in their efforts to contain the disease (13). Most believed that the disease was spread through the air, and the suggestion of Goryanikov and Malakhov that the disease was caused by microbes was rejected by other physicians because of incomplete data. Yet everyone did agree that cleanliness and moderate habits contributed toward a better resistance to the disease. The first Russian cholera epidemic elicited the first therapeutic method that was based on solid scientific data: R. Hermann of the Moscow Institute for Artificial Mineral Waters analyzed the blood of a cholera patient and discovered that it had lost 28 percent of its "fluidity." This finding was interpreted to mean that the patient had lost water from the body by diarrhea and vomiting. A Dr. Jaenchen thereupon proceeded to treat a cholera patient by intravenous infusion of water. The patient first became better, but died subsequently, perhaps of septicemia (14). The 1829–1834 epidemic alone involved ½ million persons and resulted in 200,000 deaths. More recently there was a widespread cholera epidemic in Russia (1908–1910) with over 60,000 deaths. Its devastating effects were ascribed to the inadequate sewage and water treatment facilities in Russia: only 192 of the 1082 cities in Russia had water works and only 38 had "drainage" systems in 1910 (15).

Plague was rare in Russia in recent years, though several cases were being reported each year in the Astrakhan province (16). The disease usually originated in Turkey and other eastern lands, and was brought to Russia by its southeastern nomadic tribes who were used to crossing borders at will. In the earlier days, the disease was apparently brought to Russia by its troops returning from a Turkish campaign. For instance, the Moscow epidemic of 1771–1772 was thoroughly documented by Samoilowitz, a "Ukrainian" military physician (17). He traced the disease to eastern traders who mingled with Russian troops occupying the Turkish city of Giurgiu (in present-day Romania) who, in turn, brought the disease back to Moscow. A commission appointed by the tsar in more recent years to study ways of combating the plague epidemics recommended that the public health services be expanded in the afflicted areas, that housing conditions be improved, that rat extermination programs be instituted, that better isolation facilities be built, and that the public be appropriately educated. In extreme cases, it was recommended that infected dwellings be incinerated, but that this was to be done "with tact" and with compensation to the owners (18). It is said that plague was not eradicated in Russia until the 1930s, when eradication measures sometimes included execution of the afflicted persons (2).

A major proponent for improved public health facilities in Russia was Anton Chekhov, the master of drama and short story (19). Chekhov himself was a physician whose writing career began in the medical school as a means of supplementing his income. Throughout his short lifetime (1860–1904), Chekhov alternated between his practice of medicine and writing. He was a strong advocate for the strengthening of Zemstvo medical services, and he himself served for a while (1892) as a district doctor in the Moscow province where his estate was located. His ability to organize the district's resources to combat the cholera epidemic of 1892 earned him a national reputation. Unfortunately, Chekhov's health could not withstand his boundless energy, and he died in 1904 of tuberculosis in Germany, where he had gone to seek a nonexistent cure for his malady.

While Chekhov molded much of the public opinion in favor of free health care via his literary talents, the technical impetus in that direction, according to Soviet sources, was provided by Friedrich Erisman (20). He came to Russia from Switzerland in 1869 and worked in St. Petersburg as an ophthalmologist until 1879, when he was appointed to a commission inspecting health conditions in Moscow factories. His commission found the factories highly unsanitary, and his report to that effect contributed to the establishment of tighter health regulations for industry. Mainly through his efforts, a chair of hygiene and public health was created in the Moscow University in 1882, and Erisman was invited to become its first occupant. During his tenure at the university, Erisman trained numerous public health physicians and advocated a public health approach involving all aspects of clinical and laboratory medicine. He insisted that public health officials and inspectors be physicians fully trained in various public health matters. In 1890, a public health institute was built for him in accordance with his own design and specifications. Politically Erisman was a social democrat and as such, he was often at odds with the more conservative government and university bureaucracies. As student riots broke out in 1893, Erisman threw his support behind them, and his association with the university was terminated. He returned to Switzerland, where he was active in politics until his death in 1915.

In addition to the academic and literary advocates for better public health facilities in Russia were the Zemstvo doctors themselves (21). The most notable of these were Evgraf A. Osipov and Ivan I. Molleson, who dedicated their lives to Zemstvo medical work (10, 12). Molleson was born in Irkutsk, Siberia, in 1842 and obtained his medical education at Kazan University, graduating in 1865. Molleson subsequently spent his entire life as a medical officer in Zemstvo service. His most notable job was with Perm Zemstvo, which he held with a few interruptions from 1871 to 1889. While serving in Perm, Molleson made proposals for the creation of sanitary commissions in

Russian provinces, which would be concerned with public health matters. Such commissions were indeed established and functioned successfully up until the Bolshevik takeover. Molleson retired in 1911 and died in 1920 in Voronezh.

Molleson advocated a dual approach to rural medicine, namely the therapeutic and preventive varieties. The former was to be concerned with establishment of hospitals, dispensaries, and drugstores, whereas the latter—by establishing provincial sanitary commissions whose membership would consist of all the provincial Zemstvo doctors. Such commissions would be concerned with combating epidemics and infectious diseases, administering vaccination programs, and improvement of water supplies and sewage facilities. The commission would appoint county sanitary officers who would have responsibility for implementing the sanitary commission recommendations at the local level. Molleson's proposal was approved in St. Petersburg and a sanitary commission was established in the city of Perm. Molleson was appointed the sanitary commissioner of the Shadrinsk county in the Perm province in 1872, the first such job in all of Russia. As the chief sanitary officer there, Molleson established the machinery for epidemiological control, water works inspection, and bacteriological testing, which served as a model for the creation of similar organizations throughout Russia. The sanitary commissions soon after their creation established statistical bureaus, which were concerned with gathering of epidemiological, vital, and population mobility data for the provincial Zemstvo organizations.

Molleson's suggestions for the betterment of conditions in rural Russia were made from practical experience and not from the lofty heights of the academe. As such, they were much more realistic and helpful for the peasants. For instance, he noted that many villages got their drinking water from lakes where people bathed and women washed their clothes. He also noted that the presence of swamps near the villages increased the susceptibility of their inhabitants to disease. Molleson thus strongly urged the peasants to drill wells for their water supply and drain any swamps in the vicinity of their villages. Though these ideas sound quite superfluous in our modern age, to the Russian peasants of the 1870s without any knowledge of bacteriology and insect-borne disease, these suggestions were quite novel.

The problem of syphilis was also very acute at that time in Russia and affected mostly the countryside; it was estimated that as much as 0.06 percent of Russia's population was afflicted with this disease. Yet over 90 percent syphilis cases were not acquired through sexual contact but were, instead, spread by generally unhygienic living conditions and eating habits. Molleson combated this disease in his district by establishing a program whereby the syphilitic patient would undergo a three-month treatment at the Zemstvo hospital, while his farm would meanwhile be worked at Zemstvo expense. Again, this program was copied by many Russian medical districts.

During his brief service as an industrial physician, Molleson was able to observe the misery and suffering of child laborers. Due to initiative and insistence, the Russian government passed legislation in 1882 prohibiting child labor in Russia. The numerous ideas and proposals of Molleson's received a large amount of publicity. This was probably due to the fact that he lost few opportunities to participate in the various Zemstvo medical meetings and congresses, because of his service on the editorial boards of a number of medical publications, and because of his own prolific writing, which numbered some 200–300 papers.

Evgraf A. Osipov was a contemporary of Molleson and also a Kazan University graduate (1865) (10). His area of activity was the Moscow Zemstvo, where he strongly advocated against the circuit rider form of health care delivery, and like Molleson, argued for both the curative and preventive forms of medical care delivery in the

Moscow Zemstvo. The exemplary health care delivery of the Moscow Zemstvo referred to above was developed largely under Osipov's guidance.

The city dwellers of Imperial Russia, as has already been pointed out above, fared much better with respect to medical care than did the country folk. This included the availability of various medical specialists. Thus, St. Petersburg in the year of 1885 and a population of some 900,000 persons had 20 specialists in internal medicine, 12 surgeons, 36 obstetrician-gynecologists, 19 pediatricians, 7 psychiatrists, 4 respiratory disease specialists, 6 oculists, 2 otologists, 2 urologists, 6 venereal disease specialists, 6 dermatologists, and 13 dentists. This was, of course, in addition to numerous general medical practitioners.

In contrast to the Western world, medical care in Russia was initiated and maintained by the government. Though private practice did exist, most medical care in Tsarist Russia was delivered by municipal or Zemstvo organizations. Public health delivery systems enjoyed strong support from both the academic and the practicing physicians, though the supply of them was far short of demand. For this reason, much of medical case in Russia was provided by the feldshers. This situation was viewed as being of temporary nature, and plans were in the offing to both increase the number of physicians and to phase out the feldsher profession (5). It is thus clear that the medical care system developed by the Soviets was not a product of their ideology but was, instead, a version of the system developed under the tsars.

Academic Medicine

Introduction. It has already been stated that medical practice of the Western type did not reach Russia until the beginning of the eighteenth century. The first medical school with reasonable academic standards was not established until 1755 (Moscow University), and it took another 100 years or so until competently executed medical research originating in Russia began to appear in quantity in world medical literature. Because of the late start of its academic medicine, Russia cannot boast of being responsible for such far-reaching discoveries as circulation of blood, vaccination, and the existence of infectious microorganisms. Yet when Russian academic medicine did come of age in the middle of nineteenth century, its practitioners threw themselves into medical research with remarkable enthusiasm, and thus assured for themselves a respectable place in the annals of world medical history.

Much of the medical research done in Russia before the 1860s was described in pamphlet or monograph form, and in the few Russian medical journals that existed at that time. Little work was published abroad. The first Russian medical periodical is said to have been the *St. Petersburg Vrachebniya Vedomosti (St. Petersburg Medical News)*, which operated between the years 1792 and 1794. A more successful journal was the *Drug Zdraviya (Health's Companion)*, which was established before 1833 and was operating into the 1850s. Other Russian medical journals were the *Moskovskii Vrachebnii Zhurnal (Moscow Medical Journal)*, established sometime before 1848, the *Zapiski Dubovitskago (Dubovitskii's Archives)*, established before 1840, and Inozemtsev's *Moskovskaya Meditsinskaya Gazeta (Moscow Medical Gazette)*, which operated between 1856 and 1862. The *Voenno-Meditsinskii Zhurnal (Medical-Military Journal)* was an important medium for medical communications, and its pages were open to both the civilian and military personnel. It was established before 1829, possibly as early as the late 1700s. As the quality and quantity of Russian medical research improved, papers originating from Russian institutions began to appear more often in foreign publications, especially the German ones, and less often in France. At the same time, a large number of medical publications embracing all types of specialties were established in Russia. Starting with 1870s, it became a custom among Russian medical researchers to publish papers abroad in addition to having them published in their native tongue. In this fashion, Russian scientists received maximum exposure both at home and abroad.

The Evolution of Medical Philosophy and Therapeutics in Russia. Medicine was introduced to Russia on a large scale at about the time when Hermann Boerhave's influence on the medical profession was at its zenith in Western Europe. Russia was especially receptive to Boerhave's ideas with respect to their iatrophysical views and methods of caring for patients, because there was no medical establishment in Russia to resist innovations, and because Boerhave's nephews, the Kaau-Boerhave brothers, made their home in Russia. Hermann Kaau-Boerhave (1705–1753) received his medical training at Leyden and went to St. Petersburg in 1740 as a court physician. In 1748, he became the head of the Russian Medical Chancellery, a position equivalent to that of the US Surgeon General. His contribution to therapeutics was a concoction including tea and syrup to treat respiratory disorders. This "drug" was in use in Russia for the next 200 years. Abraham Kaau-Boerhave (1715–1758) studied medicine in Leyden under his famous uncle. In 1744 he was elected to membership in the St. Petersburg Academy of Sciences, and in 1746 he moved to Russia as a professor of medicine and pharmacy at the Admiralty Hospital and its medical school in St. Petersburg.

For the next 100 years, Russian theoretical medicine was nothing more than an echo of Western European medical thought. Thus, the humeral theory of Rokitansky had its day, and the theory of dynamic etiology of disease (Solidarpathologie) was especially popular for a time. According to the latter, diseases were caused through the action of sympathetic nervous system, which received ever-changing stimuli from the environment. Whereas a given disease was caused by some specific agent at one time, the same disease could be caused by a different agent a few months or years later. The main proponent of this theory in Russia was Feodor Inozemtsev (1802–1869), a professor at the Moscow University who was convinced that the etiology of all diseases had undergone a drastic change in the 1840s. The bloodletting methods of the earlier era were thus all right, since diseases of that time were caused by inflammatory processes and the bleeding was thus useful to cleanse the blood of causative evil humors. On the other hand, the diseases of the post-1840 era were the result of the nervous system emergence as the dominant character and that, therefore, other therapeutic methods had to be utilized. A universal drug used by Inozemtsev was a mixture of ammonium salts and tartar, which, in his opinion, suppressed the deleterious effects of the nervous system. In spite of his primitive concepts of disease etiology, Inozemtsev was a superb diagnostician and teacher. He was the son of a Persian prisoner of war who remained in Russia after its end. For this reason, his name means literally "the foreigner" in Russian. Inozemtsev studied medicine at Kharkov University, but because of his participation in student riots, he was expelled and forced to teach in a rural school in the Kursk province. In 1825, he resumed his medical studies and graduated in 1828. After a study abroad, notably in Berlin, Inozemtsev was appointed to a professorship in surgery at Moscow University (1835). His diarrhea medicine containing opiates was popular in Eastern Europe until quite recently, and was even used to treat cholera during the nineteenth century.

The first Russian theoretical clinician who formulated a theory of the etiology of disease without the benefit of Western European thought was Diakovsky, a professor at Moscow University from 1831 to 1835. He felt that the psychic state of the patient was often responsible for pathology and, conversely, that different emotional states could be used to treat diseases. Another medical theoretician of purely Russian vintage was Gregory Sokol'sky (1807–1886), who was a graduate of Moscow University (1828) and its professor from 1835 to 1848. He could best be classified as medical empiricist and philosophical agnostic. He saw little value in the understanding of basic medical sciences for the purpose of treating patients, and advocated practical experience as the most useful approach to medicine. He wrote a superb treatise on chest disorders using the empirical approach, which was translated into a number of foreign languages. He was one of the first to describe in detail the symptoms of rheumatic heart. In his day, Sokol'sky appears to have been quite an innovator in the sense that basic medical scientific theories then in vogue were not only of little value to the patient, but more often than not proved to be harmful. Yet basic medical sciences remained an integral part of the medical curriculum in Russia and elsewhere to the great benefit of medicine that accrued only in later years.

The rather primitive state of Russian medicine of the early 1800s was drastically altered by the arrival of Rudolf Virchow on the world's scientific and medical scene (22). A number of Russian physicians were sent to his laboratory for postgraduate training and upon their return to Russia, they proceeded to develop a modern scientific and experimental approach to both the diagnostic and therapeutic branches of internal medicine. One of the more prominent disciples of Virchow was Sergey P. Botkin (1832–1889), who was also a student of Inozemtsev.

Botkin was born in Moscow in a merchant's family and was educated in Moscow University. After graduating in 1855, he took part in the Crimean War under Pirogov, the "father" of Russian surgery. The years of 1856–1860 were spent abroad with Virchow, Traube, and Hoppe. Upon his return to Russia, Botkin was appointed head of the medical clinic of the Military-Medical Academy in St. Petersburg. At the clinic, Botkin established a large research laboratory where many Russian doctors, including Ivan Pavlov, did their doctoral research. His laboratory served as a model for the later establishment of the Institute of Experimental Medicine. Much of the research originating from Botkin's laboratory was published in *Botkin's Clinical Gazette*, a medical periodical edited by Botkin himself. He was instrumental in persuading the St. Petersburg Medical Society to open a free hospital (the Alexander Hospital), which served as an example for the establishment of similar institutions by other local medical societies. Botkin served as court physician for Tsar Alexander II.

Botkin, like Virchow, advocated scientific approach to medicine and the incorporation of basic sciences into the arsenal of the clinician. His research laboratory was principally concerned with application of physiology and chemistry to clinical practice. Botkin's ideas and lectures were published annually in the form of *Clinical Lectures*, the volumes of which were in great demand by both the medical students and practicing physicians. The Soviets have claimed that Botkin had established the infectious nature of hepatitis, and this disease is still referred to as Botkin's disease. The Soviets of the Stalinist era (*Large Soviet Encyclopedia,* 1951) referred to Botkin as a leader of a school of physicians that toppled the "idealistic" and "reactionary" cellular pathology of Virchow. It is not clear how pathology could be idealistic and/or reactionary, and such an allegation is surprising in view of the high respect with which the Russian and world medical community had for Virchow and his ideas (22). It is doubtful that Botkin himself considered Virchow to be a reactionary, since he spent a considerable chunk of time in Virchow's laboratory and had coauthored a number of papers with his teacher (23). Regardless of whether or not Botkin was responsible for the demise of Virchowian concepts in Russia, his memory is esteemed highly by its people and the world's medical profession. One of Botkin's descendants, Dr. Eugene Botkin, was a well-known hematologist (24), and as such was often called to treat Tsar Nicholas II's hemophiliac son. He followed the tsar and his family into exile, and was murdered along with the tsar's family by the Bolsheviks in the Ipatiev House in Ekaterinburg (Sverdlovsk) in 1918.

Whereas Botkin's approach to medicine was experimental with an eye for potentially new discoveries, another disciple of Virchow's, Gregory A. Sakharin, a Moscow professor, developed an approach with the purpose of working strictly for the benefit of the patient. He developed intricate diagnostic procedures that could pinpoint the patient's disorder with surprising degree of accuracy by today's standards. Sakharin was born in Moscow in 1829 and studied medicine in his hometown from 1847 to 1852. He then worked as an assistant in the university's medical clinic, and in 1856 went to Germany to work with Traube, Hoppe, and Virchow. In 1862, he was appointed to a professorship at Moscow University and retired in 1896. Sakharin attended Tsar Alexander III in his last illness, diagnosing Bright's disease as a complication of a bout of influenza. After the tsar's death, he was accused in some circles of having diagnosed the complication too late to save the tsar's life, and his reputation declined somewhat as a result of this (25). The charges were, however, never substantiated, and even if they would have been, it is doubtful that much could have been done for the tsar at that time.

Sakharin's accomplishments included the area of basic medical sciences in addition to his clinical fame. He was able to determine that sodium in horse blood was localized in its plasma fraction, and that little if any sodium was present in the red cells (26). Unfortunately, Sakharin was due to leave Hoppe's laboratory where his work was done, and thus had no time to investigate the distribution of potassium in the blood, even though he had developed a method for the analysis of both sodium and potassium. His work was being done for the purpose of finding a good method for the determination of blood plasma and red cell volumes. He did not realize at that time that his discovery would assume in the future such importance in the diagnosis and management of disease. Today, the determination of serum sodium and potassium is an indispensable tool in the hands of the diagnostician and the surgeon.

G.P. Sakharov (22) describes another medical theoretician, V. P. Krylov (1841–1906), who was a pathologist and professor at Kharkov University. Krylov's opinions were published in a Kharkov University yearbook *Shkol'naya Khronika 1890–1895*, which was apparently financed by Krylov himself; it is totally unknown in the West. He wrote that there were three types of connective tissue in the human organism: the loose or "areolar," tissue which, among other things, served to connect the skin with the underlying tissues; the dense or "fibrillar" tissue, exemplified by the skin, blood vessels, and various membranes; and the meshlike or "fascicular" form. Diseases, according to Krylov, were caused by imbalances among the three types of connective tissue, and individuals were susceptible to certain types of diseases because of imbalances among their connective tissues. One type of an individual, the "fibromatose" one, had extensively developed "fibrillar" and "fascicular" systems, and a poorly developed "areolar" system. The "fibromatose" constitution was characterized by a generous muscle and bone development, a small heart and vital organs, and a very large aorta and other blood vessels possessing very thick walls. Such individuals were likely to develop inflammatory and sclerotic processes, carcinoma, and infections. Krylov's ideas on the existence of the various constitutions and the accompanying predispositions to disease were *per se* nothing new for his time. What was new, however, was the definition of the different constitutions on the basis of anatomical differences, and this was, of course, a strictly Virchowian concept.

Therapeutic methodology of the pre-1850 Russia could trace its origins to both the lay persons and to physicians, and the discovery of such methods was accomplished strictly *via* empirical means. Some remedies appearing in Russia have already been mentioned in connection with Hermann Kaau-Boerhave and Inozemtsev. Another remedy was actually discovered by a field marshal, Count Alexey Petrovich Bestuzhev-Ryumin (1603–1766). In 1725 he prepared what we today call Bestuzhev's tincture. It is nothing more than a solution of ferrous chloride. It apparently made young ladies' cheeks glow and, as the story goes, Empress Catherine II was willing to pay Bestuzhev 3000 rubles for his secret. It is not known if the offer was accepted.

A popular antispasmotic and expectorant that has only recently been supplanted by more effective drugs was derived from leaves and flowering tops of the *Grindelia* species of plants. The leaves and tops of these plants were used either in the powder form, or as an alcohol extract in the treatment of asthma attack, bronchitis, whooping cough, and other respiratory disorders. The plant's medicinal value was recognized by David Hieronymus Grindel (1777–1836), a chemist, pharmacist, physician, and entrepreneur. Grindel was born in Riga, the capital of the Livonian province of Russia (today, capital of Latvia), and obtained his training in the natural sciences at Jena University in Germany. Upon his return to Russia in 1798, he established a chemical company and a pharmacy. In 1804, he went to Dorpat University and remained there as a chemistry professor until 1814. He then went back to his commercial enterprises, but returned to Dorpat in 1820 as a medical student. He graduated in 1822 and worked as a doctor for the rest of his life. His main research work and writings were concerned with botany and medicinal chemistry. He published the *Russian Yearbook of Pharmacy* from 1803 to 1810.

An interesting semiempirical therapeutic measure popular both in Russia and abroad was Karel's milk diet (Milchkur, as it is referred to in the original literature). Philip Karel (1806–1886) was a highly respected physician, the founder of the Russian Red Cross (1867), personal physician to Tsar Nicholas I, and a personal friend of Ernst von Baer and Pirogov. Karel was born in Tallin, the capital of the Estonian province in Russia and today, the capital of Estonia. He graduated from the Dorpat University in 1832. In his travels to the Asiatic holdings of the Russian Empire, Karel noted that the native tribes used to treat typhoid fever and "fever epidemics" with milk. Moreover, he could not have overlooked the fact that the ancients and the not-so-ancients have used milk diets to treat gout and other maladies. In addition, Inozemtsev had reported in 1857 that he was able to cure 1000 severely ill patients by feeding them milk. In 1865, Karel published in *St. Petersburg Medizinische Wochenschrift* his experiences with the milk diet that he had devised to treat patients in whom all conventional methods of treatment had failed. His method involved the initial administration of ½ to 1 cup of skim milk four times a day. Water, but no other foods were permitted. In case of constipation, castor oil or prunes were given. On the second or third week, other foods were added to the diet. Karel used his diet successfully in various edematous conditions, heart diseases, especially as follow-ups after heart attacks, liver disease, dyspepsia, and some nerve disorders. His basic milk diet was later modified by others to improve the caloric intake of the patient: the skim milk intake was increased to 800 ml per day, or, alternately, condensed milk was administered. The various milk diets have been described by Ebstein (27), who devoted a large portion of his article to Karel's method.

The final stage in the evolution of Russian therapeutics from the empirical to the scientific is exemplified by the contributions of Alexander Jarotzky (b. 1866). He was a graduate of the Military-Medical Academy (1899) and worked for three years with Kernig at the Obukhov Hospital in St. Petersburg. He was subsequently on the faculties of the Women's Medical College in St. Petersburg, Dorpat University, and Moscow University. Jarotzky is best known for invention of a diet to treat acute gastric and duodenal ulcers. His diet was especially popular in the United States at the beginning of the twentieth century. In contrast to Karel's diet, Jarotzky's ulcer diet was firmly based on scientifically established facts, namely Pavlov's work on the physiology of the GI tract. Jarotzky pointed out that a successful ulcer diet should elicit as little stomach juice secretion as possible, should pass into the duodenum as fast as possible, should have a neutral or basic pH, should not be particularly appetizing to prevent the secretion of appetite juice, and should have a sufficient caloric value (28). He felt that the milk diet was too crude to handle cases of bleeding ulcers, and instead suggested a diet of raw egg white and fat. Eggs were found to elicit very little appetite juice and gastric juice formation, and were passed very rapidly into the duodenum. Fat was shown to inhibit stomach juice formation, and elicited the secretion of large amounts of bile and pancreatic juice. These, due to their weakly alkaline nature, served to neutralize the acidic gastric juice. The diet was begun with two raw egg whites per day and 20 g. of fat in the form of butter or vegetable oil. However, the two foods had to be administered separately some 4 hours apart, otherwise a copious gastric juice secretion would take place. The ration was gradually increased to 8–12 egg whites and 100 g. of fat per day. The diet lasted for 8–10 days, at which time it was supplemented with carbohydrate in the form of mashed potatoes. Salt and water were to be avoided, and if the patient showed signs of dehydration, water was administered *via* an enema. It is not known if biotin deficiency developed as a result of this diet. Jarotzky later summarized his work and that of others in this field in a very extensive review article published after the Bolshevik coup (29).

Jarotzky also contributed to the fundamental knowledge of the anatomy and physiology of pancreas (30). In what was apparently his thesis work for the doctorate degree in medicine, he maintained mice on protein, carbohydrate, and fat diets, and noted the effects thereof on the pancreas. In the starch and lipid-fed mice, the secretory cells but not their nuclei were smaller than normal. There were fewer zymogen granules in the pancreas of mice maintained on carbohydrate diets, or in those not receiving food at all. The zymogen granules were smaller in the lipid-fed mice. In some cases, the carbohydrate diets were accompanied by a cirrhosis of the pancreas. The Islets of Langerhans appeared to be somewhat smaller in cases where an extensive atrophy of the secretory cells had taken place. Jarotzky concluded that the Islets of Langerhans were a tissue completely independent of the rest of pancreatic cells.

One of the most frequently used procedures in physical examinations is the measurement of blood pressure, and the most popular method of performing this in the United States is the ausculatory technique. It was invented by Nicholas S. Korotkov (b. 1874), an affiliate of the Military-Medical Academy in St. Petersburg. The sounds heard through the stethoscope during such a procedure are termed Korotkov sounds. He published these important findings in a Military-Medical Academy journal in 1905 (31), which was not indexed by the *Index Medicus* of that time. Korotkov also contributed to the management of arterial aneurisms through what we today call the Korotkov test. In this procedure, the artery is compressed above the suspected aneurism, and the peripheral blood pressure is measured. If the latter is within normal range, the patient is considered to have developed collateral circulation to bypass the aneurism. Korotkov's test is published in a book, a copy of which is apparently owned by the National Medical Library in Bethesda, MD (32).

Pathology. Russian medical investigators were responsible for establishing the relationship between high fat diets and circulatory disorders. As early as in 1868, V. C. Krylov working in Rudnev's laboratory at the Military-Medical Academy reported a correlation between fat content of heart and heart "degeneration" (33). In 1909, Ignatovsky reported that rabbits developed damage to the liver and aorta if they were maintained on animal-based diets (meat, eggs, milk) as opposed to vegetable-based diets (34). He did these experiments because E. Mechnikov, a Nobel Prize-winning scientist, had earlier proposed that eating too much protein was toxic. Ignatovski's results seemed to confirm Mechnikov's views. But in 1912, Stuckey excluded protein as the causative agent of tissue damage because feeding rabbits egg white and yolks did not produce the same results. He proposed that fat was the toxic agent, though maintenance of rabbits on neutral fat diet did not result in toxicity. Vaselkin's experiments later excluded lecithin.

It was only a matter of time until cholesterol became implicated, and this was shown by Anichkov and Khalatov (35, 36). Anichkov (1885–1964) was a graduate and later a professor at the Military-Medical Academy at St. Petersburg, whereas Khalatov (b. 1884) was appointed professor of pathology at Moscow University after his tenure at the Military- Medical Academy from 1908 to 1922. In their initial paper (35), Anichkov and Khalatov reported the results of feeding cholesterol to rabbits. In 4 to 8 weeks, they noted fat droplets in the liver parenchyma, spleen, and the aorta. The pathology observed was identical to that seen by Stuckey in his egg yolk experiments. Blood cholesterol values were also very high in these animals. Although guinea pigs reacted much like the rabbits in these experiments, rats were immune to tissue fat infiltration. In a follow-up on this work (36), Anichkov made extensive histological studies on the affected tissues. He noted tissue infiltration by fat in the form of cholesterol ester "liquid crystals," and observed that such "crystals" caused cell destruction in the liver and formation of scar tissue in its place. In the aorta, fat droplets were accumulated in the extracellular areas, and were localized along the elastin fibers. The latter were eventually altered into a fine meshwork, which sealed to entrap phagocytes. The middle and inner layers of the aorta wall thus became indistinguishable. Aortic muscle fibers were also damaged. Anichkov noted that the histologic picture seen in rabbit aortae was similar to that seen in aortae of sclerotic human beings, and he proposed that this disorder was caused in the human beings by excessive amounts of cholesterol in their diets. A. N. Klimov, Anichkov's successor in the St. Petersburg laboratory, showed that atherosclerosis was directly caused by cholesterol associated with blood serum LDL and VLDL lipoproteins (37). Prof. Daniel Steinberg wrote the following: "If the full significance of his (Anichkov' s) findings had been appreciated at the time, we might have saved more than thirty years in the long struggle to settle the cholesterol controversy and Anichkov might have won a Nobel Prize..." (38).

The first instance of the use of radiation to treat cancer is ascribed to two Russian medical researchers, London and Goldberg. London (1869–1939) was born in the Lithuanian city of Suvalki, was graduated from the Warsaw University in 1894, then joined the Institute of Experimental Medicine in St. Petersburg. In 1919, London became the professor of pathological anatomy in St. Petersburg University. His main interests centered around the effects of radiation on living systems. London began his investigations on radiation by testing the effects of RaBr on mice (39). He noted that mice died within four to five days of exposure to the chemical. Before death, their hair had fallen out, they had hemorrhages under their skins, and their spleens were smaller than normal. London and his collaborator Goldman then tested the effects of RaBr on themselves (40). London noted that simple handling of the compound resulted in burns to his hands, whereas Goldberg went so far as to place 75 mg of RaBr against his skin for 3 hours. The result was that there appeared a red spot on the place of exposure, which soon became inflamed, blistered, and necrosed. It had to be handled surgically. London and Goldberg felt that radiation could possibly be used against skin disorders. They obtained two patients with facial *ulcus rodens,* which is now known as basal cell carcinoma. They irradiated the lesions with 75 mg of RaBr. One male patient was treated for 1 ½ hours, the treatment then being repeated weekly for 2 ½ months with irradiation time of ½ hour per session. The total time of irradiation was 7 hours. The second patient, a female, was treated with total irradiation time of 4 1/2 hours. The ulcers of both patients healed completely. The researchers felt that the radiation was not of the x-ray type, and that it had a very promising therapeutic potential.

One of the most distinguished tumor specialists in Russia was A. I. Abrikosov (1875–1955), who was born in Moscow and received his medical education there. From 1914–1918, he was a medical officer at the Moscow Children's Hospital, then joined the faculty of Moscow University. Abrikosov was interested mainly in muscle tumors, and one of these, the myoblastoma, is also called Abrikosov's tumor. However, other types of tumors also received his attention. For instance, he was interested in myeloma and described a case that involved the bone marrow of a number of bones (clavicle, sternum, ribs, and vertebrae) (41). On the basis of these studies, he proposed that the myelomas, which had previously been separated into the primary multiple myeloma and the lymphosarcoma of the bone marrow, were one and the same disease. The development of both, he pointed out, was identical to the classical lymphosarcoma, and he proposed the name of *primary multiple myelosarcoma* for both bone marrow disorders. That term apparently never caught on.

Abrikosov's study of the muscle tumors included the rhabdomyomas (42). In one such case, found in the heart of a three-year-old child, the tumor was composed of primitive muscle cells. In addition to the heart tumor, Abrikosov also found sclerosislike damage in the child's brain. The brain was also characterized by the presence of large cells (80–100 u in diameter). He concluded that the rhabdomyomas and the accompanying disorders of other tissues were caused by a systematic disturbance during the development of the embryo or fetus, and that this disturbance then prevented the differentiation of primitive muscle, brain, and other cells into normal adult cells.

Abrikosov's most important discovery, the myoblastoma tumor, was described in 1926 (43)—that is, after the fall of the empire. However, he had apparently observed his first case in 1915, well before the Bolshevik revolution, and the description of this accomplishment thus seems to be within the scope of this book. The first myoblastoma tumor seen by Abrikosov was removed from the patient's tongue, and he subsequently saw four more cases, two on the tongue, one on the calf of the leg, and one on the upper lip. These tumors were characterized by the presence of large round cells (20–25 u in diameter) with light protoplasm and no fat. The cells were separated from each other by a fine pericellular network. He felt that this tumor was of the blastoma type (myoblast origin), and that it was basically benign. He proposed the name of *Myoblasten Myom* for this tumor.

There does not seem to be any definite record as to where and by whom a malignant tumor was first transplanted. It seems that Novinsky of St. Petersburg was one of the first investigators to have accomplished this in 1875 (44). Two millimeter pieces of canine carcinoma were implanted under the skins of 14 normal dogs; the tumors grew in 2 animals. After about 4 months, the tumor of one had a diameter of 3 ½ cm; it was histologically identical to the implanted tumor. It had also metastasized into the lymph nodes. Novinsky concluded that cancer was a transmissible disease.

A noted kidney pathologist and physiologist was Casimir A. Buinevich, who left Russia after the Bolshevik coup and settled in the then independent Lithuania. He was born in 1872 in Chernigov in Little Russia (Ukraine), and obtained his doctorate in medicine at Moscow University in 1902. He then worked in the university's medical clinics as an assistant (1898–1902), a privat-docent (1903–1911), and professor (1911–1918). From 1918 to 1922, he was on the faculty of a medical institute at Ekaterinoslav, then went to Kaunas, Lithuania, as the head of its university medical clinics. Buinevich made many contributions to medical literature, but his most important accomplishment was probably his theory on the function of the kidney (45). He proposed that there were two types of kidney disease: that involving glomerular damage and characterized by uremia, polyuria, and loss of electrolytes; and that involving tubular damage and characterized by edema but normal nitrogen excretion. He pointed out that both the simple filtration theory of Ludwig (1844) and the active process theories of Bowman, Haidenhein (1879), and Koranyi (1897) were not in agreement with either his view of kidney disease, or the clinical facts of life. As a result of glomerular damage, such as is seen in certain sclerotic processes, Koranyi's theory would predict that increased amounts of water and salt would be excreted, and that nitrogen would be retained. In fact, nephrotic patients often suffered from edema but excreted normal amounts of nitrogen.

As an alternative to Koranyi's theory, Buinevich proposed one of his own that was consistent with clinical observations stated above: water and salts were to be transferred from the bloodstream into the tubules, whereas urea and uric acid did so *via* the glomeruli. Water and electrolytes were to be reabsorbed into the bloodstream in the glomeruli in exchange for urea and uric acid. Thus, in case of glomerular damage, as is the case in glomerular nephritis, one would expect to find an impairment in the excretion of urea and uric acid, which would result in uremia. Water and electrolytes would not be reabsorbed in the glomeruli and this would, in turn, result in polyuria and increased loss of electrolytes. On the other hand, if the tubules were damaged, Buinevich's theory would call for lowered water and electrolyte excretion, but a normal exchange of water and salts for urea and uric acid. Clinically, this would result in a normal nitrogen excretion with a concomitant edematous condition. Though beautifully logical, Buinevich's theory was not entirely correct. It is known that the glomerular filtrate consists of all the nonprotein components of plasma, and that upon the passage of the filtrate through the tubules, water, glucose, electrolytes, but not urea and uric acid are reabsorbed into the bloodstream. Buinevich's error probably lay in an insufficient amount of experimental evidence on renal function that was available at his time, plus his oversimplification of the nature of kidney disease. Be as it may, Buinevich's contribution was an important milestone on the road toward an elucidation of the mechanics of urine formation.

The diabetes syndrome was an extremely puzzling disease in the nineteenth century, since no obvious disorder could be observed upon autopsy of deceased diabetics. The consensus of opinion was that either the brain was somehow involved in the disease, or, more likely, the liver was overproducing glucose through nerve impulses originating in the brain. One very amusing theory explained the synthesis of glucose by condensation of four glycine molecules to form one molecule of sugar and two urea molecules. At no time did anyone suspect that the utilization of sugar may actually be diminished. Vinogradov (1831–1885), a Russian investigator was one of the first to suggest this possibility. He was born in Nizhnii Novgorod and studied medicine at the Moscow University. He defended his doctorate thesis successfully at Warsaw University in 1858 and then worked with Botkin, Virchow, Kuehne, and others. In 1863, he was appointed to the faculty of the Kazan University, where he remained until his death.

The problem of the diabetes syndrome was approached by Vinogradov through an induction of artificial diabetes with curare poisoning in frogs and rabbits (46, 47). He demonstrated that he could not induce the disease in hepatectomized animals, and this seemingly supported the theory of liver involvement. However, Vinogradov then proceeded to measure the amounts of glycogen and glucose in the livers of diabetic animals, and found that the levels of these compounds therein were no different from those of normal animals. He then concluded that the increased amounts of sugar in the blood and urine of diabetic animals was not due to overproduction of sugar by the liver, and raised the possibility that sugar may be underutilized in such animals. The most likely reason for the underutilized sugar in curare-induced diabetes, Vinogradov reasoned, was the inactivity of the muscles, which under the influence of curare were partially paralyzed. Curare diabetes was, of course, a special form of the disease, and Vinogradov was careful not to equate it with diabetes mellitus. Yet his work undoubtedly gave further impetus to the idea that clinical diabetes may be caused by the underutilization of sugar rather than its overproduction.

The work of Minkovski demonstrating the importance of pancreas in carbohydrate metabolism stimulated further research into the mechanism whereby pancreas regulates carbohydrate metabolism, including the role of the various cell types found in the pancreas. The Islets of Langerhans were suspected by some to be in some way involved in the regulation of carbohydrate metabolism, but most researchers ascribed other functions to these structures. Langerhans himself believed that these cells were of nervous origin. It was, however, another Russian investigator who was able to demonstrate conclusively the role of the islets in carbohydrate metabolism. L.W. Sobolev (1876–1919), working in the pathology laboratory of the Military-Medical Academy in St. Petersburg on his doctoral project, tied the pancreatic ducts of rabbits, cat, and dogs, and watched for the development of glycosuria. None occurred even 400 days after the operation. After the animals were killed and their pancreatic tissues were examined microscopically, all such tissues except for the Islets of Langerhans were degenerated or invaded by fatty deposits or scar tissue. Sobolev noted that the previous work of others had demonstrated that the islets were supplied by capillaries, and proposed that the factor responsible for control of carbohydrate metabolism was probably secreted by the Islets of Langerhans into the bloodstream. Sobolev also examined 15 pancreatic tissue samples obtained by autopsies of deceased diabetics. In 13 of these he found a dramatic decline in the number of Islets of Langerhans, and in some these structures were totally absent. Sobolev published his results first in 1900 as a preliminary note, and a full account of his work was published in 1902 (48).

One of the most versatile Russian pathologists was Ivan I. Shirokogorov. He was born in 1869 in Vladimir and studied medicine at Dorpat University, graduating with a degree of Doctor of Medicine in 1907. He remained at Dorpat as an assistant, and in 1919 became a professor of pathology at the newly established Baku University. Shirokogorov's contribution to microscopic anatomy included his discovery of mitochondria in the central nervous system (49). He used both the *in vivo* and *in vitro* techniques to show the existence of mitochondria in cerebral hemispheres, the spinal cord, and the medulla. The latter organ was especially rich in these subcellular structures. Shirokogorov also did an extensive study on the effects of adrenaline on the development of arteriosclerosis in rabbits (50). He found that the animals showed considerable degrees of individual variation with respect to the adrenaline dosages they could tolerate, so that the animals succumbed to the injections anywhere from one day to several months following initiation of the experiments. The most prominent damage seen in the aortae was the degeneration of the muscular layers, the thinning of the elastic fibers, and the formation of fatty plaques with calcium deposits. There was also a hyaline layer in the subendothelial layer of the intima in the damaged aortae. Numerous mitotic spindles were also seen in the cells of the damaged blood vessels. In addition to the aortic pathology, adrenaline also caused liver cirrhosis and the formation of brain cysts. The older animals were apparently much more susceptible to adrenaline-induced disorders.

Oncology was another area of interest to Shirokogorov. He was one of the first investigators to provide clear documentation on the occurrence of primary sarcoma of the pancreas (51). Upon autopsy of a fifty-two-year-old man who had succumbed to pneumonia, Shirokogorov noted that the pancreas had a size three times larger than normal. Microscopic examination revealed that the entire organ was infiltrated by small round cells with the exception of the Islets of Langerhans, where such infiltration was minimal. Metastases were found in the mucous membranes of the stomach and the intestines, in the bone marrow, and in the spinal cord. He noted that his findings were consistent with the idea that the Islets of Langerhans were involved in carbohydrate metabolism, since his patient had in effect lost the entire pancreas with the exception of the Islets. The patient showed no signs of diabetes when he was alive. In addition, Shirokogorov pointed out that pancreatic tumors may elicit neurological disorders because he had found metastatic nodules in the spinal cord of his patient.

There were several noted pathologists who were connected with Imperial Russia either by birth, training, or professional association and whose contributions were made either after the fall of the empire or in some foreign land where they chose to reside. One such individual was Bernard Naunyn (1839–1925), who was born and educated in Berlin, and who occupied a professorship at Dorpat University from 1869 to 1871. He is known for the introduction of the concept of acidosis, especially as a complication of diabetes mellitus. Another scientist in this category was Oscar Minkovski (1858–1931), a coworker of Naunyn. He was born near Kovno, Russia, but obtained his medical education in Konigsberg and spent his entire scientific life abroad, principally in Germany. He is known for his induction of diabetes *via* pancreatectomy, the demonstration of the pituitary gland involvement in the etiology of acromegaly, and the description of familial jaundice characterized by accelerated breakdown of hemoglobin (Chauffard-Minkovski syndrome). Alexander A. Bogomolets (1881–1946) was an outstanding immunologist. He is best known for his introduction of the antireticular cytotoxic serum in the early 1940s. Though much of this was done during the Soviet era of Russian history, Bogomolets was already a professor at Odessa University (1911–1913) and Saratov University (1913–1925), well before the fall of the empire. Another physician trained under the pre-Soviet system who later became a distinguished hematologist was Mikhail Arinkin (b. 1876), a graduate of the Military-Medical Academy. In 1927, he developed a method for extracting the bone marrow from the sternum for hematological analysis, which is used today for the diagnosis of a variety of anemias as well as multiple myeloma.

Neurology and Neuroanatomy. Neurology was an especially popular branch of medicine in Russia. It was dominated by essentially two schools of approach, one based in St. Petersburg and the other in

Moscow. The former was under the influence of Pavlov, Bekhterev, and, in part, of the reflexology teaching of Sechenov. Its practitioners, thus, took a more theoretical approach to neurology. The Moscow school was of the more practical and pragmatic nature, and was under the influence of Alexey Kozhevnikov, the famous clinician. The Moscow neurologists were mainly interested in diagnosis and treatment of nervous disorders and in finding anatomical bases for such disorders. A number of neurological diseases was discovered by this group of physicians. They were organized into the *Moscow Society of Neuropathologists and Psychiatrists,* a very active organization with numerous meetings during the year and its own publication. In addition, its activities were regularly reported by the German periodical *Neurologisches Centralblatt.* American medical libraries thus have an excellent account of the Russian neurologists and neuroanatomists of the tsarist era. In addition, an excellent collection of neurologist biographies is available, and it includes a number of Russian practitioners (52). Most Russian neurologists combined basic research with clinical practice, thus making significant contributions to both areas of endeavor. For this reason, it was necessary in this section to combine the clinical science of neurology with the basic medical science of neuroanatomy and, in part, with neurophysiology.

The most accomplished clinician-neuroanatomist and neuro-physiologist in Russia was probably Vladimir Mikhailovich Bekhterev (53). He was born in 1857 in the Vyatka province of Russia and received his medical degree in 1878 from the Military Medical Academy. After graduation, he worked with I. P. Merzheyevsky in the psychiatric and neurologic clinics of the academy. He was interested mainly in the function of the brain and spinal cord. In 1884, Bekhterev went abroad to work with Wundt and Flechsig in Leipzig, and attended lectures given by Charcot. The ideas of the latter were apparently most instrumental in the development of Bekhterev's future activities. In

1885, he went to Kazan University as professor of psychiatry. At first, he experimented in N. O. Kovalevsky's physiological laboratory, then organized his own psychophysiological lab. In Kazan, Bekhterev founded the *Society of Neuropathologists and Psychiatrists,* and established a periodical *Nevrologicheskii Vestnik (Neurological Herald).* For some time, he served as the president of the society and editor of the journal. In 1893, Bekhterev was appointed to occupy the department chair of psychiatric and neurologic diseases at the Military Medical Academy, the job formerly held by Merzheyevsky. He remained at the Academy for twenty years, during which time he completely reorganized the teaching department and clinic, founded the *Russian Society of Normal and Pathologic Psychology* and the *Psychiatric Society,* edited the periodicals *Review of Psychiatry, Neurology, and Experimental Psychology,* the *Journal of Psychology, Pedology, and Criminal Anthropology,* and established the first in Russia neurosurgical unit at the academy.

Through private donations and his own unending energy, Bekhterev was able to acquire land and build thereon a large laboratory complex, which he called *Psychoneurological Institute.* This included a neurosurgical unit, an alcoholism treatment center, a child psychology institute, and psychological and neurological clinics. The institute soon evolved into a medical school and, during World War I, into a full-fledged university, the first and only private university to exist in Russia. After the Bolshevik coup, the pedagogical complex was merged with the St. Petersburg University, and the research laboratories became the *Psychoneurological Academy and Brain Research Institute.* Bekhterev remained its director until his death in 1927. He was a remarkably versatile scientist who contributed profusely to the areas of neuroanatomy, physiology, psychology, psychiatry, neurosurgery, and neurology. Frequently the research under his nominal direction was carried out by neophyte students who, due to the unavailability of their chief, committed blunders in the planning, execution, and interpretation of experiments. Through them, Bekhterev not infrequently came under fire from his scientific colleagues, most notably Pavlov, who was Bekhterev's competitor in the field of conditioned reflexes.

In the area of clinical neurology, Bekhterev was responsible for the discovery of about a dozen reflexes, some of which are still useful in neurologic examinations. The most important one, the Bekhterev or Mendel-Bekhterev reflex, is in normal individuals the flexing of toes upon tapping of the back of the foot. In individuals with corticospinal tract disease, the sole of the foot is instead flexed upon stimulation of the back of the foot. A similar response of the fingers to the tap of back of the hand is called the Mendel-Bekhterev reflex of the hand. An abdominal reflex discovered by Bekhterev is called the hypogastric reflex, where in normal individuals the lower abdominal muscles contract upon stimulation of the inner surface of the thigh. Absence of the reflex is indicative of pathology. Bekhterev's nasal reflex involves the stimulation of nasal mucosa and the accompanying contraction of facial muscles. Another facial reflex discovered by him is the orbicular one, where the eye is closed in response to tapping of the forehead and the nasal bone. This reflex is absent whenever the facial nerve is damaged. The scapular (interscapular) reflex represents the contraction of scapular muscles upon the stimulation of the skin over the scapulas or between them. Bekhterev also described a reflex obtained in patients with general paralysis of tabes dorsales, where the pupils of the eyes dilate in response to light. Finally, Bekhterev described a sacrolumbar reflex, which is obtained upon tapping of the sacrolumbar segment of the spinal column in patients with spastic paralysis (53).

Among other diagnostic aids discovered by Bekhterev were a number of signs. He found that the loss of a painful sensation by the gastrochnemius muscle was an early indication of tabes dorsalis, a condition due to sclerosis of the spinal cord. He also discovered that the intense pain in the sole of the foot in polyneuritis patients served to distinguish this disease from poliomyelitis. The same sign was also used as an early indicator of pregnancy. Patients with sciatica were unable to lift the affected limb to the same height as the unaffected lower extremity. In addition, such patients were found to experience a sharp pain under the knee joint upon stretching of the affected limb. Intercranial diseases originating at the base of the brain were characterized by a painful reaction to a tapping of the cheekbone. Sensitivity to the tapping of the skull and the spinal column were also found to be indicative of some brain disorders. Bekhterev was apparently one of the first clinicians to note the shuffling gait of patients with Parkinsonism. Techniques such as those described above seem to be primitive indeed compared to what is available today. Yet at the end of the nineteenth and beginning of the twentieth centuries, these were the only tools the physician possessed when dealing with neurological disorders. And Bekhterev appears to have been a major contributor to the arsenal of signs and reflexes available to pioneer neurologists.

Bekhterev was the discoverer or codiscoverer of a number of neurologic and psychiatric disorders. The best known is the so-called Bekhterev's Disease, which is characterized by a stiffness of the spinal column, a sharp pain in that area, and an atrophy of muscles. Bekhterev first described this disorder in the Russian-language journal *Vrach* in 1892, and in the German journal *Neurologisches Zentralblatt* in the following year. In 1897, he published an extensive treatise on this disease in *Deutsche Zeitschrift fuer Nervenheilkunde,* a journal edited by Adolf V. Strumpell (54). The latter then published a paper immediately following Bekhterev's one, claiming that he had observed Bekhterev's disease in 1884, and had included a short paragraph thereon in a neurology textbook that he had written. Strumpell's disease involved not only the spinal column, but also the hip joints. In 1899, Bekhterev published two papers (55, 56), where he presented three case reports clearly showing that the diseases described by Strumpell (and later in 1898 by Marie) were different, though similar to that described by himself. Thus, whereas the Strumpell-Marie disease (ankylosing spondylitis) often started with the extremities and progressed to the spinal column, Bekhterev's disease always began with the stiffness of the neck and the upper spinal column, and did not involve the extremities. Bekhterev believed that the Strumpell-Marie disease was a generalized ossification and was of arthritic origin, whereas his disease was of neurologic origin. Upon the autopsy of one of his patients, Bekhterev did find a degeneration of cartilage and nerve roots of the spinal cord and a thickening of the pia mater.

A form of epilepsy similar to myoclonic epilepsy was described and named *Epilepsia choreica* by Bekhterev. It was characterized by cramplike twitching movements of the extremities, the face, and often the entire body (57). Bekhterev also discovered a disorder that usually followed a stroke in childhood. Instead of a paralysis, the patient developed a muscular tension in one half of the body. The extremities were especially affected, whereas the skin sensitivity and mental functions remained unimpaired. The affected muscles often developed a pronounced hypertrophy. He called this disease *Hemitonia apoplectica* (58). A complication of syphilis was also discovered by Bekhterev, where the patient's symptoms were similar to those seen in multiple sclerosis. He called this disorder *Disseminated syphilitic sclerosis.* The disease was characterized by the destruction of spinal cord and brain tissue, slurring of speech, and weakness (paresis) of limbs and facial muscles.

A large number of psychiatric disorders was studied by Bekhterev, some being of more importance than others. An interesting disease described by him was what he called *Paranoia suggestio delira.* The afflicted patients imagined themselves to be under the hypnotic spell of others, and were thus compelled to do all sorts of things against their wills. The disease did not have a tendency to be transformed into megalomania. Bekhterev also studied several forms of hallucinations, such as those associated with hypersensitivity to various odors, sounds, and the imagination of the presence of reptiles in the stomach (Reptiliophrenia).

Bekhterev's anatomical investigations (53) involved the entire central nervous system. In the area of reticular formation of the medulla he described several nuclei, which he named the *Nucleus reticularis, Nucleus medianus, Central nucleus, and Nucleus innominatus.* The best-known structure discovered by him is the Bekhterev nucleus, or the superior vestibular nucleus, which is situated at the edge of the fourth ventricle. It represents the terminus of the vestibular nerve, and is apparently involved in the central facilitation function of the central nervous system. He was also interested in the various nerve tracts of the spinal cord, the brain peduncle, the medulla, and the connections between the various nuclei of the medulla. The corticospinal tract was shown by him to degenerate in the descending direction, indicating that, contrary to Charcot's findings, the tract could be considered to be of the afferent type. In the cerebral hemispheres Bekhterev noted a strium, which carries his name, and in the spinal cord, at the interface of the white and gray matters, he noted a group of cells that also carry his name in some literature. Bekhterev's anatomical experiences were summarized in a two-volume treatise *Leitungsbahnen,* which first appeared in the Russian language in 1896, then was translated into German (1899) and French (1900).

Neurophysiology was a field which was probably closest to Bekhterev's heart. One of such areas studied in his laboratory was equilibrium. He cut the acoustic nerve of experimental animals and noticed that they moved in a circular fashion in the direction of the side with the cut nerve. He postulated that the abnormal movements observed were due to the cessation of impulses on the side of the cut nerve and a facilitation of impulses in the intact nerve. Aberrations in equilibrium also occurred when he cut the base of the cerebellum. He proposed the existence in the brain of a center that controlled equilibrium in addition to the semicircular canals.

The functions of several ganglia were elucidated in Bekhterev's laboratory. The *globus pallidus* (part of the lenticular nucleus) was adjudged to be a motor ganglion, because its stimulation elicited cramplike movements in the animal. *Substantia nigra, a* group of cells near the hypothalamus, was shown to contain the chewing and swallowing centers. And the *corpus mammillare,* a portion of the hypothalamus, was shown to be concerned with respiration.

The functions of the various portions of the spinal cord were studied by Bekhterev's group *via* its partial resection. When Borovikov, a student, cut the dorsal portion of the spinal cord in the dorsal region of a dog, he noted paresis and ataxia of the legs, whereas the skin sensitivity remained intact. It was concluded that the motor tracts carrying impulses to the posterior portion of the body were situated in the dorsal portion of the chord. When the dorsal portion of the chord was cut in the neck region, the animal developed not only ataxia, but also equilibrium problems. Another student named Holzinger discovered that a portion of the corticospinal tract (situated in the lateral portion of the spinal cord) carried the skin sensitivity signals.

Bekhterev's study of the cerebral cortex resulted in 1900 in the discovery of the taste center near the opercular region of the lateral central fissure (59). He also showed that the skin and muscle sensitivity center was not located exclusively in the sigmoid gyrus but was, instead, spread out over a much larger area. The sigmoid gyrus was also found to be involved in motor function of the organism and with associative reflexes of the animal. The latter finding was disputed by Pavlov. Bekhterev felt that the canine sigmoid gyrus corresponded to a portion of the human brain's central sulcus, and was thus an excellent model system for its further study. His group also showed that the vision center of the dog's brain was located in the occipital lobe, a location similar to that of the human brain (60). All the brain surfaces, except perhaps those of the occipital and frontal lobes, had effects on the blood pressure, rate of heartbeat, and vasodilation when stimulated with a very weak electric current. Involuntary muscle contraction could be elicited in the small intestine, stomach, and bladder upon the stimulation of the sigmoid gyrus. In addition, centers were found that controlled the sex organs, contraction of the spleen, prostate, and uterus, and the production of milk. These findings were thus useful in explaining the involvement of the inner organs during epileptic seizures of afflicted individuals.

Bekhterev's work on the brain centers controlling motor functions involved the use of trained dogs, and prompted him to engage in the study of conditioned reflexes. However, instead of using the latter term, which was coined by Pavlov, Bekhterev named his phenomena *associative reflexology,* a term quite descriptive though a bit awkward. In his study of conditioned reflexes, he used voluntary muscle movements as his assay system in dogs. Perhaps following Pavlov's example, Bekhterev advocated the use of his reflexology in the treatment of personality and psychiatric disorders, as well as its use to solve other practical problems. In the latter category, he designed a method that would distinguish an authentic deafness or blindness from faked conditions by either making a noise or flash a light before a person, and following this up with a weak electric shock immediately afterward. The subject would, of course, cringe or otherwise react to the electric shock. If during this cycle of events the light would be flashed or noise sounded without being followed with the electric shock, the truly blind or deaf individual would not cringe in anticipation, whereas the pretender would do so. It is not clear if this technique was ever applied; perhaps it was designed to ferret out the reluctant army draftees, though it is not known whether or not the Russian army utilized Bekhterev's methods. He also advocated using his principles of reflexology to train retarded children, the blind, and the deaf-blind individuals.

Among the pioneer neurohistologists of Russia, the names of Vladimir Alekseyevich Bets (Betz) and Alexander Dogiel are perhaps best known. Bets (1834–1894) obtained his medical degree from Kiev University in 1860. In 1868 he was appointed chair of the anatomy department at his alma mater, where he remained until his death. Bets made a thorough microscopic examination of the brain using over 1300 sections in this study. In the vicinity of the central fissure, he discovered a group of cells that he named *Giant pyramidal cells* (61). These cells were located in the fourth layer of the cortex and were some 0.05 to 0.06 mm wide and 0.04 to 0.12 mm long. They had two major and 7 to 15 minor protrusions. The pyramidal cells were more numerous in the right hemisphere than the left. Persons of all ages, as well as apes, were found to have these cells. In addition, the same type of cells was found in the corresponding location of dog brains. Bets called the area of these cells *Lobulus paracentralis.* In his paper, Bets noted that physiologically, the pyramidal cells apparently had both afferent and efferent functions concerned with muscular activity, since electrical stimulation of this area by Fritsch and Hitzig had previously elicited muscular contraction in dogs. In later years, a syndrome was described involving a sclerosis of the Bets cells and resulting paralysis of the extremities. Bets also made a very detailed study of the five layers of the brain and noted the peculiarities of each layer (62).

Alexander Stanislavovich Dogiel (1852–1922) was born in Panevezys, a town in the Kovno province of the empire (in modem

Lithuania today). He studied medicine in Kazan and graduated in 1879. He defended his doctoral dissertation in 1883, and in 1888 was appointed to the chair of histology in the newly-created Tomsk University. In 1895, he moved to St. Petersburg University, where he also lectured in the Women's Medical College. He remained in St. Petersburg until his death. Dogiel is credited with the founding of the *Russian Archive of Histology, Embryology, and Anatomy* in 1915.

Dogiel is considered by many to be the most accomplished histologist in Russia. He specialized almost entirely in the field of neurohistology, and his papers in this area are indeed masterpieces of thoroughness and attention to the minutest detail. His drawings were superb. Dogiel studied the afferent nerve endings in a number of mammalian organs, especially in the heart of the cat and dog, and popularized the term "end-plate." He developed his own staining techniques for his studies (63). Dogiel's main contribution to medical sciences is believed by some to be his classification of cells in sympathetic ganglia (64). Using methylene blue, a relatively simple dye by modem standards, he was able to distinguish three types of cells in intestinal and gall bladder ganglia: (a) The motor cells that had a stellar shape. These cells had certain morphological similarities with connective tissue cells and were usually situated in the periphery of ganglia. The motor cells were most plentiful in the myenteric plexus (Auerbach's plexus), and were less so in the submucous plexus (Meissner's plexus). Each motor cell had 4–20 dendrites, which made contact with dendrites of other cells. In addition to the numerous dendrites, Dogiel found each cell to give off an extremely long process, which is called *Nervenfortsatz* and which we today call the axon. It was found to make contact with muscle fibers of the intestine. (b) Cells of the second type were somewhat larger, rectangular in shape, and with a small round nucleus. These cells were more numerous in the submucous plexus than in the myenteric plexus.

Dendrites of these cells usually protruded beyond the ganglion, entering the interganglionic nerve fibers. Frequently, the dendrites were observed to reach the mucosa of the intestine and to enter the crypts of Lieberkuhn. The axons were seen to enter other ganglia, where their further whereabouts could not be established. The cells of the second type were believed by Dogiel to have a sensory function and a possible connection with intestinal secretory functions. (c) The cells of the third type were similar in size and shape to those of the second type. They were present in the larger ganglia only, and were distributed between both the periphery and center of the ganglion. Their dendrites originated at the poles of the cells. They were very thick at their central ends, but proceeded to taper off into thin fibers at the peripheral end. The dendrites were much larger than those of the motor cells, and formed an extensive intraganglionic network without leaving the constraints of the ganglion. The axon was very thin, and Dogiel was not very successful in tracing its fate.

The types of nerves entering ganglia of the plexus were also studied by Dogiel. He distinguished two types of nerve fibers: (a) Very thin ones whose endings inside the ganglion formed an extensive network without making contact with intracellular dendrites; and (b) Thick nerve fibers, which, upon entering the ganglion, gave off numerous nodular branches that then surrounded the ganglionic cells and made an effective pericellular network. The fibers of the second type cells were frequently seen to pass through several ganglia. Dogiel believed that the nerve fibers of the first type were typical sympathetic fibers. The nerves of the second type were believed to be a general class of sensory cerebrospinal fibers. Dogiel observed such fibers in the heart and other tissues.

An investigator who successfully bridged the fields of neuro-anatomy and neurophysiology was Liverii Osipovich Darkshevich (1858–1925). He received his medical degree from the Moscow

University in 1882 then went abroad to study in Strassburg and Leipzig among other places. His association with Flechsig was probably most influential in shaping his future research activities. While abroad, Darkshevich also collaborated with Freud. In 1888, Darkshevich passed his doctorate examination and was appointed to the faculty of Moscow University. In 1892 he *moved* to Kazan University, where he organized the first in Russia alcoholism clinic. In 1917, Darkshevich was appointed to the faculty of the Moscow Women's Medical College, where he remained until his death.

Darkshevich's research specialty was the posterior commisure of the brain, a network of nerve fibers associated with the thalamus and several structures contiguous to it. He discovered that the nerve fibers of the posterior commissure consisted of two types and called these the *dorsal* and the *ventral* tracts (65). The ventral tract was the more prominent one and thus easy to follow. It ran along the central canal from the oculomotor nucleus and the posterior longitudinal tract to the thalamus. The dorsal fibers originated in the corpora quadrigemina and eventually ended up in the cerebral cortex. Darkshevich also discovered a pair of nerve fiber tracts that connect the optic tracts with the habenular nucleus. These tracts are commonly called *Darkshevich's fibers.* The nucleus of the posterior commissure was also discovered by Darkshevich and this structure is referred to in some literature as *Darkshevich's nucleus.* This structure is connected with the posterior commissure and the posterior longitudinal tract (fasciculus).

Anatomical investigations of Darkshevich were followed by attempts to elucidate the physiological function of the ventral fibers of the posterior commissure (66). He accomplished this by sectioning the posterior commissure in various places in rabbits and observing the effects thereof on the pupils of the eyes. The thirteen rabbits in which the operation was successful fell into two groups: those whose maximum pupil constriction upon exposure of the eyes to sun- or lamplight was the same before and after the operation, and those whose pupils were constricted to a smaller extent after the operation under the same conditions. Postmortem microscopic examination of the posterior commissure area revealed that the structure had not been really cut in the first group of rabbits and that other structures were damaged instead. On the other hand, the rabbits of the second group all had their posterior commissures severed. The first group thus served as an effective control. It was concluded that the posterior commissure was involved in the transmission of stimuli to the iris of the eye.

Darkshevich also wrote a number of clinically oriented papers. As an example, one may mention his study of muscle atrophy in patients with hemiplegia of cerebral origin (67). He found that the atrophy started some 1–8 weeks after the onset of paralysis at the center of the limb, then progressing to the periphery. The extent of paralysis and atrophy were independent of each other, and the irritability of the atrophying muscle to electric stimulation was unimpaired. Microscopic examination of autopsy material showed no abnormality in spinal cord cells, and a simple rather than degenerative form of atrophy in the affected muscles. Darkshevich pointed out that cerebral hemiplegia with its early atrophy must be distinguished from hemiplegia of spinal cord origin with its late onset atrophy. He speculated that cerebral hemiplegia may be due to the severance of connections between the brain centers and the appropriate nerve tracts in the spinal cord.

Among the neurophysiologists of some accomplishment one can mention Vladimir Muratov and Joseph Kupressov. Vladimir Alexandrovich Muratov (1865–1916) was a medical graduate of the Moscow University (1889) and remained there as Kozhevnikov's student for further study. By 1893, we find him in the position of chief medical officer of the Moscow Mental Hospital, and in 1908 in the same position at a Saratov hospital. From 1911 to 1916, he was a professor of neurology at Moscow University. Muratov is credited with establishing the field of child neurology in Russia and published a number of papers on the occurrence of neurological diseases in children, thus calling attention of his colleagues to this relatively neglected field (68). Muratov did much of the work to elucidate the function of corpus callosum (69). In many texts credit for this is given to Mingazzini, who in 1922 wrote an excellent treatise on this brain structure. Muratov began his studies on corpus callosum *(Balken,* as it was called in German literature) by distinguishing two types of fibers in that structure. He called them the *traverse* and the *longitudinal* (fasciculus subcallosum) tracts. He was able to trace these fibers in the brain and noted that they began and ended in the hemispheres. He proposed that the fasciculus subcallosus was identical to the fibers discovered by Onufrovich in the brains of children born in the absence of corpus callosum. Physiologic role of corpus callosum was studied by Muratov by sectioning it in dogs, and examining it for signs of deterioration. Usually, cutting of the corpus callosum resulted in the degeneration of fasciculus subcallosum. The latter was also found to be degenerated when the motor centers of the brain cortex were damaged. Muratov concluded that the corpus callosum fibers served to connect the two cerebral hemispheres. He called the corpus callosum a true commissure.

Brain damage is in most cases associated with a paralysis of motor functions of the organism. There are occasions, however, where such damage is accompanied by involuntary movements of the muscles. Muratov was a pioneer in describing such disorders (70). He felt that the division of involuntary movements into the irregular choreatic and the more regular athetotic motions was not real, that both types of movements were seen in the same patient and thus were interchangeable. He was, in this sense, in agreement with Charcot and in disagreement with von Monakow. The reason for this condition, according to Bonhoefer, was a degeneration of the superior peduncles of the cerebellum linking it with the corpora quadrigemina and the thalamus. Muratov, however, proposed on the basis of a number of experiments that the involuntary movements were due to a damaged anterior portion of the cerebellum. Yet, he did not completely discount Bonhoefer's proposal, stating that certain types of involuntary movements may very well be due to damaged superior peduncles of the cerebellum.

Pioneering work on the nerve control of bladder function was done by Joseph Kupressov, who submitted his findings to the Military-Medical Academy as a dissertation for his doctoral degree in 1879 (71). The work itself was apparently done at Warsaw University in Navrotsky's laboratory. His experiments involved the introduction of water into the bladders of rabbits *via* the cannulation of ureters, and the subsequent observation of conditions required to empty the bladders *via* the urethra. Water was introduced into the bladder at a predetermined temperature at various pressures obtained by varying the hydrostatic head of the water column. Under normal conditions, Kupressov found that the males emptied their bladders at a water column height of 42.9 cm, whereas the females did so at 45.6 cm. When the urethrae were cut, the pressure required to affect bladder emptying was dropped to 25.2 and 34.9 cm in males and females, respectively. The spinal cord was then severed at various levels, and the amount of pressure necessary to empty the bladders was measured. It was found that when the chord was cut at the level of the first to fourth lumbar vertebrae (counting from the posterior end of the column), there was absolutely no effect. However, cutting the chord at the level of the sixth to seventh lumbar vertebrae resulted in a decrease of pressure necessary to empty the bladder to an average of 14 cm in both males and females. The results of cutting the chord at the level of fifth vertebra were not reproducible. The amount of water pressure to elicit bladder opening in dead animals was also at 14 cm. Kupressov thus came to a conclusion that the tone of the bladder is controlled by two factors: the urethral sphincter, which played a much larger role in males than in females, and the musculature of the bladder wall. Both mechanisms were apparently controlled by a spinal cord center, which, Kupressov proposed, was located somewhere above the sixth lumbar vertebra.

A number of Russian neurologists were responsible for the discovery of various neurological disorders and tools for the diagnosis of nervous system aberrations. Among the latter group of doctors one can mention Kernig and Rossolimo, who discovered respectively the Kernig sign and the Rossolimo sign or reflex. Vladimir Mikhailovich Kernig (1840–1917) was a graduate of the Dorpat University Medical School and was later the chief medical officer at the Obukhov Hospital in Moscow. As early as in 1882, he reported in a Russian journal that diseases affecting the meninges (e.g., meningitis, hemorrhage, edema) are accompanied by an inability of the patient to stretch his or her lower extremity at the knee when the thigh is 90 degrees with respect to the trunk (sitting position). If a severe case involving the meninges was present, the thigh and the lower extremity could not make an angle of more than 90 degrees, and in less severe cases, an angle between 90 and 100 degrees could be obtained. This maneuver was apparently painless to the patient, and is now commonly referred to as the *Kernig sign.* It was positive in 70–100% of patients suffering from spinal meningitis, though in children the rate was only 50–70%. The sign was obtained in all adult cases with secondary, tubercular, and epidemic meningitis, and remained so several months after convalescence. Kernig reported a 50% mortality rate in cases of epidemic meningitis, and 100% rates in secondary and tubercular meningitis. Where paralysis had set in because of meningeal disease, the Kemig sign was always negative. It was also negative in all nonmeningeal diseases except in leptomeningitis. Kernig recommended that three criteria be used in the diagnosis of meningitis: the presence of stiffness in the neck, a positive Kernig sign, and a positive examination of lumbar puncture fluid (72).

Kernig also contributed to the understanding and management of heart diseases. Soviet sources claim that his observations on angina pectoris were instrumental to the development of rationale for the management of myocardial infarcts. Kernig noted that many angina attacks were followed by sudden deaths. In a talk before the St. Petersburg Medical Society and later in a German publication (73), he pointed out that such angina attacks were frequently complicated by pericarditis. He divided the post-attack cases into three categories: patients with a slight fever (up to 38°C), a high pulse rate, and an enlarged heart; patients with no fever but a very rapid enlargement of sections of the heart; and patients with acute pericarditis. The complications described by Kernig did not constitute new findings; enlargement of the heart had previously been observed upon autopsy of angina patients, but no connection was made between such observations and the clinical management of the disease prior to Kernig's recommendations. His advocacy of a close follow-up of patients following angina attacks and absolute bed rest wherever necessary contributed significantly to more successful disease outcomes.

Grigorii Ivanovich Rossolimo (1860–1928) graduated from Moscow University along with Anton Chekhov (the author) in 1884 and remained at the university as a member of the faculty until 1911, when he resigned in protest of some educational policy developed by L. Kasso, the minister of education. He returned to the university in 1917, and remained there until his death. Rossolimo is best known for his discovery of the so-called Rossolimo reflex (74). He found that in patients with pyramidal tract damage, the Babinski reflex was positive in 50 percent of the cases. However, seventeen out of twenty patients that he studied had a positive Rossolimo reflex: when the large toe was bent dorsally, in normal patients it returned to the original position when released. In afflicted patients, a flexion of the toe took place. He speculated that this was due to a contraction of the plantar muscle. He is also credited with a formalization of a disorder characterized by the inability to swallow food, both solid and liquid. He described eight such cases where there was no physical cause for this a disorder. Instead, he proposed that psychological factors were involved, especially fears that could be traced to childhood. He called this condition *Amiotactic disphagia* (75).

A number of neurologic diseases were discovered by Russian clinicians. The elder statesman of this group was Aleksey Iakovlevich Kozhevnikov (1836–1902), a native of Riazan and graduate of Moscow University. He worked with Charcot in 1866, and upon his return to Russia was appointed to the faculty of Moscow University, where he established a neurological clinic and provided training for a number of Russian neurologists. Kozhevnikov was interested in a number of neurological issues, such as the nature of fibers linking cerebral and cerebellar cortices, lathyrism, spastic diplegia, and ophthalmoplegia. His most important contribution was the discovery of a form of epilepsy, which he called *Epilepsia partialis continua* (76), and which today is known as either *continuous epilepsy* or *Kozhevnikov's epilepsy.* He described four patients suffering from this disorder. All had cortical (Jacksonian) epilepsylike seizures. The attacks began with a specific group of muscles, then spreading to other parts of the body. There were no remissions between seizures, and there was, instead, a constant twitching of the affected muscles. The attacks often involved only one half of the body. The affected muscles became weak and atrophied without losing their sensitivity. Kozhevnikov believed that this disease was due to encephalitis that was followed up by a sclerosis or scarring of the inflamed brain membranes. He felt that the only effective treatment could be of surgical nature.

Another member of the Moscow group of neurologists was Sergei Sergeevich Korsakov (1853–1900), the discoverer of the so-called *Korsakov syndrome.* He was born in the Vladimir province of Russia, where his father was a factory manager. He studied medicine at Moscow University and in 1888 was appointed to head its psychiatric clinic. With his teacher Aleksey Kozhevnikov, Korsakov founded the Moscow Society of Neuropathologists and Psychiatrists in 1890. In addition to his teaching and research activities, Korsakov was an active crusader for the betterment of conditions at the insane asylums and introduced the concept of family therapy for mental disease patients.

Korsakov's observations on the Korsakov syndrome were first published in 1887 in the Russian periodical *Vestnik psikhiatrii,* and later in a German periodical (77). The syndrome was often accompanied by polyneuritis and was found frequently in alcoholics, though it was not restricted to them. He thus diagnosed this disorder in a woman who had given birth to a stillborn child, a woman with a pregnancy complicated by puerperal sepsis, a woman who had recovered from typhoid fever, a man who had a form of leukemia, a woman with a tumor, and an apparently healthy woman who claimed not to have touched alcohol. The symptoms of the syndrome were not uniform, but most often patients had hallucinations, confusion, amnesia, and disorientation. The psychic disorders were accompanied by pain, paralysis of the extremities, degeneration of the muscles, cardiac irregularity, and loss of patellar reflex. The disorder was found to last for up to a year, and was usually followed by a slow period of convalescence. The mental and physical functions of the patient were regained within several years' time. Korsakov felt that the syndrome he had described was caused by toxins acting on the brain. He felt that such toxins could be either of exogenous type, such as alcohol, or of the endogenous type, such as toxins of bacterial or tumor origin. On that basis, he named this disease *Cerebropathia psychica toxaemica.*

Among additional and perhaps lesser known neurologic clinicians of the Moscow school were Lazar Solomonovich Minor (1855–1942) and Vladimir Karlovich Rot (Roth) (1848–1916). Minor was a graduate of Moscow University (1879), and in the 1890s he was there on the faculty as a privat-dozent, working in the university's neurologic clinic. In 1910, he moved to the Moscow Women's Medical College and remained there until 1932. Minor's main accomplishment was the discovery of *hematomyelia.* He had first observed the disease in 1886 in a small boy, and at first thought that he was faced with the Brown-Sequard syndrome. Additional observations on other patients convinced him that he was dealing with a theretofore new disorder, which he named *Central hematomyelia* (78). He originally described five cases in persons who had suffered trauma to their spinal cord: a man upon whom fell a glass-filled sack from a second floor of a building; two who had fallen down some stairs; one who slipped on ice and had hurt his back; and one who had suffered a barn accident. Common symptoms were paralysis on one side, paralysis of the bowels and bladder. There was no evidence of any fractures or dislocations, atrophy, loss of feeling, or loss of pain sensation in the muscles. One of Minor's patients expired, and autopsy revealed that there was hemorrhage in the spinal cord. In a later paper (79), Minor described additional cases of the disease and classified his patients into three groups: a. typical Brown-Sequard symptoms with a one-sided loss of feeling and paralysis with or without muscular atrophy; b. loss of heat and pain sensitivity in the trunk and legs, with paralysis of legs, but without atrophy of the leg muscles. The arms may both be paralyzed and their muscles atrophied. In the latter case, the disease involved the gray matter of the spinal cord on both sides; and c. paralysis and muscle atrophy on both sides. Hematomyelia is also known as Minor's disease in some literature.

Minor made a number of other discoveries and innovations. He discovered a familial tremor that begins in adolescence and increases in severity toward old age. It was reported in a Russian journal called *Russkaia Klinika* in 1929, and is known as *Minor's tremor* or *essential tremor.* Minor was also an enthusiastic advocate of bedside photography, and published a number of papers on that subject. A new form of thermoesthesiometer was constructed by Minor (80); it became commercially available soon thereafter. It was a two-chamber device, one containing cold water, the other hot water. A suction bulb attached to both chambers was used to draw water simultaneously from both chambers, mix the two samples, and return the mixture to both chambers at the same time. With sufficient number of aspirations and returns, the temperature in both chambers was equalized. One of the chambers was placed in contact with the patient's skin, and the temperature of its contents was either lowered or raised. The patient's sensitivity to the temperature difference was indicative of either presence or absence of particular neurologic disorder. Minor's paper describing the thermoesthesiometer was accompanied by a detailed mathematical analysis of its operation, which was prepared by a Moscow University physics professor.

Rot was also a graduate of Moscow University, and worked in a number of Moscow hospitals and clinics until 1902, when he was appointed to the faculty of the university. For a number of years he was editor of the *Zhurnal Nevropatologii I Psikhiatrii* journal. Several sources credit Rot with the invention of the thermoesthesiometer, but no account of this seems to have appeared in the *Index Medicus.* Yet it is known that the thermoesthesiometer credited to Rot was in wide use in Russia in 1911 (80). It can be surmised that Rot had designed and constructed that device, though whether or not he was the first to do so remains to be determined. In the Western literature, Rot is best known for his description of *Meralgia paresthetica* (81), also known as *Roth's disease,* or the *Bernhardt-Roth disease.* Rot first noted this disease in 1885 and, at the time of publication of his findings in 1895, he had observed some fourteen cases thereof. He found that usually only one hip was affected, with paresthesia in all cases, and partial anesthesia and pain in some. Males were affected more often than females. *Post mortem* examination of the most affected areas showed that the external cutaneous femoral nerve was present in their vicinity, and Rot proposed that damage to this nerve was responsible for the clinical symptoms of the disease. Rot also noted that a similar condition was described by Bernhardt in patients convalescing from typhoid fever *(Neurologisches Centralblatt,* 1895), but felt that Bernhardt was merely observing complications arising from the typhoid fever.

A number of neurologists and psychiatrists had connections with Russia in one way or another, though they were active primarily in other European countries. Among these, Kraepelin, von Struempell, and von Monakow are best known. Emil Kraepelin (1856–1926) was a professor of psychiatry at Dorpat University from 1886 to 1891. He is known for the modern classification of psychiatric disorders. Ernst Adolf von Struempell (1853–1925) was born in the Kurland province of Russia and studied medicine at Dorpat University. When his father moved to Leipzig to accept a faculty position at its university, his son moved with him and graduated from Leipzig University in 1875. He then taught at Leipzig, Erlangen, and Breslau, and never returned to Russia. Von Struempell is known for his discovery of spinal spastic paralysis, polioencephalomyelitis, pseudosclerosis, and many reflexes. Constantin Von Monakow (1853–1930) was born near Moscow, where his father was in the Russian civil service. In 1863, his family moved to Switzerland, where von Monakow obtained his medical degree in

1877. He is known for his discovery of rubrospinal tract, elucidation of the function of the red nucleus, and development of the theory of diaschesis, according to which neurological disorders are caused by damage to the central nervous system. The Soviets have labeled this view as "idealistic," in the same sense as the Virchowian theory of cellular origin of disease. In his later years, von Monakow became deeply religious, though without embracing any organized faith. He was especially fond of the more philosophical Russian classics (Dostoevsky?).

Anatomy and Histology. The first Russian anatomists were foreigners who were called to occupy positions in the newly established St. Petersburg Academy of Sciences. The best known of them were Johann G. Duvernoy and Josias Weitbrecht. Duvernoy was born in 1691 in France and studied medicine in Paris and Tuebingen. In 1725 he went to St. Petersburg but returned to Tuebingen and died in 1759. Duvernoy's work was published mostly in the St. Petersburg Academy proceedings, and concerned the structure of blood vessels and the thymus gland. He was apparently responsible for establishing that the large bones unearthed in Siberia from time to time belonged to the prehistoric mammoths rather than to elephants, as had previously been believed by scientists of that era.

Josias Weitbrecht was born in 1702 and came to St. Petersburg as a student of Duvernoy. He eventually also became a member of the St. Petersburg Academy. In addition to his duties at the academy, Weitbrecht also practiced medicine. He died in 1747 in St. Petersburg. Weitbrecht's first job was apparently to inventory Ruysch's anatomical collection purchased abroad by Peter the Great. Later he occupied himself with the various connective tissue structures of the human body, and a number of such structures still carry his name. One, the so called Weitbrecht's ligament, connects the coronoid process of the ulna with the radius; Weitbrecht's cartilage is situated in the acromioclavicular joint (scapula-clavicle joint); Weitbrecht's fibers *(retinaculum capsulae articularis coxae)* are portions of the ligaments of the hip joint; and Weitbrecht's foramen is an opening in the capsule of the shoulder joint. Weitbrecht's book on ligaments, published in 1742 in Russia, has been translated by Emanuel B. Kaplan and published by W. B. Saunder's Publishing House in Philadelphia, PA.

Joseph Jacob von Mohrenheim (1759–1799) was another foreign anatomist and physician who took advantage of opportunities offered by Russia. He was a distinguished teacher and researcher in Vienna, and edited a couple of publications in the field of anatomy and medicine. In 1783 he migrated to St. Petersburg, where he attended to members of the court and practiced surgery and ophthalmology. He described the *Fossa infraclavicularis,* often called Mohrenheim's fossa, which is a skin depression below the clavicle.

Many of the early native anatomists were of German extraction and whose homes were in the Baltic region of the Russian Empire. The most famous of these was probably Justus Christian Loder, who is rightfully honored as the father of Russian anatomy. Loder was born in 1753 in Riga and studied medicine in Goettingen. He then taught in Jena, Halle, and Koenigsberg, and returned to Russia in 1809 upon the collapse of German states before Napoleon's onslaught. During the French invasion of Russia in 1812, Loder asked for and received permission to establish a hospital for the military wounded. This he did at Kassimov in the Riazan province. Following the bloody battles of Smolensk and Vyazma, his hospital handled as many as 30,000 casualties. Loder wrote that 1700 of his patients had died, but the rest were returned to the active army or assigned to garrison duty. For his efforts, Loder was awarded the order of St. Ann by Tsar Alexander I. Following the conclusion of the war, Loder planned to leave the government service and establish a medical practice somewhere on the shores of the Black Sea (82). However, he never accomplished this and remained in Moscow to teach anatomy at its university. To his credit is the establishment of the Anatomical Institute at the University of Moscow and the organization of anatomy into the strongest subject in the university's medical faculty. Loder died in Moscow in 1832. His main scientific contribution was the systematization of all anatomical knowledge of his time into a unified anatomical treatise, which he accomplished between the years 1797 and 1803.

Alexander Friedrich Hueck was born in 1802 in Reval (Estonian Province of the Russian Empire), and studied medicine at Dorpat, Munich, Berlin, Goettingen, and Paris. He then went back to Dorpat as professor of anatomy, and died there in 1842. To his credit is the discovery of *Ligamentum pectinatum,* which is often referred to as Hueck's ligament. He also wrote a number of anatomical texts and executed several works in the fields of physiology and anthropology. Another Dorpat anatomist with worldwide reputation was Ernst Reissner. He was born in Riga in 1824 and studied medicine at Dorpat from 1845 to 1850. He obtained his doctorate in 1851. Reissner then remained as a "prosector" at the medical clinic in Dorpat until 1853, when he was appointed to a professorship in anatomy at the same institution. He remained there until his retirement in 1875, and died on his Kurland estate in 1878. Reissner was a superb teacher and researcher, and served as an inspiration to a number of future anatomists and clinicians such as Ludwig Christian Stiede. Reissner is credited with the discovery of the thin fiber, which occurs free in the central canal of the spinal cord (1864) (frequently called Reissner's fiber), and of the vestibular membrane (Reissner's membrane) that separates the cochlear duct from the vestibule in the inner ear (1851).

Ludwig Christian Stieda, a student of Reissner, was born in1837 in Riga and graduated from Dorpat University in 1861 (83). He then went to Germany to study under Leuckart, Garlach, and Herz, and returned to Dorpat in 1864 as a junior member of the anatomy faculty. Upon the retirement of Reissner in 1875, Stieda became the head of the anatomy department. In 1885, he accepted a position at Koenigsberg and remained there until his retirement in 1912. He died in 1918 in Giessen, Germany. Stieda's interests involved many scientific disciplines and included zoology, anthropology, archeology, and medical history beside his main interest in anatomy. In 1866 he described an individual in whom, upon autopsy, he was able to see two ribs connected to the 7th cervical vertebra (84). The left rib was fused with the vertebra, whereas the right one was normally flexible. Such ribs are today called cervical ribs, and are occasionally found in autopsy materials. Stieda is best known in connection with the anatomy of the foot, where he investigated in detail the posterior process of the talus (85). This structure is commonly referred to as Stieda's process. The talus, as Stieda reported, has two nodules, the *tuberculum mediale* and the *tuberculum laterale.* In some 6 percent of individuals the latter is missing and is replaced by a small bone called *os trigonum.* Stieda observed this abnormality in fourteen skeletons and none had any signs of fractures in that region of the foot. Stieda felt that the abnormality was brought about through an embryonic defect.

Other mid-nineteenth century anatomists in Russia without the Dorpat connection included men like Konstantinovich and Buyalsky. Konstantinovich is credited with the discovery of the *Arteria dorsalis recti* and the marginal vein of the anus. These are sometimes referred to as Konstantinovich's artery and vein, respectively. Elias Buyalsky was born in 1789 in Chernigov and studied medicine in the Military-Medical Academy in St. Petersburg. He obtained his doctorate in 1823 and remained in St. Petersburg as professor of anatomy until his death in 1864. Beside his activities as an anatomy instructor, Buyalsky was also an accomplished surgeon who wrote several texts on the subject. He was especially adept at removing stones from the genital-urinary tract. His anatomical contributions included a study of aneurisms, congenital disorders, and his description of a unique case involving a woman with two reproductive systems who experienced pregnancies at different times.

It is difficult to select the most accomplished anatomist of the Russian Imperial era, though it would appear that Wenzel Leopold Gruber would be most suitable for this honor. Gruber was born is 1814 in the Austro-Hungarian Empire and studied medicine in Prague. From 1842 to 1847 he occupied a junior position at Prague University, but he failed to win an appointment to a professorship. Gruber then followed Pirogov's suggestion that he move to Russia, and so he joined the faculty of the Military-Medical Academy in St. Petersburg as an anatomist. He was soon appointed to a senior faculty position and spent the rest of his professional life at the academy. He was instrumental in establishing the academy's Anatomical-Pathological Institute. Gruber retired in 1888 and moved back to Vienna where he died in 1890.

Gruber is credited with as many as 500 publications, including several books on anatomy and pathological anatomy. His research interests included the structure of muscles, congenital malformations, genital abnormalities, and the gastrointestinal tract. His name is associated with a pocket in the suprasternal space near the clavicle. It is commonly known as Gruber's fossa or Gruber's cul-de-sac. Gruber is perhaps best known for his investigations on the gastrointestinal tract. He made an extensive study on the malformations of the mesenterium and the resultant occurrence of various types of internal hernias. In one of the more unusual cases, Gruber reported on the autopsy of a twenty-five-year-old victim of typhoid fever who had an enormous scrotal hernia and a *hernia mesogastrica dextra* (86). The abdominal cavity contained only the sigmoid flexure (portion of the descending colon and the rectum) and a portion of the duodenum. The mesogastric hernia consisted of a portion of the duodenum and a 3-foot section of the jejunum. The other parts of the intestinal tract were located in the scrotal sack. The mesenterium proper was attached to the small intestine and the entire large intestine up to the sigmoid flexure. In the same publication, Gruber offered a system for the classification of the various protrusions of the peritoneum into the peritoneal cavity: *retroversione epigastrica, r. mesogastrica, r. hypogastrica dextra,* and *r. hypogastrica sinistra.* The respective hernias occurring in the fossae of these protrusions as a result of mesenteric malformations were *hernia interna epigastrica, h. interna mesogastrica (dextra and sinistra), h. interna hypogastrica dextra, and h. interna hypogastrica sinistra.* Hernia interna mesogastrica is also called Gruber's hernia.

The internal mesogastric (mesenteric) hernia was investigated by Gruber as early as in 1857, when he published his first paper on the subject in *St Petersburg Medizinische Zeitschrift.* In 1863, he described a typical hernia of this type in a thirty-eight-year-old soldier who had died of peritonitis (87). Upon autopsy, an obstruction of the ileum by omentum was recognized as the cause of death. In addition, Gruber found that the patient had an apparently symptomless internal mesogastric hernia that was of sufficient size to accommodate the entire length of jejunum (4 ft. 8 in.) when the latter was in an empty state. When Gruber injected water into the jejunum via the duodenum, the jejunum was forced to back up into the peritoneal cavity, and as soon as it was emptied, it returned to its previous position in the hernia. No collapse of the hernia sack or narrowing of the opening was detectable during this cycle of events. This hernia had apparently existed in the patient for a number of years without causing him any discomfort.

An anatomist-pathologist who made a significant contribution to the technique of anatomical specimen preservation was N. Mel'nikov-

Rasvedenkov (1866–1937), an 1889 graduate of the Moscow University's medical school. Until 1901 he was an assistant at the university's anatomical institute, then went to Kharkov University as a professor of pathological anatomy. Mel'nikov wanted to devise a preservative that would maintain organs with their original color. Alcohol, he noted, caused organ discoloration, and felt that formalin, first used by Blum in 1893, might work better. He proceeded to devise the following method (88): the organ was immersed in formalin for 24 hours, then transferred to 95% ethanol for 6–8 hours (or even up to 2 to 5 days), then was finally transferred to a solution containing 30 g potassium acetate, 60 g glycerol, and 100 g water. He was even able to embalm a corpse with this method for a later dissection by students.

By the end of nineteenth century, gross anatomy had lost its former attraction for new talent and had become a stagnant part of medical research and medical school curriculum. Few discoveries were made at that time in gross anatomy, and most of its practitioners had turned to histology, embryology, and physiology, or to pathologic anatomy. At this time, in fact, physiology had begun to replace anatomy as the basic science foundation of medical practice. As a result, a new school of anatomists had emerged, which sought to develop a theoretical foundation for anatomy and to transform the mere collection of terms that anatomy had become into a dynamic scientific discipline. At the forefront of this new school of thought was a Russian anatomist, Peter F. Lesshaft (1837–1909). He was a graduate of the Military-Medical Academy in St. Petersburg, where he studied anatomy under Gruber. In 1868, he was appointed to the faculty of Kazan University, though in 1871 he lost this job as a result of some antigovernment agitation. He then worked for Gruber, but was soon appointed director of physical education programs for Russian military schools. He held this position until 1886, when he was appointed to the faculty of St. Petersburg University. He retired in 1897. At his last job, Lesshaft was instrumental in establishing a large biological research laboratory, which for some time served as home base for a number of young scientists (e.g., Tswett) who awaited appointments for permanent professorial positions.

Lesshaft was an enthusiastic innovator not only in the classroom, but also in philosophy of science as it pertained to anatomy. He complained that anatomy, as it was taught at that time, was a study of the deceased and irrelevant for the living (89). It was, according to Lesshaft, a subject concerned with nomenclature, and as such, served only to entrap students during examinations. He predicted that anatomy would remain at a primitive level as a science if its empirical character was not transformed into something with a theoretical basis. He proposed to accomplish this by viewing the human organism as a mechanical structure whereby mathematical equations could be developed for the functioning of muscles, bones, and various types of support and connective tissues. He felt that the principles of mechanical engineering might be most applicable to such an approach to anatomy.

As an example of Lesshaft's theories, one may mention his work on the forces of holding intestines in their places in the abdominal cavity (90). He considered the abdominal cavity to be analogous to the joint of an extremity, where the joint's contents are held in place by air pressure, adhesion, and muscles. This is balanced out by an opposite force consisting of the weight of the joint, friction, and elasticity. The air pressure acting on a joint was given as $\pi r^2 hd/1000^2$, where h is atmospheric pressure in mm, d is the density of mercury, and r is the radius of the joint. This figure came out to 21.65 kg. Adhesion was estimated to contribute 35 g, and the muscles were considered to exert a force of 600 kg. Going back to the abdominal wall, air pressure was calculated to exert a force of 336 kg upon the abdominal area, and adhesion was estimated to contribute 9915 g. Lesshaft did extensive measurements of the muscular dimensions surrounding the abdominal cavity and calculated their contribution to maintenance of the abdominal cavity contents in their place. He proposed that changes of pressure from the inside of the abdominal cavity were compensated for by changes in muscular activity, whereas pressures upon the abdomen from the outside were compensated for by changes in the elasticity of the abdominal wall. In a similar treatise on the structure of the pelvis (91), Lesshaft noted that this structure, being a spherical arch in nature, was the strongest anatomical entity in the whole organism. From pelvic dimensions he derived a formula that could predict the capacity of a given pelvis to withstand stress: $P=2Q\sin^2\emptyset$, where P is the force exerted on the pelvis, Q is the capacity of the articulate area of the pelvis to withstand stress, and $\emptyset$ is an angle subtended by the spinal column and crest of the ilium. Lesshaft found the average P to be 1254 kg (range of 500 to 2338 kg). Lesshaft's name is also associated with the so-called Lesshaft's triangle or rhombus, which is a space bounded by the obliquus internus, latissimus dorsi, and serratus posticus inferior muscles. This is the location of the lumbar hernia. Lesshaft was a vigorous advocate of physical exercise among children and young adults, and wrote a number of works on this subject.

Research in microscopic anatomy was represented in Russia by many able investigators; the most notable of whom was probably A. Dogiel, whose work is described above. Seraphima Shakhova (1830–1905) was another important contributor to microscopic anatomy, and much of our knowledge of the peculiarities of the kidney tubules can be traced directly to her work. She was born in Ekaterinoslav, and it is unknown where her medical training was accomplished. We know that she was awarded her doctorate by the Bern University in Switzerland in 1876, and her now classical work on the histology of the nephron represented her dissertation (92).

Shakhova looked at the cells of the entire nephron of dogs, starting with the glomerulus and ending with the collecting tubules. However, she was mostly interested in that portion of the nephron which is present in the medullary rays. She called this section the "spiraligen Kanaelchen," and it is situated between the proximal convoluted tubule ("gewundenen Kanaelchen") and Henle's loop. She found two types of epithelial cells in this portion of the nephron, which she called the columnar ("saeulenformig") and fungiform (stellar?) ("pilzformig"). The columnar cells had 5–7 folds at their bases, which disappeared toward tubular lumen. The other type of cell had a stellar shape at the base, and the cell tapered off toward the lumen of the tubule. The nucleus was located close to the base of the cell. Both types of cells possessed processes that protruded into the tubular lumen. Today we call these protrusions the brush borders, and we have to thank Shakhova for the discovery of these structures of the proximal convoluted tubules.

The epithelial cells of the proximal tubules proper were described as being cylindrical, with a base wider than the apex, and having a spreading type of a base similar to that of the fungiform cells above. These cells also had brush borders, though they were smaller and apparently simpler than the above described structures. These cells had deep fissures that connected with one another in neighboring cells giving the impression of folds running continuously throughout the entire structure. Such folds frequently served to keep the protoplasm in only a small portion of the cell, and they became more numerous toward the glomerulus. Shakhova also described the appearance of Henle's loop. She noted that the canal as it left the proximal convoluted tubule became very narrow, then became wider in the cortical area of the kidney, and as it reentered the medullary rays, it became narrower again. It was some 4-fold wider at its widest point than its narrowest point. At the latter, the entire lumen was lined by only 4 cells. The cells lining Henle's loop were smaller than those of the proximal convoluted tubule, and they had no brush borders. The distal convoluted and collecting tubules had epithelial cells that Shakhova called cylindrical.

Hematology. The field of hematology emerged as a separate discipline toward the end of the nineteenth and beginning of the twentieth century. Before that, blood was studied by various specialists such as pathologists, chemists, and physiologists. One of the first Russian investigators who studied the properties of blood was Sakharin, whose contributions were mentioned above. He was soon followed by Alexey S. Shklarevskii (1839–1906), who was educated at Moscow-University (1856–1862) and eventually became a professor at Kiev University. He was responsible for the initiation of studies on hemodynamics, which he carried out during his postdoctoral exposure in Germany under Recklinghausen and Helmholz (93). One of the first objects of his investigation was the properties of sedimentation of blood corpuscles of both mammalian and reptilian origins. He recognized that there was a centrally located and rapidly sedimenting red cell stream, the peripherally ascending stream, and individually sedimenting red cells at the top of the tube in which the sedimentation was carried out. The descending red cell stream moved 10 times faster than the individually-settling red cells in a capillary of a medium diameter (0.36 mm), whereas there was a twentyfold difference in a capillary with a large diameter (0.72 mm). The white cells entered the rapidly-sedimenting red cell stream but were soon swept up into the ascending stream of red cells and thus never reaching the bottom of the tube. This process was continuously repeated so that in the end, the white cells had positioned themselves at the top of the red cell column.

Shklarevskii then proceeded to investigate non-gravity blood flow, for which purpose he constructed a special apparatus. Like in case of sedimenting red cells, Shklarevskii found that there was a centrally-located rapidly-moving red cell stream and a slower migrating peripheral layer. At all rates of flow examined, the two layers behaved quite differently. Thus, whereas the movement of the centrally-located blood cells was parallel to the axis of the capillary and all the cells were aligned in the same direction, the cells of the peripheral layer moved in an erratic fashion, rolling, tumbling, and frequently colliding with each other, so that these cells appeared to be quite stationary. There was, however, no accumulation of red cells in any given area, so that the number of red cells at a given point in the capillary was always constant. With an increasing rate of blood flow, the centrally-located stream became narrower, and the cells were diverted toward the peripheral layer to a large extent. Because of the weight of the red cells, the centrally-located red cell stream had an axis somewhat below that of the capillary tube. The white cells moved at a speed of only 1/20 that of the red cells, and were diverted to a large extent to the peripheral layer.

Blood flow through a capillary of an uneven diameter was also investigated by Shklarevskii. The tube diameter was altered either gradually or suddenly. With gradual broadening of the tube both cell layers were slowed down, though the effect was greater with peripherally-located cells. With the gradually narrowing capillary, both cell layers moved faster, with the centrally-located stream showing the greater change. When the vessel diameter was narrowed and broadened in quick succession, the centrally-located cell stream showed hardly any change in its velocity, whereas the peripherally-located cells had stopped completely in the wider sections of the capillary. This is especially true with the white cells, which showed heavy accumulation in the wide sections of the vessel. Shklarevskii felt that this represented a model system in regard to the vascular bed of inflamed tissue. He felt that the white cells show a similarly heavy accumulations in the blood vessels of the inflamed tissue, where the cells on the bottom of the capillary and in contact with the blood vessel wall would experience a considerable pressure from cells in the upper layers, and are thus forced out through the vessel walls into the extracellular space. The ability of white cells to change their shape is helpful in accomplishing this feat (94).

The phenomenon of blood clotting received some attention in Russia, where much work on fibrinogen and fibrin was originally done. This will be discussed in the chapter dealing with biochemistry. Microscopic examination of thrombi frequently revealed the presence of epithelial cells in addition to the red and white blood cells. Bubnov, a St. Petersburg investigator, studied the origin of epithelial cells in blood clots as follows: he exposed the jugular vein of either rabbits or dogs and painted some zinc-containing dye on the outside wall of the blood vessel. In other animals, he gave the dye intravenously. He then induced the formation of a thrombus in the jugular veins and examined the presence of the dye granules in the epithelial cells of the clots. He found that such granules were present in the blood cells where the dye was painted unto the outside wall of the blood vessel, and there were no dye granules in the clots of animals that had received the dye intravenously. Bubnov concluded that the epithelial cells found in the blood clots originated outside the vein and crept into the lumen through the venous wall (95).

Much work was also done in Russia on the technique of preparing and staining blood smears for microscopic examination, and the work of two such contributors is described below in this chapter. Mikhail N. Nikiforov (1858–1915) is mentioned in most medical dictionaries as a dermatologist, though he appears to have been a classical pathologist with an interest in microscopic anatomy. Nikiforov is best known for his method of fixing blood smears unto a microscope slide (96). He noted that the theretofore utilized Ehrlich fixation procedure required a heating of the blood smear to 120°C under carefully controlled conditions, and that such a procedure could not be carried out successfully without considerable practice. He, instead, proposed to fix blood smears with organic solvents, where the smear was first air dried and then immersed in a 1:1 (v/v) mixture of ether and anhydrous ethanol for 1–2 hours. The preparation was then stained as usual. Nikiforov reported that such fixation procedure produced specimens with a quality equal to that of heat-fixed smears; leukocyte granules were distinctly seen, and the spirochetes, if present, were especially well stained with basic aniline dyes following this fixing procedure.

Nikiforov also published on thrombosis and fat emboli (97). In one such report, he described performing an autopsy on a girl who had succumbed following an operation to correct *talipes varus.* Microscopic examination of her tissues revealed that the blood vessels of the brain, liver, and lungs were blocked by various emboli and thrombi: fat droplets, bone marrow cells, and in the case of the brain, clumps of leukocytes were all involved. This case was unique in that emboli and thrombi carried such a diverse nature in the same individual patient.

Soviet sources ascribe to Nikiforov the discovery of the true nature of granulation tissue and the refutation of Ziegler's theory that granulation tissue consists of cells of leukocyte origin. Nikiforov did indeed create an inflamed area in the skin of dogs and, upon microscopic examination thereof, discovered that the granulation tissue formed in the locus of the inflammation was of epithelial origin from the surrounding tissue (98). However, since this work was done in Ziegler's laboratory, apparently with his full knowledge if not under his direction, it is only fair that Ziegler, in addition to Nikiforov, be acknowledged as the true discoverers of granulation tissue cell origins.

Perhaps the greatest Russian innovator in the area of histologic staining was Dimitrii L. Romanovsky (1861–1921). He was born in Pskov and graduated from the Military-Medical Academy in 1866. In

1891, he was awarded the degree of Doctor of Medicine for his discovery of what we today call the Romanovsky stain. In 1899, he joined the Helena Pavlovna Medical Institute and remained there his entire professional life. Romanovsky was principally interested in the etiology of malaria, a disease extremely common in certain parts of Russia. He developed the eosin-methylene blue staining procedure for the purpose of studying the life cycle of the malarial parasite in blood smears (99). His stain was prepared by titrating a saturated solution of methylene blue with a 1% aqueous solution eosin until a violet precipitate appeared. This was generally accomplished when 1 part of the methylene blue solution was mixed with 2 parts of the eosin solution. Romanovsky's dye stained all nuclei dark-purple, whereas red cells were stained pink and the platelets appeared dark-red to violet. Eosinophile leukocyte protoplasm stained intensely pink, whereas that of the polynuclear neutrophilic leukocytes stained light-violet with dark-violet coloration in the granules. The protoplasm of both the lymphocytes and the mast cells stained intensely blue. Spirochetes in blood also stained blue.

The malarial parasite itself showed remarkable detail when stained with Romanovsky's dye mixture. He found two types of structures in the cell: a blue-staining area and a colorless area, which contained purple-staining granules. He recognized the latter as being the parasite nuclei. On the basis of his microscopic examination, Romanovsky was able to divide the life cycle of the malarial parasite into four stages. The first stage involves the extracorpuscular parasites, which are soon seen to send out protrusions that puncture holes in the red cells. After the parasite enters the red cell through the hole, the second stage begins with the parasite growing to a size almost as large as that of the red cell. The mature intracorpuscular parasite represents the third stage of development. Here it begins to form chromatin network in its nucleus, and the parasite cell divides into as many as 20 daughter parasites per single red cell. The red cell then bursts and releases the parasites, which then search out new hosts. Rarely sporulating forms of the malarial parasites were observed by Romanovsky, and to these he assigned the fourth stage of development. The editors of the *St. Petersburger Medizinische Wochenschrift,* where Romanovsky's paper was published, prefaced his article with an editorial note predicting that his method would prove to be extremely useful for the recognition of malaria and its stage of development in clinical practice. This proved to be an understatement, since Romanovsky's method, or variations thereof, are today used not only for the staining of blood smears, but also for the microscopic examination of a number of other tissues.

The most distinguished Russian hematologist was undoubtedly Alexander A. Maximov (100). He was born in 1874 in St. Petersburg and graduated from the Military-Medical Academy in 1896. He obtained his Doctor of Medicine degree in 1899, then went to Germany to study under Hertwig in Berlin and Ziegler in Freiburg. Upon his return to Russia in 1902, he was appointed to the faculty of the Military-Medical Academy and remained there until 1922, when he left Russia to join the University of Chicago. He died in Chicago in 1928. Maximov was a classical pathologist by training, and his professional activities are divided by Bloom into four stages: 1. work on the histology of blood, the placenta, and pancreas; 2. the study of inflammation; 3. the study of the formation of blood; and 4. work on tissue culture. As an example of Maximov's early contributions, one can mention his paper on the origin of blood platelets (101). He stained his blood smears first with eosin, then with methylene blue, and claimed to have seen nucleuslike particles within the red cells. He concluded that the platelets were formed from such red cell nuclei, since both structures stained identically with his stain. It is not clear exactly what Maximov saw in the red cells that he considered to be nucleuslike material (nucleoid, as Maximov called these structures). He also followed the formation of mature red cells from the stage of erythroblast and demonstrated that the nucleus was ejected therefrom in the process of mature erythrocyte formation. He was violently attacked for this view by A. Pappenheim, who held that the nuclei of erythroblasts were resorbed rather than ejected.

Maximov's research of classical pathology nature can be exemplified by his paper on the artificial induction of parenchymal cell embolism in rabbits (102). Upon inducing pregnancy in rabbits, he attempted to find parenchymal cell emboli in the lungs, but found none under normal conditions. However, if he manipulated the uteri of rabbits for 3–5 min., placental cells were found in the lung arteries along with multiple thrombi. These were the giant cells of the "periplacenta" (placental sack), rather than the cells of the endovascular plasmodium, as what he had expected. In addition, some glycogen-containing cells were also seen. Maximov was also successful in inducing the appearance of bone marrow cells and fat emboli in pulmonary arteries by damaging bones that contained bone marrow, and induced liver cell emboli in the lungs by manhandling the liver. The thrombi observed were in many cases very poor in fibrin, but were rich in platelets and were surrounded by leukocytes.

Maximov's greatest contribution to hematology was undoubtedly his proposal that there was but one precursor for all circulating blood elements (103). Up to that time it was believed that the red and white blood cells originated from two separate types of primitive cells, the myeloblast and the lymphoblast, respectively. Maximov investigated the problem, and came to the conclusion that both types of blood cells originated from an "endothelial" cell, which separated itself from the tissue and swam freely in plasma. This "endothelial" cell then evolved into a "large lymphocyte." This, as Maximov had found, was the cell that was one of the first to appear in embryonic circulation. The red and white cells then evolved from this "large lymphocyte." He examined a number of hemopoietic organs of the embryo and fetus, including the bone marrow, mesenchyme, liver, lymph nodes, and the thymus, and found the development of blood to be identical in all of these. In the postnatal organism, Maximov found the hemopoietic situation to be essentially the same as that of the embryo, i.e., the development of both the red and white cells came from the "large lymphocyte." In addition, he found that the small lymphocyte could transform itself into the "large lymphocyte." Though Maximov's scheme had to be modified as more data became available, his concepts of hemopoiesis are still remarkably close to modern views on the subject. Maximov's proposal is summarized in this chapter's illustration pages.

A large portion of Maximov's work after leaving Russia was concerned with tissue culture methodology, a field which Maximov helped to develop. His first attempts at culturing tissues were concerned with lymphocytes, where he used plasma and bone marrow in the growth medium. In later years, he successfully cultured rabbit mammary gland cells and other tissues using methodology developed with leukocytes (104). He found that the epithelial cells of the mammary tissue were transformed starting with the third day of culturing into atypical or undifferentiated tissue, with multiple mitotic spindles and an invasion of the nearby connective tissue. He was not able to distinguish such cells from those of breast carcinoma. He felt that the alteration of normal epithelial cells into cancerlike cells was due to at least three factors: mechanical stimulus due to the process of extirpation; the growth-promoting factors of bone marrow; and the characteristics of the mammary gland itself. Maximov's view on cancer in general was that cancer was a systemic disease, and probably was the result of a breakdown of the immune system.

Infectious and Parasitic Diseases. Russian scientists made respectable but not extraordinary contributions to the understanding of causation and mode of transmission if infectious diseases. This is to some extent surprising and disappointing, since Russia was frequently visited by epidemics of cholera, typhoid fever, typhus, recurrent fever, and even plague. One would have expected that much effort would have been devoted in Russia to the study of these diseases. Yet Russian scientists did not discover the causative agents for any of the major infectious diseases, and during epidemics, most of the experts in such fields limited themselves to visiting the locations of the epidemics and recommending measures for the containment of them. In the opinion of this author, this defect was due to lack of attention by the academe on the causes of infectious diseases, microbiology, bacteriology, and parasitology. Chapter 3 describes some significant Russian contributions to the basics of these sciences, yet application to practical uses thereof in medicine was lacking. There were, nevertheless exceptions to the above general rule, and it is more than worthwhile to review Russian contributions, such as they were, to the understanding of infectious and parasitic diseases.

It has already been mentioned that cholera was endemic in Russia since the 1820s and that several devastating epidemics visited that land in the next 100 years. Among the early physicians who attempted to explain the etiology of cholera was Carl Rosenberger (1805–1866), a native of Dorpat and a naval physician. He is credited with putting down a mutiny on a vessel of the Black Sea flotilla, where a plague epidemic had broken out in 1830. In 1836, he left the navy and joined the Russian civil service. He directed the first successful battle against a cholera epidemic in Russia emphasizing preventive measures to contain the disease. Another early cholera expert was Leonty Rklitzky (1815–1867), a professor of the Military-Medical Academy. In 1843, he published in the Military-Medical Journal a method for the treatment of cholera patients with salts. It is today known that cholera causes a dramatic loss of body electrolytes, and the successful treatment of the disease must include intravenous replenishment of them.

The most famous cholera fighter of Russian origin was Waldemar Haffkine (Vladimir Aronovich Chavkin) (105). He was born in 1860 in Odessa in a Jewish merchant's family and graduated from Odessa University in 1883 with a degree in zoology. He then worked in a zoological laboratory in the Odessa Zoological Museum, and in 1888 left Russia for Geneva, Switzerland. A year later he joined the Pasteur Institute in Paris, where he worked on the etiology of infectious diseases. In 1882 he was able to report his production of a cholera vaccine, and petitioned the Russian government for permission to test it in Russia. His request was denied, apparently because of the sad results of Ferran's experiments with the "viriolization" of cholera in 1885 (104). Instead, Haffkine was invited to test his vaccine in Calcutta, India, where he was able to reduce cholera mortality rates by 72 percent in vaccinated populations.

In 1886 Haffkine moved to Bombay, India, to see what could be done to halt the plague epidemic that at that time was raging in Bombay. Using the newly discovered *Bacillus pestis* (Yersin and Kitasato, 1894), he was able to prepare an effective plague vaccine that was able to decrease the mortality rate in vaccinated population by 80–90 percent. During an inoculation program in Malkowal, a few of the vaccinated villagers died from tetanus whose microorganism had apparently contaminated a vaccine bottle. Haffkine was severely attacked by his critics for the so-called "Malkowal disaster," and was even accused of being a Russian spy. Haffkine retired in 1914 and settled in Paris. He dedicated his remaining years to religious studies and the promotion of Jewish cultural awareness among its youth. In 1928, he moved to Switzerland, where he died in 1930.

Waksman's biography of Haffkine (105) is entitled "The brilliant and tragic life of W. M. W. Haffkine, bacteriologist." His life was indeed brilliant, but the term "tragic" presumably means that there was not sufficient opportunity for him to engage in scientific work in his native land, Russia. Both Waksman and Hart (107) in their biographies imply that this lack of opportunity was due to Haffkine's Jewish nationality. This may partially have been the case, though Waksman and other sources (108) have mentioned other variables in this connection—the alleged membership of Haffkine in the underground Narodnaya Volya organization. This was a party consisting mainly of student and intellectual members whose aims, in contrast to its seemingly idealistic name, were the overthrow of the existing Russian governance by violence and terrorism. Its apostles were the exiled anarchist theoreticians, Bakunin and Herzen, who operated from abroad and advocated the replacement of the tsarist autocracy not by a parliamentary democracy but by a total absence of authority and "total freedom," i.e., anarchy. Members of Narodnaya Volya made at least fourteen attempts on the life of Tsar Alexander II, the last one being successful (March, 1880). The tsar's assassination took place only a few days before "a constitution" was to be announced to the Russian people. The new tsar, Alexander III, did not go through with such reforms and instead instituted severe repressive measures against all dissidents. The pogroms of 1880s followed, and Haffkine had apparently to concern himself and his community against the mobs rather than with the overthrow of the tsar's government. He was arrested for being a member of a Jewish "self-defense" organization, but was soon released through the intercession of Elie Mechnikov. It is thus reasonable to suppose that Haffkine's difficulties in Russia may have been due to his participation in anarchist activities instead of, or in addition to, his Jewish nationality. It is interesting to note that Haffkine's nephew was the director of the epidemiological division of the St. Petersburg Department of Health in 1917. He had some seventy physicians and a multitude of lesser-caliber medical personnel under his supervision (4). And he, like Haffkine, was also Jewish.

Haffkine's anticholera vaccine was prepared according to the principles enumerated by Jenner, and was administered in the form of two subcutaneous injections. The first injection consisted of attenuated organisms, i.e., organisms grown at 39°C under aeration with sterile air. The second one consisted of microorganisms whose virulence was decreased twentyfold by the passage through guinea pigs. The first injection didn't provide a significant degree of immunity, nor did both injections when given in small doses. However, both injections given in maximal amounts resulted in a seventeenfold lower mortality—and nineteenfold lower morbidity rate in vaccinated populations. Protective effects of the vaccine lasted from the eleventh day following vaccination until at least the 459th day (106). The Haffkine-type vaccine was in use throughout the world until recently when a more effective vaccine based on the vibrio toxin was developed (109).

In preparing the plague vaccine, Haffkine attempted to combine ingredients that would protect the patient against both the invasion by the microorganism and against its toxin (110). However, he apparently settled for a vaccine against the bacillus invasion only. He grew the plague bacillus suspended from a layer of butter or oil on top of a nutrient broth. The bacilli grew down into the broth in the form of "stalactites" and were periodically shaken to collect at the bottom of the flask. After a 6-week growth period, the bacteria were collected and inactivated by heating at 50–65°C. The vaccine was then tested in several Indian jails (!), then in larger populations of the Portuguese colony of Damoan, and/or the domains of Aga Khan. Mortality rates were reduced by 80–90 percent in the vaccinated populations.

Many narratives of Haffkine's life make the impression that the tsarist government was not interested in controlling cholera in Russia and, for that reason, Haffkine had to go to India to test his vaccines. In fact it appears that the Russian government wasn't interested in supporting the person of Haffkine, a member of revolutionary groups, rather than in his scientific activities. It is a fact that vaccination against cholera was widespread in Russia, though by far not sufficient, especially during the 1906–1908 epidemic. One vaccine prepared in Russia was that of Korwacki from Warsaw Department of Health (111). He treated his vibrios first with heat at 56°C for one hour, then with 0.5% phenol. The vaccine was administered in 3 doses at 5-day intervals, and the sera of the vaccinated persons showed vibrio agglutinating activity at dilutions of 1:200 to 1:400, whereas convalescents had titers of 1:200. Korwacki also mentions that the standard Russian "Kharkov" vaccine gave similar results. Zabolotny (112) also mentions that vaccinations were administered *en masse* in St. Petersburg during the 1906–1908 epidemic. Among a group of 30,000 vaccinated persons, only 12 contracted cholera and 4 of them died. The incidence of the disease among nonvaccinated persons was 68/10,000 with about 50% mortality rate. Such a vaccine was apparently even more successful than that of Haffkine.

The possibility of passive immunization against cholera was investigated by Fedorov from the Pathological-Anatomical Institute of Moscow (113). He was able to protect rabbits against cholera by injecting them with the serum of immunized rabbits. The animals were protected against the disease even if injected during the incubation period. It is not known if Fedorov tried his method on human patients.

The nature of immunity against cholera in immunized animals was investigated by Isayev (1854–1911) in cooperation with Pfeiffer. Isayev was a navy physician, first with the active fleet, then as a chief physician of a naval hospital in Kronstadt. His research work was done during his visit with the Institute for Infectious Diseases in Berlin, Germany. Isayev was able to produce temporary immunity to cholera in guinea pigs by injecting into the peritoneal cavity various substances such as serum from healthy human beings, bacterial growth media, urine, and even physiological saline. He showed that the irritation produced in this fashion attracted macrophages to the locus of irritation, resulting in a 4–5-day immunity to a challenging dose of virulent cholera bacilli (114). In the same paper, Isayev reported on the active and passive immunization of guinea pigs against cholera. Vaccinated guinea pigs were resistant against the disease, though this resistance was directed against the whole organism rather than the bacterial toxin. Serums from convalescent human cholera patients also protected the guinea pigs against the disease, and the immunity lasted much longer than did that produced by the nonspecific irritants. Such immunity was not entirely due to the presence of macrophages, according to Isayev, though he failed to speculate on the nature of such additional immunizing factors.

The mechanism of inactivation of the cholera bacilli by immunized guinea pigs was elucidated by Isayev and Pfeiffer (115). They observed that when vibrios were injected into the peritoneal cavity of the animals, they were immediately inactivated by lysis and the dissolution of particulate matter therein. The peritoneal cavities became sterile within a short period of time. It was moreover observed that macrophages and other white blood cells did not take part in the immune reaction, and concluded that the immunity was carried out by as yet unknown blood serum factor. Soviet sources indicate that bacteriolysis was discovered by Isayev independently of Pfeiffer. As was pointed out above, this was hardly the case, as both investigators worked on this problem together. However, the lysis of bacteria by immune serums is often referred to in the West as Pfeiffer's phenomenon. In all fairness then, the process should be referred to as the Pfeiffer-Isayev phenomenon.

The most active Russian cholera researcher was probably Simeon I. Zlatogorov (1873–1931). He was born in Berlin but went back to Russia to study medicine at the Military-Medical Academy, graduating in 1897. He was then variously associated with his *alma mater,* the Women's Medical College in St. Petersburg, and Zabolotny's laboratory in the Institute of Experimental Medicine. From 1911 to 1917 he worked in Bekhterev's Institute, and during the Soviet era he was once again associated with the Military-Medical Academy, and later—with Kharkov University. During the Lyssenkoist purges era, he was accused of promoting the concepts of Mendelian genetics as they apply to microbial genetics.

Zlatogorov became interested in the etiology of cholera as a result of the 1906–1908 epidemic. He traveled to the most heavily infested area of Russia, the Volga basin, and studied the mode of cholera transmission. Taking 89 water samples from the Volga River in the vicinity of Saratov, Zlatogorov found vibrios in 19 of them. He also noted that no sewage treatment facilities existed in Saratov and Tsaritsin (later Stalingrad, now Volgograd), the raw sewage being first collected in special ravines then pumped into the river. In addition, the drinking water, taken from the Volga River, was in some instances treated very poorly (using "American filters"); vibrios were found in samples of such river water. Water that was treated with the more expensive "British filters" contained no vibrios. Zlatogorov concluded that contamination of the water supply was the main mode of cholera transmission. His conclusions were substantiated by the fact that the epidemic was confined to those areas that obtained their drinking water from the Volga River. Cholera cases further inland were rare. Zlatogorov further wrote that such epidemics could be controlled only when the city and local governments were willing to appropriate funds for the construction of adequate water works and sewage treatment plants, and this could come about only through a democratization of the local power structures (116).

Having returned to his laboratory in St. Petersburg, Zlatogorov proceeded to investigate some of the fundamental properties of the cholera vibrio. He noted that some of the cultures he had brought back had lost their power to agglutinate with the cholera antiserum. Investigating this further, he found that the vibrios in water could assume a saprophytic character, and would be neither virulent nor would agglutinate with the specific antiserum. These cultures could, however, be transformed into the virulent variety by either *in vivo* or *in vitro* culturing and transfers. On the other hand, he could transform the virulent vibrios into the nonvirulent type by permitting them to stand for 7 days in water at room temperature, followed by repeated washing with water. He concluded that the lack of agglutination does not necessarily indicate the absence of cholera microorganisms in the medium tested (117).

The duration of the carrier state in convalescent cholera patients was also investigated by Zlatogorov (118). He examined fecal material of 255 cholera patients during and after their illness, and showed that the largest group (132) showed the persistence of the vibrios for 14–17 days following the onset of symptoms. In one person, the vibrios were found 56 days after onset of the disease. Experiments with rabbits showed that the long survival of vibrios in the alimentary tract was due to the infiltration of the gall bladder and the liver. The survival of the vibrios *in vitro* was limited by the presence of other microorganisms, whereas in the absence of such microorganisms the vibrios were able to survive from 7 to 9 months in fecal material (e.g., sewage). Vibrios found in the excrement of patients at various times after the disease was gone, but they were frequently not agglutinable by specific antisera. Yet at a later time, vibrios isolated from the same patient resumed their capacity to be agglutinated. Thus, whereas the specific antiserum agglutinated vibrios obtained from the patient on the third day of illness at a dilution of 1:5000 to 1:15,000, the vibrios obtained from the same patient on the 17–37th day after the onset of the illness failed to agglutinate altogether. At a later date, the vibrios were once again agglutinated at serum dilutions of 1:200 to 1:5000. The agglutinating properties of the inactive vibrios could be restored by a variety of *in vitro* methods, such as several passages over nutrient agar, by freezing-thawing, or by *in vivo* passage of the organisms through guinea pigs. He thus confirmed his earlier observations on the inactivation of the vibrios by water, and showed that human beings could be carriers of the vibrios even when their fecal material reacted negatively with the vibrio-agglutinating antiserum.

Zlatogorov's contributions to science were not limited to cholera, and he is credited with providing the first comprehensive description of the bubonic plague bacillus after its discovery by Yersin and Kitasato in 1894. Zlatogorov investigated 22 different *B. pestis* cultures obtained from either animals or human beings from different parts of the world, including Russia, South Africa, India, South America, France, England, and Mongolia (119). No significant differences among these cultures were observed, where all were gram-negative, indole-negative, and all were able to form capsules. All had bipolar staining properties. The bacilli were found to exist in several forms, the most common of which was the oval-shaped type 0.8–1.5 u in length and 0.8 u in width. They were not motile. The optimum growth temperature was 30°C, and the bacilli grew best in the presence of 0.5% NaC1 and a slightly alkaline medium. The best growth medium proved to be calf meat broth. Zlatogorov noted the characteristic "stalactite" growth pattern of the bacilli previously observed by Haffkine, and also saw that the bacilli were arranged into long chains. He called them "streptobacilli" for this reason. In older cultures, the "stalactite" growth pattern was seen to give way to growth at the bottom of the flask with a thin layer of bacteria remaining on the surface of the medium as an oily layer. Biochemically, *B. pestis* cultures did not liquefy gelatin, did not convert fructose and lactose to alcohol, and produced alkali during their growth period. The bacilli did not form spores, and were completely inactivated by heating at 60°C. Their growth was much more rapid on the semisolid agar than in liquid culture, where a tough membrane was formed over the agar surface. Zlatogorov noted that the pseudotuberculosis bacillus *(B. pseudotuberculosis rodentium Pf)* was in most respects identical to the plague bacillus. This included all the staining and biochemical properties, and even the ability to agglutinate with serum of patients recovered from the plague disease. The only difference observed was that the antiserum to the plague bacilli was able to form a precipitate with the *B. pestis* culture filtrates (toxin-antitoxin precipitin?), whereas no such precipitate was formed with the filtrate of the pseudotuberculosis organism.

Zlatogorov's interest in infectious disease was undoubtedly stimulated by Daniil K. Zabolotny, the director of the laboratory of the St. Petersburg Institute of Experimental Medicine where Zlatogorov worked. Zabolotny (1866–1929) was an active investigator of the cholera, plague, syphilis, and a number of other infectious diseases, for which purpose he participated in expeditions to India, Manchuria, and Arabia. He was a graduate of the Odessa and Kiev Universities, and in 1898 became simultaneously professor of bacteriology at St. Petersburg University, the St. Petersburg Women's College, and the director of the syphilis laboratory at the Institute of Experimental Medicine. While still a student, Zabolotny is said to have prepared, under the supervision of a Professor Savchenko, an anticholera vaccine that he used to immunize himself. He then proceeded to inject himself with live vibrios, which failed to affect him. His expeditions to Asia were mostly concerned with the study of endemic loci of the plague.

During that time, he demonstrated that bubonic and pulmonary plagues were caused by the same microorganism in the monkeys. He also showed in 1910–1911 that the persistence of plague in certain areas of Asiatic Russia was due to a small wild rodent that was carrier of the disease.

Syphilis was produced in baboons by Zabolotny independently of Mechnikov in 1903, though Zabolotny published his results a year later than Mechnikov did in the *Russian Journal of Skin and Venereal Diseases* (1904), and did not receive the publicity that his competitor received. Soviet sources make the claim that Zabolotny had observed the spirochete in syphilitic patients some two years before Schaudinn and Hoffman announced their discovery in 1905. Yet, in describing the agglutination of spirochetes by serum obtained from syphilitic patients, Zabolotny never made any claims as to the discovery of the microorganism (120).

The search for effective means to treat syphilis led another Russian medical scientist, Alexey Polotebnev (1838–1907), to demonstrate the curative powers of Penicillium mold. He successfully treated syphilitic ulcers by applying the mold locally, and his work was published in 1872 in *Meditsinskii Vestnk,* a journal hardly known in the West. Polotebnev is also considered to be the father of Russian dermatology, where he was instrumental in establishing departments of dermatology in Russian universities, and the inclusion of dermatology as a major component of the medical curriculum. In true Russian tradition, Polotebnev advocated a systematic approach to skin disease, and believed that such diseases were caused by disturbances in the nervous system. He spent the majority of his professional life at the Military-Medical Academy in St. Petersburg as professor of dermatology and the director of the academy's dermatology clinic.

The presence of spirochetes in the blood of animals was first demonstrated by M. N. Sakharov of Tiflis in the Caucasus, who was in the employ of the Russian railroad system (121). As part of his studies on the etiology of malaria, he noted that every summer there appeared a disease among geese near certain railroad stations, which would cause them to lose weight, cause diarrhea, and would eventually kill all of them. No microorganisms could be identified in the blood and organs of the deceased animals, but Sakharov was able to show the presence of spirochetes in the blood of the animals up to one day before their death. He was able to transfer the disease from one goose to another, but not from goose to other birds. The spirochetes were very fragile, and were easily crushed between the microscope slide and its cover glass if care was not taken to prevent this. Sakharov was not able to culture the spirochetes *in vitro*. Since the microorganism attacked geese only, he named it *Spirochaeta anserina*.

Another spirochetal disease that assumed major importance in Russia was recurrent fever (*Febris recurrens*). It made its appearance in Moscow and St. Petersburg in 1864–1865, and possibly earlier in the more remote areas of Russia. The clinical progress of the disease was described by several St. Petersburg and Moscow physicians of that time, most notably by Zorn (122), who proposed that the classical and biliary forms of recurrent fever were one and the same disease; and by Botkin (123) who, on the basis of certain similarities between malaria and recurrent fever, treated his patients with quinine. The most accomplished investigator of the recurrent fever in Russia was probably Osip O. Matchutkovsky (1845–1903), who was a graduate of the Kiev University (1869), a staff physician of the Odessa Municipal Hospital (to 1893), and finally, a professor in the Postgraduate Medical Institute in St. Petersburg. Using human volunteers, Motchutkovsky was able to transmit recurrent fever from the ill to the healthy individuals by injecting infectious blood (124). He thus showed that insects could function as carriers of diseases. He could not transmit typhoid fever by this route, nor could he infect the animals with human blood. Blood of patients undergoing the acute phase (pyrexia) only was infective, whereas apyretic blood from the same patients was not.

Motchutkovsky also investigated various means for predicting the recurrence of acute phases in patients with recurrent fever. He looked at the urine and fecal properties, the size of spleen and liver, and blood microscopy as possible prognostic aids, but found that the most reliable indicator in this respect was the temperature chart of the patient (125). He found several typical temperature curves in his patients:

1. During the remission period, the temperature steadily rises with a difference of 1.5 to 2.5°C between the lowest and highest temperature of the remission period. In such cases, a recurrence of pyrexia happened in 9 out of 10 cases;

2. During the remission period, the temperature declined gradually. In such instances, no relapse occurred in 10 out of 12 cases;

3. A very slowly rising temperature (1 degree in 7 days) signified no relapse;

4. Irregular changes in temperature during the remission period, which did not fit any of the categories above, meant that a relapse would occur in 50% of the cases.

In one of his patients, a little girl, Motchutkovsky noted relapses every 13–21 days, whereas in a normally progressive disease these occurred once in 4–7 days. He concluded that in this case, he was observing a *de novo* reinfection of the disease. And since each episode was less severe than the preceding one, it was clear that the patient was building up a resistance to the disease, though Motchutkovsky concluded that immunity was very difficult to elicit.

In addition to his work on recurrent fever, Motchutkovsky made significant contributions to the field of neurology. In 1883, he published in the Russian journal Vratch an article on the treatment of *Tabes dorsalis* patients by suspension (modern traction?), and in 1890 this therapeutic method was described in detail in a British journal (126). The patients whom Matchutkovsky saw were usually in the final stages of the disease: most had total or partial paralysis of the extremities, interference with bowel movements, micturition, impotence, shooting pains and numbness of the legs, and atrophy of muscles. If no complications were present, such as a heart condition, the patients were treated by suspension from a specially designed harness that would stretch their bodies by 3–5 cm. There was a carefully worked out suspension plan, starting with a suspension period of 1–2 min., then increasing it gradually to 10 min. The treatment was carried out once every 2–3 days for 50 to 100 treatments. Of the 12 *Tabes dorsalis* patients under his care, ten showed marked improvements, whereas the other two showed some. Patients who were helped were once more able to walk 1–2 miles, the shooting pains were gone, sexual activities had returned, and their limbs felt much stronger. Pain reflexes were almost normal, though the patellar reflex was still absent. Motchutkovsky felt that the beneficial effect of the treatment was due to a change in the relationship of the spinal cord to the position of the fifth and sixth spinal nerves. Three of the patients treated by Motchtkovsky were seen 5 to 7 years later, and in no case had the disease returned to its severe original state. The pain reflexes were measured by Motchutkovsky with a special algesimeter of his own design (127).

Returning to the theme of recurrent fever, there were apparently no therapeutic measures that one could invoke once the disease had set in. Gabrichevsky was probably one of the first investigators who made an attempt at preparing a vaccine against the disease by taking blood of a patient during his pyrexia episode, heating the blood to 56°C and reinjecting the blood into the patient during the apyretic period (128).

The experiment was not successful. It may be noted that at that time (circa 1905), no one had succeeded culturing the spirochete *in vitro*. Gabrichevsky believed that spirochetal infections in man, apes, and geese gave rise to humoral antibodies that appeared to be both bacteriostatic and lytic (129). He was severely criticized for his view by Mechnikov, who, of course, did not like anyone who believed in the humoral theory of immunity. Not having succeeded in the active vaccination experiment described above, Gabrichevsky attempted to treat his patients by passive immunization. He injected into horses spirochete-containing human blood, and after an appropriate period of time, prepared the antiserum. He then gave the antiserum to his recurrent fever patients during their first apyretic period and got the following results:

No. of relapses	Treated patients (%)	Untreated patients (%)
None	47.0	12.8
2	37.3	32.9
3	13.1	46.5
4	1.3	7.1
5	1.3	0.7

Clearly, the horse antiserum was able to significantly reduce the number of recurrent fever relapses in Gabrichevsky's patients.

Gabrichevsky was Russia's most versatile investigator of infectious diseases and their therapy via immunological means. He was born in 1860 and graduated from Moscow University with a degree in medicine in 1886. He then worked with Koch, Ehrlich, Mechnikov, and Roux, and was appointed to the faculty of Moscow University in 1892. In 1895, Gabrichevsky established the Bacteriological Institute at Moscow University with the mission to produce diphtheria and scarlet fever vaccines. He remained on the faculty of the university and as director of the institute until his death in 1907. In the West, Gabrichevsky is best known for his development of the scarlet fever vaccine (130). He cultured the streptococci isolated from patients that had succumbed to the disease, heated the culture to 60°C, then treated it with 0.5% phenol. The vaccine, containing 0.02 to 0.03 ml. of packed cells per ml suspension, was administered initially in a dose of 0.5 ml, then the inoculation was repeated twice with a dose of 0.75–1 ml at 7–10 day intervals. Gabrichevsky also prepared a vaccine against glanders in horses. The streptococci were isolated from diseased animals, grown in a calf meat broth at 37°C for 48 hours in a weakly alkaline medium, and inactivated by either heating to 60°C or by 0.5% phenol. When injected into horses, the vaccine resulted in a temporary rise in temperature, and protected the horses against effects of massive doses of live organisms. The scarlet fever vaccine of Gabrichevsky was tested by Langovoy at the St. Vladimir's Children's Hospital, who reported that among the vaccinated children in a surgery ward, only 1 in 120 came down with the disease, whereas the general rate of infection in the surgical wards was 3% (131).

In addition to his work on spirochetal and streptococcal diseases, Gabrichevsky was very much concerned with the malarial problem in Russia. He participated in expeditions to some remote areas of Russia that were infected with the malarial parasite to recommend procedures for the eradication of the disease. The commission in which Gabrichevsky participated was able to recommend the prophylactic use of quinine and the use of mosquito nets. Where these simple recommendations were implemented, the incidence of malaria was reduced some two- or fourfold (132).

It was already stated that Gabrichevsky's institute was established for the purpose of preparing diphtheria and other disease antitoxins. He was able to raise 30,000 rubles from private sources for this purpose. Since the introduction of immunological methodology for treating of diphtheria, Gabrichevsky noted (133, 134) that the mortality rate from that disease had declined from 43% in 1890 to 25%

in 1898 in the city of Moscow, and in the city of St. Petersburg, the same rate of decline was noted. To aid in the diagnosis of the disease, Gabrichevsky advocated the routine use of throat cultures and strict quarantine procedures to arrest the spread of the disease. He was especially impressed with the New York Dept. of Health procedures, and urged their adoption in Russia.

Russian scientists were responsible for much of the early work in the field of parasitic diseases and their modes of transmission. Melnikov, working with Leuckart, was one of the first to show the role of insects in the transmission of parasitic diseases. He found that canine tapeworms were introduced into the animals via the dog louse (135). A major contributor to our understanding of insect-borne diseases was Alexey P. Fedschenko (1844–1873), a natural sciences graduate of the Moscow University. He was particularly interested in two-winged insects and compiled an extensive encyclopedia on the subject. He was also an enthusiastic traveler and explorer, having visited several Siberian and Asiatic regions of Russia for the purpose of gathering specimens. He met an early death in an attempt to climb Mt. Blanc in the Swiss Alps. In 1869, Fedchenko investigated the life cycle of a filarian, which he named *Filaria medidensis,* and which today is known as *Dracunculus medinensis.* He showed that the parasite invaded the cyclops, a small ticklike arthropod whose infected larvae, in turn, found their way into a human or canine body *via* drinking water. This work was published in 1871 in a local Russian journal of naturalists (136). In 1872, Fedchenko described the life cycle of the nematode parasite *Gnathostoma hispidum,* which he found mostly in mammals feeding on cold-blooded vertebrates.

Animal parasites were investigated by several Russian scientists, among whom one could mention Wrublevsky and Danilevsky. A trypanosome that invaded the bloodstream of the bison was discovered by Wrublevsky in 1909 (137). The protozoan was some 30–50 u long, i.e., of intermediate size compared to other known protozoans, it had a well-defined nucleus, centrosomes, and flagella, and moved rapidly when examined in aqueous suspension under a microscope. Vladimirov and Yakimov confirmed Wrublevsky's discovery and proposed that the new protozoan be named *Tryponosoma wrublevskii.* Danilevsky, a Kharkov University professor, is credited with the discovery of nonpathological blood parasites in birds and reptiles, which he called *Haematozoa* (138). He also noted the similarity between these animal parasites and the malarial parasite of man, and today these are grouped into a single subclass, the *Haemosporina.*

Several years before the appearance of the work by Leischman, Donovan, and Wright on kala-azar and the Oriental sore in 1903, a Russian military surgeon, Peter F. Borovsky (1863–1932) had already established the causative agent of the Oriental sore *(Leischmania tropica).* Borovsky's work appeared in the *Military-Medical Journal, and* was not known to the West until Hoare (1939) endeavored to write about it in 1938. Borovsky was a graduate of the Military-Medical Academy and, upon graduation in 1891, became a staff surgeon and director of bacteriological laboratories at a military hospital in Tashkent. He remained there for the rest of his life. Borovsky's studies on the Oriental sore were started in 1894. He was able to isolate many different microorganisms from the ulcers of affected patients, yet he felt that all of them, except a protozoan 1.5-2 u in diameter, were there because of secondary infection. The protozoan that Borovsky had observed had a distinct nucleus and multiplied by simple fission or budding. The protozoa were found in the skin papules of infected individuals and in the early stages of ulcer development. However, the protozoa were absent from older ulcers, where, instead, a multitude of microorganisms was found. He definitely concluded that Oriental sore was not caused by bacteria but was instead caused by a higher-order organism, the protozoan. E. J. Marzinovsky and S. L. Bogrov, in a

Russian publication in 1904, classified Borovsky's organism as a member of the *Trypanosoma family.*

Borovsky's work was later extended by his Tashkent colleague, K. Shulgin, who in 1902, in a Russian publication, reported on the mode of transmission of the Oriental sore. In 8 cases, he was able to establish that the development of the sore was preceded by mosquito bites. He also noted that the disease made its appearance during late summer and early fall, and that the disease among army personnel was confined to enlisted men and was practically absent from officers, who were used to sleeping under mosquito nets. All these facts indicated that the disease was transmitted *via* insect bites.

There was some interest among Russian scientists in fungal diseases as, for instance, the work of Berestnev (140). He is credited with distinguishing actinomyces-induced actinomycosis from other diseases that were symptomatically similar to actinomycosis. True actinomycosis, Berestnev wrote, should be a disease caused by any of the organisms that carried the names of actinomycetes, oospora, nocardia, streptothrix, and cladothrix. Diseases resembling actinomycosis that were actually caused by Gram-positive or negative bacteria were termed by Berestnev *pseudoactinomycosis.* Another Russian scientist, Katharina Kastalsky, was able to obtain actinomycetes in a pure culture, and her contributions are described more fully in a later chapter.

Mycobacterial diseases received relatively little attention in Russia from the point of view of their fundamental characteristics. The possibility of vaccinating patients against tuberculosis was brought up as early as in 1869 by A. Petrov of Kazan University (141) who sprayed a tubercular nodule extract into the pleural cavity of a guinea pig. The animal acquired immunity against tuberculosis. Petrov did not follow up on this with human experimentation. A very ingenious proposal to treat tuberculosis in human beings was put forth by S.

Metalnikov from the zoological laboratories of the St. Petersburg Academy of Sciences (142, 143). He had noted that bee larvae required beeswax for proper development, and reasoned that the waxy cell walls of the Mycobacteria might also be digestible by the bee larvae, and that they might thus be immune to tubercle bacilli. When he injected these bacilli into the larvae, he noted that they were rapidly taken up by the phagocytes and had completely disappeared from the animals after 60 min. He then purified the mycobacterial wax and upon injecting it into the larvae found that it was completely and immediately destroyed. He felt that there was a specific lipase in the blood cells of the larvae that digested both the beeswax and mycobacterial wax. Metalnikov then proposed that tuberculosis patients may be treatable by the injection of larval blood. To test this hypothesis, he infected guinea pigs with a massive dose of tubercle bacilli and followed this up with an injection of larval blood. All animals getting the larval blood survived, whereas those that didn't died within one month. It is not known whether or not attempts were ever made to treat human beings with this method. Metalnikov was a zoologist, and his work on tuberculosis was undoubtedly inspired by another zoologist, i.e., Elie Mechnikov, in whose laboratory Metalnikov had spent some time. As is well known, Mechnikov was able to make a major contribution to medicine at an earlier time by working entirely with insects.

Leprosy was initially studied in Russia by Gregory H. Minkh (Muench) (1836–1896), a professor at Kiev University. Soviet sources have claimed that Minkh had established the infectious nature of this disease, but it appears that his contribution was confined to epidemiological studies in southern Russia. A major contribution towards understanding of leprosy was made by V. J. Kedrovsky (1865–1937) of the Pathologic-Anatomical Institute of the Moscow University (144). He was one of the first to grow the leprosy bacillus in a pure culture. He devised a culture medium where human placenta was homogenized with distilled water, the suspension was filtered through a bacterial filter, and either mixed with peptone broth or with agar. Such a medium served well for the culturing of tubercle bacilli, gonococci, and influenza bacilli. He then inoculated agar slants containing this medium with the blood and the extract of a skin nodule of a leprosy patient. Growth was seen in both cases within 2–5 days. The organisms grown were morphologically identical to authentic leprosy bacilli, though they proved not to be acid fast. He scrupulously excluded any possibility of contamination, and concluded that the difference he had observed between the *in vivo* and *in vitro* cultured *Micobacteria leprae* were due to changes that occurred during the *in vitro* culturing. He felt that the characteristics of the leprosy bacilli were situated between those of the tubercle bacillus and the diphtheria bacillus.

An energetic advocate of culturing of blood for diagnostic purposes was N. Klodnitsky, a bacteriologist with the Interior Ministry's Bacteriological Laboratories in Astrakhan (145, 146). He was convinced that red cells had bacteriocidal properties, and proposed that for the purposes of bacteriological examination, the drawn blood should be immediately mixed with distilled water in a ratio of 0.5 to 4.5 ml to lyse the red cells and dilute the humoral antibodies. This dilution was then to be inoculated into agar or nutrient broth. Using this method, Klodnitsky was able to obtain cultures from the bloods of typhoid, gonorrhea, paratyphoid A and B fevers, pneumococcal pneumonia, and streptococcal and staphylococcal septicemia.

The causative agent of erysipelas was established in 1887 in St. Petersburg by Meierovich (147). In his doctoral dissertation, he reported the isolation of erysipelas streptococcus from the skin and blood of several patients. These microorganisms grew rapidly in conventional broth media either at room temperature or at 37°C, and such cultures were then used to transmit the disease to rabbits. Those animals that did not survive the disease showed microorganisms in their blood and many organs. Those that did survive were immune to reinfection for 1–2 months following convalescence. Meierovich called attention to the fact that several types of microorganisms could cause erysipelas, though he appeared to consider *Streptococcus pyogenes* as the principal offender. This organism was identified as the causative agent of erysipelas also by A. Pavlovsky (148), who pointed out that it was responsible for the less serious forms of the disease.

Fundamental work on the pathogenesis of *Enterobacteriaceae* may be exemplified by the contributions of L. Rosenthal of Gabrichevsky's institute in Moscow. Shortly after the discovery of the dysentery bacillus by Shiga, Rosenthal was able to isolate the bacillus from his patients in Moscow and to show that they did not generally invade the bloodstream. He found agglutinating antibodies in the serums of convalescing patients on the tenth to twelfth day of the illness. He even tried to immunize guinea pigs with the microorganism, though this did not meet with much success (149). Rosenthal was successful in preparing the dysentery bacillus toxin by growing the microorganisms in a conventional broth at a slightly alkaline pH-value for 3 weeks. A 0.1 ml. aliquot of the culture filtrate was able to kill a 2 kg rabbit. The toxin was precipitable with ethanol, and could be redissolved in a weak salt solution. It was heat and acid stable (150). Rosenthal then used the toxin to produce an antiserum in horses, and used this to treat his dysentery patients. Of the 157 patients treated, only 8 (4.5%) did not survive. The death rate among the untreated patients was 12.2 to 17.7%. In addition, the hospital stay of the treated patients was on the average 10 days, whereas that of the untreated ones was 16 days (151).

An important contribution toward the understanding of spirochetal jaundice or Weil's disease was made by N. P Vasiliev, a privatdozent and an attending physician at the Alexander Hospital in St. Petersburg (152). Between the years 1883 and 1888, he observed 11 cases of an acute disease characterized by a sudden onset of chills, headache, weakness, excruciating muscular pains, and jaundice. In addition, albumin and various types of kidney cells were observed in his patients' urines. The disease lasted from 2 to 4 weeks, and the patients experienced complete recovery. At no time was Vasiliev able to observe the presence of microorganisms in his patients' blood. Vasiliev reviewed the observations of all investigators that had described similar symptoms, starting with Landouzy in 1883, Weil in 1886, and his own work. He noted that a total of 48 cases had been described, and his investigation further showed that the disease was most prevalent during the months of June to August, and struck mostly men aged 16 to 24. Vasiliev's paper was also valuable for distinguishing this disease from other similar disorders. He argued against Weil's proposal that the disease was a form of typhoid fever, and rejected the idea that it was a variant of relapsing fever, as had been proposed by Griesinger. The latter, in 1853, reported on the occurrence of a disease in Egypt with symptoms quite similar to those seen by Weil and Vasiliev; he had named the disease *Febris recurrens biliosa.* Vasiliev proposed that the disease observed by himself, Weil, and Griesinger be named *Typhus biliosus,* and it should not be confused with any intestinal disorder. Today, it is known as Weil's disease, and is caused by a unique microorganism, namely *Leptospira icterohaemorrhagiae,* discovered in 1914 in Japan. Vasiliev's contribution to the etiology of Weil's disease is recognized in some literature as Vasiliev's (Vasilieff's) disease.

Pediatrics. Perhaps the best-known aspect of pediatrics in Imperial Russia was its extraordinarily high child mortality rate. It was already stated previously that child mortality in Russia averaged from 200–300 per 1000 live births, which was higher than that of most European countries of that time. From this, it is commonly assumed that few if any pediatric services were available in Russia. Though such an assumption is certainly valid in part because of a general shortage of physicians at that time, it nevertheless is also true that pediatric services and knowledge of child diseases among Russia's general practitioners were at the same level as those of other medical specialties. In a remarkably detailed analysis of child mortality in Russia, Wladislav Hubert, a St. Petersburg pediatrician, stated that the problem was not so much with medical practitioners as with the customs of Russians who did not believe in breastfeeding (153). Thus, among the Russians, during the years 1895 and 1899, child mortality was 275 per 1000 births during the first *year postpartum.* As many as 75 percent of the deaths were due to gastrointestinal problems. From the first week of life onwards, the children were fed black bread with salt, sauerkraut, borsht, fermented bread drink (kvas), etc. in spite of physicians' warnings not to do so. The Russian minority groups, which did not follow the Russian peasant custom and whose medical care was certainly not much better than that in Russian villages and cities, showed drastically lower child mortality rates: for the same period, the Moslems (Caucasus and eastern part of the Empire), had a mortality rate of 163/1000, the Jews—128/1000, the Catholics—150/1000, and the Protestants—179/1000. According to Hubert, improvements were possible only upon a massive educational effort among members of the Russian peasantry.

There were some 30 special pediatric hospitals in Russia by 1912, the oldest ones being the St. Petersburg Children's Hospital (1855), the Prince Oldenburg Children's Hospital in St. Petersburg (1869), and the St. Olga Children's Hospital in Moscow (1887). Incidentally, the first children's hospital in the United States was established in 1855 in Philadelphia. Every university in Russia had a children's clinic (hospital), where patients could be referred to by local doctors. In some 55 of the 78 Russian provinces there were foundlings' hospitals with conditions ranging all the way from poor to excellent. The Simbirsk Foundlings' Hospital was said to have a child mortality rate of only 25% of that in general population. In order to combat the high rate of gastrointestinal disorders among Russia's infants, a network of milk stations was established beginning in 1901. The latter effort was financed primarily by private funds, and was of limited benefit to the Russian population as a whole (only 20,000 sterile milk bottles were distributed yearly to the poor); however, the idea and organization of such stations was expected to serve as a model for the expected entry of the government into the field of public health. Governmental help was, however, lagging, because of chronic shortage of funds. Russia, contrary to popular thought, was a poor country, with a relatively small percentage of productive individuals; of the 125 million persons in European Russia in 1908, only 33 million were employed. The rest were either children (some 50% of the population), or retired, or in some way incapacitated. For every working individual, there were thus 3–4 persons that were not employed. Though Russia was first in the world to form establishments to care for homeless infants and children (early eighteenth century), by the beginning of the twentieth century it was behind most of the European nations in this regard.

Russian academic pediatricians were mostly concerned with ways to improve the morbidity and mortality from the well-known childhood diseases such as scarlet fever, diphtheria, etc. For this reason, a large portion of Russian pediatric journal papers was concerned with the effectiveness of various serums used to treat acute childhood diseases, complications resulting therefrom, and the effectiveness of vaccination programs. Another high proportion of papers dealt with rare congenital disorders and diseases primarily observed among adults. Russia thus produced only a handful of truly great pediatricians who are still remembered in treatises on the history of medicine. One such individual was Nil F. Filatov (1847–1902) (154), who is often considered to be the founder of Russian pediatrics. After graduating from Moscow University in 1869, he worked as a Zemstvo physician for a year then went abroad to Prague, Vienna, Heidelberg, and Berlin. Upon his return to Russia, he was associated with the Moscow Children's Hospital and the Moscow University until his death. He was instrumental in establishing the Moscow Pediatric Society (1892) and served as its president for ten years.

Filatov is credited with discovering the *fourth disease* and *infectious mononucleosis,* though for some reason he does not appear to have published papers dealing exclusively with these diseases. Instead, Filatov described both disorders either in papers dealing with the therapy of children's diseases in general, or in his pediatric texts. Perhaps, for this reason, his discoveries went unnoticed by the outside world until both diseases were rediscovered by Dukes in case of the fourth disease, and by Pfeiffer in case of infectious mononucleosis. Thus, the fourth disease, also called *Rubeola scarlatinosa* by Filatov, was first described by him in 1885 in the Russian publication *Russkaya Meditsina (No.* 48), and later in his book *Vorlesungen ueber akute Infektionskrankheiten im Kindesalter* (1897, p. 382) (155). It was a scarlet feverlike disease, which lasted for a few days and resulted in full recovery. However, many such patients came down later with a *bona fide* scarlet fever. Dukes described the disease in 1900, terming it the "fourth disease," and today, it is generally known as the Filatov-Dukes disease.

Infectious mononucleosis was described by Filatov in 1887 (156) as follows: *"Ich habe hier nicht die Adenitis im Auge, welche Entzuendungsprozesse der tiefen liegenden Theile begleited, z. B., der Muendhoehle, des Rachens, u. s. w., wobei nicht die Druesen, sondern die zu Grunde liegende Krankheit das entscheidene Moment abgeben, sondern ich wuensche Ihre Aufmarksamkeit auf ein Leiden des*

Kindesalters zu lenken, welches freilich nicht in Buechern beschrieben wird und daher angehenden Aertzten gaenzlich unbekannt ist, aber nicht destoweniger ziemlich of im Leben vorkommt; ich meine namentlich die idiopathische Entzuendung der lymphatischen Druesen, welche am oberen Ende des Musc. sterno-cleido-mastoidens, d. h., unter dem Ohr und dem Proces mastoidens, und hinter dem Angulus maxill, inferior liegen." He called the disease idiopathic inflammation of the lymph glands. It was observed in children between the ages of 2 and 4, and less frequently in adults. The fever remained 5–10 days, though the swelling persisted for 2–3 weeks. In 1889 Pfeiffer described the same disease, and it is sometimes referred to as Pfeiffer's disease in Western literature.

Filatov wrote numerous articles and books whose purpose appears to have been educational, though as noted above, most such publications were replete with original observations. One such paper contains his description of influenza in children (157), where he considered the fever, sniffles, and earache as the three most characteristic symptoms. In many cases, he notes, lesions (herpes) were present in the mouth and nasal mucosa. Pneumonia was seen to be the most serious complication of influenza. Filatov's *Klinische Vorlesungen ueber Kinderkrankheiten* (Leipzig, 1901–1902) described each disease in terms of specific case histories, and this work served as a standard pediatric reference throughout the world for many years.

Karl Adreevich Rauchfuss (1835–1915) was another Russian pediatrician with worldwide reputation, and the high esteem in which he was held by foreigners was expressed in his obituary written in Germany by Heubner in 1916, a time when Russia and Germany were at war with each other (158) (A. B.'s note: Rauchfuss' German name was likely helpful). Heubner described Rauchfuss as a self-made man who belonged to no specific medical "school" of thought but who made inestimable contributions to the field of pediatrics. Rauchfuss was born in St. Petersburg, was a graduate of the Military-Medical Academy (1859), and spent some time in Virchov's laboratory. He designed and helped to establish the Prince Oldenburg Children's Hospital in St. Petersburg (1867–1869) and the St. Vladimir Children's Hospital in Moscow (1874–1876). In 1876 he was appointed to the position of *Leibpediater* or pediatrician to the court, and was frequently called to treat the hemophiliac son of Tsar Nicholas II. In addition to his busy medical practice and administrative prowess, Rauchfuss contributed significantly to the fundamental knowledge, diagnosis, and methods of treatment of many children's diseases. He is credited with the discovery of subglottic laryngitis (158), a number of congenital cardiovascular disorders, the so-called Rauchfuss triangle for the diagnosis of exudative pleurisy, and the so-called Rauchfuss sling. The latter is a device that permits changing the position of a patient without actually moving him. It is extremely useful in postsurgical cases and wherever absolute immobility of the patient is necessary. Among the congenital disorders described by Rauchfuss were the various cases of thrombosis involving the ductus arteriosus, pulmonary artery, or inferior vena cava (159). He also reported seeing a case where the diameter of the aorta was only 1/3 of a normal aorta. Rauchfuss also described 25 cases in which the aorta originated from the right ventricle instead of the left one, and where the pulmonary artery originated in the left ventricle. The symptoms of this disorder were cyanosis of the skin and mucous membranes and a hypertrophy of the heart without murmurs (160). The so-called Rauchfuss triangle was described by its discoverer in 1903 at a medical meeting, but Rauchfuss claimed to having used it for the diagnosis of pleurisy in children for a number of years before that (161). He noted that in children with exudative pleurisy, there was on their backs a triangular area, opposite to the area of effusion, which was dull to percussion, i.e., the percussion sounds were muted in that area. Rauchfuss believed that the triangular area of dullness was caused by displacement of the ava mediastinum by the diffused lung into the healthy lung cavity. It should be mentioned that an area similar to the Rauchfuss triangle was at the same time observed in adults by Grocco, and, thus, the Rauchfuss triangle is often referred to as the Grocco sign.

Pleurisy was investigated by a number of other Russian physicians. S. Levashev, a University of Kazan professor, developed an operative method for the draining of the exudate from the pleural cavity (162). He noted that simple drainage of the abscess frequently led to complications with fatal outcomes, and proposed to replace the fluid drained with a solution of 0.7% NaCl that had been carefully sterilized. The 0.7% concentration was chosen because tests in animals had indicated that NaCl concentrations of 0.2 to 0.6% and over 0.75% caused tetanization of the muscles. The fluid temperature was 30°C, but if inflammation was present, it was infused at a temperature of 12 to 15°C with beneficial effects. The operation, which appeared to cause no pain for the patient, was performed repeatedly draining the abscess and replacing the exudate with saline in small portions. A single series of treatments was sufficient to cure exudative pleurisy, and purulent pleurisy if treatment was started sufficiently early. Replacement of the exudate by saline was accompanied by an immediate drop in temperature, which later rose again for a short period of time. At the same time, vital capacity was observed to return to normal. Levashev's method was applied to children by Alexander Kissel (1859–1938), a graduate of Kiev University and student of Botkin. Of 8 pleurisy cases, Kissel was able to cure 5, whereas the other 3 showed considerable improvements. He suspected lesions of tubercular origin in resistant cases (163). Kissel was also one of the first pediatricians to recognize the connection between rheumatic disorders and cardiac complications. He reported on a case of a boy who had developed a heart valve disease two years after recovery from a condition involving rheumatic joints (164). Purulent pleurisy was investigated by A. N. Shkarin, a St. Petersburg pediatrician who was able to isolate pneumococci either in pure culture, or a mix with streptococci, staphylococci, or tubercle bacilli from his patients (165). In St. Petersburg, for some reason, pleurisy was always found to coexist with pneumonia or bronchitis.

The nineteenth century concepts of the development of pleurisy were modified by Dimitry Sokolov, a pediatrics professor at the Women's Medical College in St. Petersburg. Using the newly available X-ray technology, he discovered that the exudate developed at the bottom of the lung and in its fully developed form had the shape of a triangle with the base at bottom of the lung and apex in the axillary region. Position of the fluid was not changed by changing the position of the patient. It had previously been believed that such exudates had triangular shapes with the apex somewhere along the spinal column. Sokolov tested his findings by introducing water into pleural cavities of cadavers, then taking X-ray films. They showed the same images as those of his live pleurisy patients (166). Sokolov also constructed an apparatus that he called the differential pneumograph that was designed to measure the degree of chest expansion during breathing on both sides of the chest (167). In a normal individual, both sides of the chest expanded and contracted in a perfectly parallel fashion during normal or forced breathing. Significant deviations from normal patterns were observed in patients with various chest diseases, such as the lateral transposition of thoracic and abdominal viscera, stenosis of the larynx, cardiac hypertrophy, chorea, mediastinal tumors, pleurisy, and pneumonia. Many of the pneumographic patterns were correlated with chest X-ray films and were found to be in perfect agreement. Sokolov was also interested in a number of other pediatric problems, such as inflammation of the various types of glands (168). In addition to the submaxillary and cervical glands, the lymphatics of the thoracic and peritoneal cavities were also often involved, according to Sokolov. The most common infective agent found was the tubercle bacillus, followed by the streptococci, staphylococci, pneumococci, and even *E. coli* and the typhoid bacilli. The infections were found to be acute or chronic, and some were found to have lasted for several years.

Bronchiectasis, an unusual chest disorder in children, was studied in detail by V. Chernov of Kiev University (169). He noted that few if any researchers had bothered to maintain an adequate history of the patients suffering from this disorder, hence the causes and eventual outcomes of these diseases were frequently unknown. Chernov distinguished the acute and chronic types of the disease. The former always ended lethally, whereas 2 out of 3 chronic cases proceeded to recovery. One such chronic case lasted for 6 years before full recovery took place. Chernov noted that all bronchiectasis cases were preceded by some acute lung disorder such as pleurisy or pneumonia. The disease itself was characterized by a high fever, weakness, and most typically by the expectoration of a large quantity of foul-smelling sputum. The lungs of the deceased patients had numerous lesions, their air passages were 2 to 3 times larger than normal, and there was a complete absence of elastic tissue. Chernov felt that the exudate contained an elastase that digested the elastic components of the lungs and promoted the formation of lesions characteristic of the disease.

Chernov also published his views on the diagnosis and mode of treatment of spinal meningitis (170), where he strongly advocated the use of bacteriological techniques to characterize the disease. He did not believe in the use of lumbar puncture technique as a therapeutic tool, which had been advocated by Western authorities, and instead used the cerebrospinal fluid to identify the infective agent. In advocating the bacteriological culturing of cerebrospinal fluids, Chernov made an excursion into the area of law, where he related a case of a boy being brought to the clinic after being hit on the head by a man. The boy soon developed a high fever and delirium. His cerebrospinal fluid was cultured and was found to contain diplococci. Though the boy eventually recovered, had he died, the man who had hit him would have been unjustly accused of murder, Chernov theorized.

The common childhood diseases, measles, scarlet fever, whooping cough, and diphtheria were investigated in connection with complications arising therefrom, and the mode of action of their toxins by several Russian pediatricians. The effects of scarlet fever on heart action were studied by J. V. Troitsky, a professor of pediatrics at Kharkov University (171). He sought to alleviate the cardiac symptoms and other complications by administering streptococcal antiserum (Moser's serum) to his patients. Instead of disappearing, the cardiovascular symptoms were actually enhanced; upon the subcutaneous injection of the antiserum, Troitsky noted the appearance of systolic murmur at the tip of the heart, cardiac arrhythmia, and a change in the pulse wave from monocrotic-dicrotic-hypertropic to exclusively monocrotic with a rise in blood pressure. Upon resorption of the antiserum, the cardiac symptoms disappeared. Scarlet fever was investigated by a number of other investigators, and generally involved the search for methods of its treatment. Gabrichevsky thus developed a vaccine for immunization against the disease. Vinokurov of Odessa reported on the effectiveness of passive immunization, and Vladimirov of the St. Vladimir Children's Hospital in Moscow wrote on the means of diagnosing the disease (172). He lamented the fact that doctors were depending too much on laboratory results for the diagnosis of childhood diseases, and were not interested in developing skills in the area of physical diagnosis. As an example, he mentioned the strawberry tongue symptom characteristic of scarlet fever. He considered this to be the most reliable symptom of the disease, and noted that the tongue has the typical appearance even in the absence of skin eruptions, as does happen in rare instances. Vladimirov also reported on an apparently uncomplicated case of the measles, where the child suddenly expired from cardiac arrest (173). He decided that the measles microorganism elaborated a toxin that served to paralyze the vagus nerves. Sotov was another investigator who was interested in measles complications (174). One of his patients developed a tremor of the extremities that lasted for 6 weeks after the onset of the disease, another developed a psychosis with epileptic seizures that continued for 2 months after the illness began, and a third child who showed the presence of retinitis albuminurica, though no kidney disease was present as revealed by laboratory tests. All patients eventually recovered.

The involvement of adrenal glands in the course of diphtheria was studied by V. Molchanov, a professor of Moscow University (175). He noted that at the early stages of the disease the adrenals were hyperactive, but under the influence of the diphtheria toxin its medullary cells became atrophied, and the circulation within the gland was impaired due to the adverse effects of the toxin upon the blood vessel walls. The result of the damage to the adrenal gland was that, according to Molchanov, a paralysis of the gland occurred with insufficient production of adrenalin, which, in turn, resulted in cardiac arrest in many cases. He considered it a possibility to treat diphtheria by administration of adrenalin. Molchanov also wrote on the limitations of the tuberculin test and, during the Soviet era, he is said to have discovered a connection between scarlet fever and development of heart valve disease.

The microorganism causing whooping cough was supposedly isolated by M. I. Afanassiev in 1887, and reported on by Russian journals. He cultured a spore-forming organism from the sputums of whooping cough patients, which were able to produce pneumonialike symptoms in dogs. After the discovery of the whooping cough bacillus by Bordet and Gangou in 1906, Klimenko, a researcher at the Institute of Experimental Medicine in St. Petersburg, attempted to repeat Afanassiev's work, but isolated the Bordet-Gangou bacillus instead (176). When dogs were infected with the microorganism, they developed symptoms much closer to those of *bona fide* whooping cough than did Afanassiev's' dogs. It is not clear what organism Afanassiev had isolated.

As stated above, numerous Russian investigators described a variety of congenital disorders and deformities found in newborns, most of them seen upon autopsy. A prolific writer in this respect was V. P. Zhukovsky (Shukovsky), who described a number of digestive tract malformations. One of the more interesting cases was that of an infant who, while alive, kept regurgitating his food (177). The child expired soon after birth and upon autopsy, it was found that his esophagus ended in a blind sac slightly above the tracheal bifurcation point. It did not communicate with the stomach. Yet there was a structure corresponding to the lower segment of the esophagus, which led from the stomach to the trachea. Zhukovsky was also quite interested in malignant tumors both in the newborn (congenital) and in older children. In one case, he noted a swelling on one side of a newborn infant, which grew rapidly until the child expired on the eighth day of life (178). The immediate cause of death was hemorrhage, the source of which could not be pinpointed in autopsy. There was, however, a tumor the size of an egg on one of the adrenal glands, which, upon microscopic examination, proved to be a sarcoma *(Sarcoma globocellulare)*. Zhukovsky's interest in pediatric oncology was not restricted to specific cases, but also involved the gathering of literature on the subject that had appeared in Russian publications. In an extensive review of the area (179), Zhukovsky mentioned such investigators as K. Raeumer, N. Filatov, W. Niessen, A. Lazarev, D. Sokolov, A. Muratov, and others, attesting to a great degree of interest in childhood tumors among Russia's pediatric researchers. Lastly, one may mention J. Vinokurov, the chief of pediatric services at the Odessa Jewish Hospital. He described a four-year-old patient with a pleural effusion, exophthalmus, and numerous knots about the neck (180). The patient expired and upon autopsy, a large yellow mass was found behind the eye affected by the exophthalmus, and smaller tumors were distributed throughout the inner surface of the skull, on the ribs, the spinal column, and the kidneys. Microscopic examination indicated that the tumors were of the round cell sarcoma type, and the diagnosis of chloroma was thus made. Vinokurov held that the origin of such tumors were the blood-forming organs.

Rickets was studied in Russia to a rather great extent, because as many as 50% of the children in some localities were affected by it. The causes of the disease proposed by Russian investigators ranged from environmental factors to malnutrition to aberrations in nitrogen metabolism. H. Kovarsky of Vilna found that some 33% of the city's children were affected by the disease, and that Jewish children were more prone to develop the disease than were gentiles (181). He explained this difference, perhaps not too incorrectly, by the fact that the Jewish mothers did not allow their children to play outside as much as gentile mothers. He also noted that the disease was more prevalent among those who lived in dark and damp quarters and especially if they were malnourished. Kovarsky thus advocated that bad living conditions contributed to the development of the disease. He discovered a complication of rickets called *disuria spastica,* i.e., difficulty in urination. At that time, the connection between rickets and vitamin D deficiency was still a *terra incognita.*

Shkarin and Kurayeff attempted to explain the beneficial effects of sunshine on the prevention of rickets by its effects on nitrogen metabolism (182). Careful measurements of nitrogen intake and excretion during sunbathing revealed nothing more than that in both

145

the normal and rachitic children, the retention of nitrogen was lowered during exposure to the sun. No definite conclusion could be based from such data. The most logical proposal on the etiology of rickets was probably offered by J. A. Shabad from St. Petersburg, who felt that an inadequate intake of calcium was the cause of the disease (183, 184). He carried out a remarkably detailed calcium, phosphorus, ash, and moisture analysis of normal and rachitic bones, and found a decrease of calcium and a normal amount of phosphorus in the latter. He then examined the mothers' milks of normal and rachitic children and found that the latter had an average of 63 mg Ca per 100 calories, whereas that of normal milks was 77 mg. He concluded that rickets was caused by an insufficient intake of Ca by the nursing infant. He excluded the possible use of cow's milk as a Ca supplement, because he recognized that the latter, though higher in Ca than human milk, does not provide enough Ca in the form that can be absorbed.

Finally, such exotic diseases as the Tay-Sachs disorder were also known in Russia. Kovarsky described several cases and noted that the disease was reported very infrequently in Russia (185). He attributed this not to a low degree of incidence (though it indeed was in the Russian Slavic population), but to the alleged fact that the Russian doctors were not familiar with it. As was the case with the English and American experiences, Kovarsky's patients were all of Jewish ancestry, since he worked in a Jewish hospital. He discovered an additional symptom of the disease—total lack of a sense of equilibrium in affected children.

Obstetrics and Gynecology. Among the medical subspecialties, obstetrics was probably the first to receive special attention in Russia. This can perhaps be explained by the special interest that the female rulers of Russia in the eighteenth century took in this field. Thus, in 1754, upon the edict of Tsarina Elisaveta Petrovna, two schools of midwifery were established, each in St. Petersburg and in Moscow.

The state of the art developed further during the reign of Catherine II, when the first obstetrician of Russian nationality, N. Ambodik, published a handbook on obstetrics at government expense (1784). It was supposed to be distributed to all practicing doctors in Russia. Catherine's role in establishing foundlings' homes has already been mentioned, and shortly after her death, a maternity hospital and an obstetric research institute were established in St. Petersburg (1797). Its first director was Joseph von Mohrenheim. The research arm of the hospital later became known as the Imperial Clinical Institute of Obstetrics and Gynecology with a worldwide reputation for excellence (186). In 1838, a school was established with the purpose of training peasant girls in the art of midwifery, and similar schools were later founded in numerous other cities (187).

With the emergence of the universities as the main institutions for the training of physicians, research and instruction in the art of obstetrics passed to the obstetrics clinics of medical schools, though the Clinical Institute of Obstetrics retained its function as a research institution and postgraduate school of obstetrics. The obstetric clinic of the Military-Medical Academy was established in 1806, and between 1835 and 1838 it delivered 2109 babies. The death rates at these early Russian maternity centers varied between 3 and 6%, and did not improve until the introduction of aseptic techniques into the birthing room.

Among the early Russian academic obstetricians were the Dorpat professors Ernst von Bidder (b. 1839) and C. F. von Deutsch, who taught there from 1804 to 1834. The more prominent early St. Petersburg obstetricians were Otto v. Gruenwalt (b. 1830), who wrote extensively on puerperal sepsis, Eduard A. Krassovsky (1821–1898), and Kronid Slavyansky (1847–1898). The latter two physicians were probably most instrumental in introducing modern obstetrics and gynecology into Russia. Krassovsky (188) was a graduate of the

Military-Medical Academy (1851) and upon further study in Germany and France, joined the St. Petersburg Maternity Hospital, whose director he became in 1870. Among his contributions are the systematization of obstetric and gynecologic instruction in Russian medical schools, and the training of numerous physicians in abdominal surgery. He is said to have performed the first successful ovariectomy in Russia in 1862. Slavyansky was also a graduate of the Military-Medical Academy, and was a student of Krassovsky in the field of gynecology. He was responsible for introducing into Russia the use of aseptic techniques in the delivery room (189), and reported that, as a result, the mortality rate from puerperal sepsis in Russian obstetric clinics had dropped to 0.3–0.5% by 1890.

Russian gynecologists were one of the rare groups of physician-scientists who publicly praised Tsar Nicholas II for his support of scientific and medical research (186, 190). The tsar had apparently donated from his personal funds a large sum of money to rebuild and expand the Imperial Clinical Institute of Obstetrics and Gynecology, whose staff was grateful for his generosity. During the 1917 takeover by the Bolsheviks, the institute was almost totally destroyed (because of the tsar's donation?), and it was not rebuilt until the late 1920s under the direction of its prerevolutionary chief, Dr. Dimitry v. Ott.

Russian obstetricians and gynecologists were organized into the Russian Society of Obstetrics and Gynecology, headquartered in St. Petersburg. Its proceedings were regularly recorded in the German publication of *Monatschrift fuer Geburtshuelfe und Gynaekologie,* and the meetings were apparently dominated by the two Russian obstetric luminaries, v. Ott and Stroganov. Other obstetric societies existed in Russia, such as the Moscow and Kiev ones, though no record of their sessions is to be found in the Western medical literature.

The nineteenth century concept of the female menstrual cycle was limited by the fact that the regulatory hormones had not yet been discovered, and by the limited knowledge of the time of ovulation with respect to the menses. As an illustration of the state of the art in this area, one can mention the fact that Ulesko-Stroganova (see below) believed that the function of the corpus luteum was in the role of a phagocytic organ. Nevertheless, numerous theories on the interconnection of menses, ovulation, and propensity of conception abounded not only in the Western world but also in Russia.

One of the more interesting theories was proposed by Feoktistov (191), a St. Petersburg gynecologist. He noted that recent work from Leopold's laboratory in Germany indicated that menstruation and ovulation were independent of each other, and that ovulation was not periodic. Leopold also proposed that menstruation was a bodily rhythm, similar to the pulse and respiration. Feoktistov disagreed and instead proposed that menstrual bleeding was brought about by ripening ovaries, which, due to their large size, exerted a mechanical stimulus upon the sympathetic nerve endings in the uterine tissues. This, in turn, initiated a vasomotor response (vasodilation) and subsequent bleeding. He felt that this process was similar to the mechanical ejaculation in the male. He then attempted to correlate the processes of ovulation, menstruation, and conception, noting that according to the then recent work of Hansen, conception occurred easiest immediately after the menstrual blood flow had ceased. He further noted that menstruation was accompanied by the loss of endometrial epithelium, and that the endometrium, prior to the commencement of blood flow, was very heavily impregnated by blood and various other secretions. From these partially erroneous data, Feoktistov reconstructed the following sequence of events: the ripe ovaries initiated menstrual bleeding and thus caused the loss of endometrial epithelium. This then permitted the fertilized ovum to be implanted in the uterine wall, since, with the epithelium in place, the ovum could have never been implanted in the endometrium. If conception did not occur during the relatively short period following the menstrual bleeding episode (1–7 days), the female was then relatively immune to conception because of the rebuilding of the epithelium. Feoktistov felt that the shedding of epithelium was the true "menstruation" in the physiological sense, and the accompanying bleeding was an ancillary phenomenon connected with the sexual cycle of the female. Feoktistov's logic was thus superb; unfortunately, the data that he had to work with (incorrect findings of Hansen and the total absence of endocrinological concepts) did not permit him to make conclusions that would stand the test of time.

The cyclical nature of female physiology in general was later recognized by v. Ott (192). He found that the peak of a woman's activity was reached sometime midway between the menses, then dropped sharply just before the bleeding. He used temperature curves, muscular strength, vital capacity, and response to reflexes to demonstrate such periodicity, and concluded that it depended on the sexual cycle of the female. Ott was also responsible for establishing the independence of fetal circulation (193). He accomplished this by replacing a portion of a pregnant dog's blood with a 0.6% NaCl solution, removing the fetus after an appropriate time interval, and analyzing for protein in both the maternal and fetal blood. He found a pronounced hydremia in the maternal blood and a normal protein concentration in the fetal blood. Prior to performing these experiments, Ott investigated the effects of 0.6% NaCl transfusions on the blood compositions in dogs (194). He was able to demonstrate that protein was not lost from blood at the same rate as the rate of bleeding. Thus, 2/3 of a dog's blood volume was replaced by 0.6% NaCl, and the blood still contained some 57% of its former protein. Complete recovery of protein occurred 30–40 days after bleeding, whereas red cell counts returned to normal after 16 days. It must be noted that 0.6%

NaCl transfusions were used in Ott's time to treat acute blood losses in human patients.

Dimitry v. Ott (1855–1929) was one of the most accomplished gynecologists in Russia. He was a graduate of the Military-Medical Academy (1879) and subsequently studied in Germany, France, Italy, and England. Upon his return to Russia, he joined the Imperial Institute of Obstetrics and Gynecology and eventually became its director. In addition to his contributions to the fundamental knowledge of the female physiology, Ott was a superb operator and surgical innovator. He described a case where a large fibromyoma was complicating a pregnancy and which he treated by a Caesarian section on the 263rd day of pregnancy (195). The tumor was situated partly in the uterus and partly in the connective tissue surrounding the uterus. He removed the uterus, but left the other portion of the tumor intact. Both the woman and her child recovered uneventfully, and the tumor portion that had remained in the peritoneal cavity after the operation had completely receded two years after surgery. Ott also devised an operation to repair damaged urethral musculature. He used the labia minora to create a new urethral opening, which eventually permitted the patient's bladder to work normally (196). A similar operation was later devised by Nicholas Volkovich, a Kiev surgeon, to treat urinary incontinence (197).

There were numerous distinguished gynecological surgeons in Russia in addition to Ott, and most of them were, for some reason, concerned almost solely with the art of hysterectomy. Alexander Gubarov (b. 1855), a professor of gynecology at Dorpat and later at Moscow Universities, exemplified this emphasis in the Russian gynecological circles. He wrote an interesting review article on the subject (198), where he placed special emphasis on the methods to ligate uterine blood vessels. He was opposed to the ligation of veins, stating that this was likely to give rise to thrombosis and embolism. He called attention to the rapid pulse rate in the absence of temperature rise as a symptom of thrombosis in gynecologic patients. Gubarov was also responsible for the development of a technique to tie the uterine artery during surgery.

The title of the most distinguished Russian gynecologic surgeon probably belongs to Vladimir F. Snegirev (1847–1916), a graduate and professor of Moscow University. He was an expert on the causes and treatments of uterine bleeding, and his book on the subject (1886 and 1895) was translated from the Russian into a number of foreign languages. Snegirev was the developer and promoter of a vaginal irrigation method for the treatment of uterine infections. The irrigation with 2% phenol or potassium permanganate was carried out continuously for 1–3 days with excellent results. He was responsible for the development of a modification of Doyen's hysterectomy procedure, where he was able to cut the mortality rates from the usual 18% to 4.5% (199). This was especially significant, since in Russia of that time uterine tumor cases were usually not seen until there was a significant abdominal swelling. The operation lasted from 40 to 90 min. With his associate, N. Altukhov, Snegirev described an improved procedure for the tying of the uterine artery during hysterectomy (200). The technique involved the retraction of the broad uterine ligament and making a 3 cm incision into the tissue binding the broad ligament to the ovarian ducts. The uterine artery was then bound just above the point of branching of the hypogastric artery, and resulted in no damage to the uterus, to the artery, or the uterine veins. Seven cases were treated by this method without a single complication. Snegirev was also responsible for the discovery of a uterine bleeding condition that he called *Endometritis dolorosa* (201). The disease was characterized by hefty abdominal pains, heavy bleeding, hysteria, and even epileptic seizures. The acute stages were usually observed during menstrual periods. The disease was also associated with female sterility. Five points of pain were described as an aid to diagnosis, with the most prominent point being just above the pubic bone. Therapy involved the application of leeches, dilation of the cervix, and irrigation of the uterus with a tincture of iodine or iodoform.

Pregnancy complications received attention as early as the 1860s from J. L. Lazarevich (1829–1902), a graduate of the Kiev University and later a professor at Kharkov University. He studied eclampsia and claimed that it was of nervous origin. The nervous system was also implicated by Lazarevich in the initiation of labor. He developed a very successful method of inducing labor by infusing fluid into the area of fundus uteri by an apparatus of his own invention (202, 203). The births occurred some 17–36 hours after the infusion, and depended on the amount of fluid infused as well as the temperature differential between the infused fluid and the uterus.

Placenta previa was studied by Gubarov (see above), who described a specimen that he obtained from an autopsy of a mother who had died of the complication (204). He called attention to the unusually prominent veins of his specimen, and especially to what he believed was the sinus circularis near the entrance to the uterus. He felt that these veins were exceptionally fragile and were responsible for the bleeding episodes seen in placenta previa conditions. In addition, the damage to the veins afforded an easy access for the bacteria into the systemic circulation, thus predisposing the patient to sepsis. Placenta previa was described at length in a widely read book by N. Pobedinsky (1861–1923), a Moscow University professor. He was also an accomplished gynecologic surgeon who reported delivering a child from a mother with a 1.6 kg tumor in her uterus without losing either the mother or the child, or even the mother's uterus (205).

Vasily V. Stroganov was best known for his method of eclampsia treatment. In the West, it is termed the "conservative method of Stroganov." He was born in 1857 in the Smolensk province of Russia and was a graduate of the Military-Medical Academy (1882). His professional life was spent almost entirely at the Imperial Obstetric and Gynecological Institute of St. Petersburg (1885 to 1926), first under Ott, later as a full professor. In the late nineteenth century, the mortality rate among those who had contracted eclampsia was some 15–30%, and the various treatments available at that time did not seem to do much good. Stroganov based his course of treatment on the premise that eclampsia was complicated by some disorder in the central nervous system, and that the body was getting insufficient amount of oxygen (206). He therefore administered oxygen to his patients, placed them in a darkened room with minimal noise level to eliminate central nervous system stimuli, and administered a combination of chloral hydrate and morphine to control cramps and pain. Within 24 to 48 hours the episode was usually over. At no time was the removal of the fetus performed, a procedure advocated by some authorities. However, children born of eclamptic mothers showed an 11–22% mortality rate, a figure well below the 25 to 40% rate seen in mothers treated by the then conventional methods. In his first report (206), Stroganov described 58 cases of eclampsia that were treated by his method and which did not result in the loss of a single patient. He later reported mortality rates of about 4%. Stroganov believed that acute eclampsia was brought about by an infection and was complicated by central nervous system stimuli. He felt that chloral hydrate acted as an antagonist to toxins elaborated by the infecting microorganisms.

Stroganov's contributions were of course not limited to the treatment of eclampsia. He did an extensive study of the bacterial flora of the female genital tract all the way from infancy to menopause (207). He found that the vagina of the newborn was essentially sterile, and that bacteria were introduced by handling of the baby such as bathing and diapering. Later in life, the vagina contained numerous organisms including bacilli, diplococci, staphylococci, and streptococci. He observed no significant change of bacterial flora during menstrual bleeding periods, pregnancy, or abortion. Contents of the cervix were sterile. Stroganov felt that the pathologic microorganisms could not survive because of the acidic medium in the vagina and because of the action of normal vaginal microorganisms. He thus concluded that the vagina was capable of cleansing itself. Lastly, Stroganov was able to realize the medical utility of the samovar (208). He reported that instruments and other delivery paraphernalia could be sterilized in the samovar by a 30-minute heating period, or by 3 periods of 10 min. each. This work was apparently meant to provide a reliable sterilization procedure for use in the countryside where hospital facilities were not available.

Puerperal sepsis was in the nineteenth and early twentieth centuries a common and dangerous complication following delivery. Many methods had been proposed for its handling, and the most popular techniques in Russia were those of drying out the uterine cavity (Matsevsky), the infusion of $NaCl$ solutions (Wernitz), and irrigation with corrosive sublimate (Shchetkin). One of the more successful methods was that of Grammatikati (209), a University of Tomsk gynecology professor who irrigated the uterus with a solution of 5% aluminol–2.5% iodine in 95% ethanol. The technique was especially useful with gonorrheal uterine complications. However, the method of Sitsinsky, a young gynecologist working under the direction of Ott at the Imperial Institute of Obstetrics and Gynecology, appeared to be most successful in eliminating the 5–10% mortality rate observed in patients with puerperal sepsis (210). Sitsinsky observed 3147 births and abortions over a two-year period at the institute and of these, 246 patients developed uterine sepsis. Treated by his procedure, none of the affected individuals succumbed to the disease. Sitsinsky's procedure involved four steps/uterine washes with solutions containing sublimate, boric acid, water, and 95% ethanol. He cautioned the gynecologists to avoid moving the position of the uterus during the washing procedures.

Uterine pharmacology was studied *in vitro* by Kurdinovsky (b. 1874), who began his studies in Kravkov's pharmacological laboratory at the Military-Medical Academy, and then continued them independently in Switzerland, Germany, and again in Russia. He first exsanguinated uteri by pumping Locke's solution through the aorta, excised the uteri, and placed them in Locke's solution at 38°C with the appropriate amount of oxygenation. The uteri were kept viable for as long as 49 hours (211). He found that the uterus acted as an autonomous organ with rhythmic contractions and peristalsis-like motions in the horns. It responded by contracting vigorously to various stimuli such as heat, cold, and mechanical manipulation. He observed several cases of apparently normal deliveries by isolated uteri, whose fetuses lived for as long as 30 min. after birth. This finding prompted Kurdinovsky to propose that the central nervous system had little to do with the initiation and completion of labor, and proposed that delivery was strictly a local reflex process. Such a rejection of the central nervous system role was indeed unusual for a Russian scientist. Kurdinovsky also tested the action of adrenaline on the uterus, and found it to cause a tetanic contraction. Such a contraction was elicited even when only a small portion of the uterus (e.g., one of the horns) came into contact with the hormone. The *in vivo* effects of adrenaline on the female reproductive system were studied by Fenomenov, a professor first at the Kazan University, later at St. Petersburg, who found its action on the vaginal tissue to be different from that on the uterus wall: whereas uterine contraction was observed in response to adrenaline administration, no effects on the vaginal wall were observed. He used adrenaline to stop uterine bleeding, noting that a marked vasoconstriction accompanied this effect (212). The actions of other biologically active compounds on the isolated uterus were also studied by Kurdinovsky (213), with the purpose of selecting drugs that would be of potential usefulness in inducing labor or abortion. Quinine was found to cause a tetanization of the uterus, whereas berberine, hydrastimine, and physostigmine brought about an increase in the intensity of normal uterine contractions. Cotaramine chloride caused a mild tetanization, whereas caffeine and strychnine had no effect.

The most accomplished investigator of the microscopic structure of the female reproductive system was Catherine Ulesko-Stroganova (1858–1943), a graduate of the St. Petersburg Women's Medical College (1886) and a member of the Imperial Institute of Obstetrics and Gynecology. She made a thorough study of the epithelium of the female genital tract (214), the decidual membrane (215), and the placenta (216). She confirmed the findings of other authors to the effect that placental villi consisted of an outer cell layer called syncytium and an inner layer called the Layer of Langerhans. The latter was found to consist of several rather than a single cell layer. She believed that both types of cells had a common origin, namely the fertilized ovum and not the uterus as had previously been thought. The uterine wall itself was found to be transformed into a heavily vascularized structure, into which villi buried themselves. All placental cells observed in the intervillar spaces by other authors were traced by Ulesko-Stroganova to the syncytium, and were thus adjudged to be of fetal origin.

Decidual tissue was studied by Ulesko-Stroganova at different times of pregnancy. At all of its stages, the decidua were found to contain 3 types of cells. The most prevalent were large oval cells with a single nucleus, or, occasionally, with 2–3 nuclei. The second type of cells was smaller with a single nucleus. The two types of typically decidual cells differed with respect to their safranin staining properties: protoplasm of the larger cells stained pale-pink, with the edges taking up a larger quantity of the dye, whereas the protoplasm of the smaller cells was completely transparent except for the heavily staining granules. The smaller cells also contained mitotic figures much more often than did the larger cells, and Ulesko-Stroganova proposed that the former were actually precursors of the latter. The third type of cells observed in decidual tissue was the mononuclear leukocytes, which were especially prevalent near the blood vessels. All decidual cells, including the white cells, were encased in a fine meshwork of connective tissue, and all cells contained large quantities of glycogen granules. During the very early stages of pregnancy (2 weeks), Ulesko-Stroganova observed a swelling of the uterine mucous tissue with swollen and widely separated connective tissue cells (an edemalike condition). However, in addition to the connective tissue cells, numerous mononuclear leukocytes were also evident, most of them being in the process of division. She concluded that the leukocytes were the chief precursors of decidual cells, with the connective tissue cells also making a certain contribution to the formation of decidua. She also felt that the function of the decidua was in the capacity of a phagocytic organ whose assignment was to inactivate the wastes elaborated by the fetus. The glycogen observed in decidual cells was, in her view, used to provide energy for the phagocytic process and not for the nourishment of the fetus, as had been previously proposed by other authors.

Ophthalmology and Otology. The practitioners of ophthalmology and otology were the first medical specialists of Russia to achieve worldwide reputations. Important papers in these fields were written by Russian scientists as early as in the 1840s, at which time other medical specialties were still in their infancies. This was not, however, the result of some progressive attitude on part of the government or educational establishments, but was rather caused by the initiative and high degree of competence on the part of Russian ophthalmologists and otologists themselves. By 1893, there were only 200 ophthalmologists in Russia and an additional 300 physicians who considered themselves competent in these areas. By that time, the population of Russia was 112 million souls and blindness and eye diseases were very common there. According to Talko (217), there were 200,000 blind persons in Russia in 1893, many of whom were cared for by a semiprivate network of asylums ("curatoriums") under the nominal protection of Empress Maria Alexandrovna. Later, under the initiative of Bellarminov, six eye clinics were established in various parts of Russia, which spearheaded blindness prevention programs in their regions.

The 200,000 figure reported by Talko may or may not have been correct, since other authors reported vastly different figures. Perhaps the most pessimistic report was written by A. J. Skrebitsky (218). He made a survey of the medical records of army recruits between the years 1879 and 1883 and found that for every 108 recruits examined, one was excused because of "blindness." This report was challenged by V. Dobrovolsky (219). He pointed out that first of all, Skrebitsky failed to include in his sample the large number of recruits who were assigned to the reserves rather than taken into the active army, and secondly, he did not mention the criteria used by the Russian army in defining "blindness." For instance, if the recruit's vision in the right eye was worse than 20/40, he was excused as being "blind," no matter how good his left eye was (apparently aiming the rifle to shoot required a good right eye: auth.). Dobrovolsky also related a method that had originated among the Caucasian Jews, whereby the right eye of their inductees was injured with a long needle so that a cataract would develop there. This was then sufficient to stay out of the army. Dobrovolsky himself presented some statistics on blindness in St. Petersburg, stating that its incidence in the city was 8.8 per 10,000. This figure, he stated, was somewhat higher than those in other parts of the country because there were several large asylums for the blind in the city. Among the 27,291 St. Petersburg recruits that he studied, thirty-four were excused from military service because of "blindness." This included those who were totally blind as well as those whose vision was worse than normal in the right eye. This came to one "blind" recruit for every 802, a number considerably lower than that of Skrebitsky.

Russian ophthalmologists were organized into the Russian Ophthalmological Society, established in 1885 in St. Petersburg by a group of 16 oculists (220). Many were also members of the German Ophthalmological Society: of its 336 members in 1886, eighteen were Russians, and of the 264 members in 1892, twenty were from Russia.

The first chairs of ophthalmology in Russian medical schools were established in 1859, one at the Military-Medical Academy and the other at Moscow University. The first occupant of the academy chair was Eduard Junge, and Gustav Braun was the first chairman at Moscow University. The most distinguished ophthalmologist before their time was most likely Vladimir Karavayev (221), who died in 1892 at the age of eighty-one with a fifty-two-year professional career behind him. He was a professor at Kiev University and was especially skillful at removing cataracts. He is said to have removed cataracts from the eyes of Empress Helena Pavlovna, sister of Tsar Nicholas I. The man who was primarily responsible for elevating ophthalmology to the dignity of an independent medical specialty in Russia was Eduard Junge (1832–1898). He was born in Riga, obtained his medical education at Moscow University, and studied abroad under Virchow, Mueller, Helmholz, and Graefe. He occupied the ophthalmology chair at the Military-Medical Academy from 1859 to 1882, when he resigned to accept the directorship of the Agricultural Academy in Moscow. Through Junge's efforts, an oculist was appointed to the medical staff of every military district in Russia to look after the vision of soldiers (222).

In addition to being an able administrator and promoter of the field of ophthalmology in Russia, Junge also contributed some original work to his specialty. He was able to demonstrate a clinical case where, in a child with general weakness, conjunctivitis, and keratitis, the cornea was inflamed because the nerve fibers of the trigeminal nerve were degenerated (fatty degeneration). The appropriate ganglia showed a heavy infiltration by connective tissue and fat droplets (223). He was also responsible for some fundamental work on the pathology of retinitis pigmentosa and managed, along the way, to accuse H. Mueller of absconding with his ideas on the subject (224). Junge discovered a striated retina in a seventy-six-year-old woman who had died of a stroke. The pigment cells that had invaded the retina had positioned themselves along the blood vessels, thus forming a characteristic pigmented network. The same pigment cells were also found in the choroid, and pathological changes were also seen in the nerve fibers. The retina had lost its function wherever the pigment cells were localized, and such atrophied areas had become extraordinarily thick. In another individual with a similar disorder, the pigment cells were distributed throughout the retina at random without any relation to the blood vessels. In this case the optic nerve disc was unrecognizable, and the ganglion layer of the retina had completely degenerated. He felt that the retinitis was caused by an inflammation of the choroid, and that the immediate causes of blindness were optic nerve atrophy and the deterioration of the blood supply.

The first chair of ophthalmology to be established at Moscow University was occupied by Gustav Braun (1827–1897), a graduate of the same university. He wrote the first Russian language ophthalmologic text, and was the author of several original articles on glaucoma (225). Simultaneously with Donders and Graefe, Braun proposed that a condition known as excavation of the optic nerve disc *(Amaurosis cum excavatione papillae nervi optici)* was a glaucoma. Today, it is known to occur as a complication of glaucoma. The most striking property of this disease, according to Braun, was the absence of acute episodes before partial blindness set in. He treated most of his patients with iridectomy, but when surgery was not consented to, he observed a development of opaqueness of the vitreous humor, the lens, the cornea, and the iris. The cornea became very hard and insensitive, and, at this point, complete blindness set in.

There were several other pioneering oculists in Russia whose primary positions were in the various Russian hospitals rather than medical schools. One of these, W. F. Froebelius (1812–1886), was born in St. Petersburg and was a graduate of Dorpat University. In 1842 he joined the St. Petersburg Eye Hospital and was also on the staff at the St. Petersburg Foundlings' Hospital. He was the first to perform ophthalmoscopic examinations in Russia (1851), the first to perform an iridectomy in Russia (1857), and was instrumental in establishing the first Russian vaccination institute at the Foundlings' Hospital. He was an accomplished eye surgeon who advocated the removal of a large portion of the iris for the purpose of treating glaucoma (226).

Robert Blessig (1830–1878) was another famous Russian oculist of the pioneering era. He was born in St. Petersburg, studied medicine at Dorpat graduating in 1855, and spent some time abroad with Virchow, Arlt, and Graefe. In 1863 he became the chief physician at the St. Petersburg Eye Hospital, where he remained until his death from typhus fever acquired from a patient during surgery (227). Blessig has been credited with being first to describe retinal cysts in 1855, which today are called Blessig's (or Ivanov's) cysts. He was also one of the first to describe the embolism of retinal artery (228). This disorder was observed in a sixty-two-year-old merchant who suffered a sudden blindness in one eye. Blessig saw a yellow spot in the area of the embolus and numerous minute hemorrhages throughout the retina. Later, a discoloration of the retina set in, which Blessig explained on the basis of insufficient circulation. Blessig's recognition of this disorder was undoubtedly aided by his experience with arterial emboli during his sojourn in Virchow's laboratory, where Blessig studied the effects of tying the renal artery in dogs and rabbits (229). The pathological changes observed in the kidney as a result of this maneuver included a marked hyperemia, hemorrhage, inflammation, fatty infiltration, and the destruction of uriniferous and Malphigian tubules.

An oculist who was honored by Junge as the first Slavic ophthalmologist was Victor F. Szokalski (1811–1891), a Pole who had considerable influence on the development of ophthalmology in all of Russia (230). Szokalski was a graduate of the Warsaw University (1827) and participated as a Polish army physician in the Polish insurrection of the 1830s against the Russians. When the rebellion did not succeed, Szokalski emigrated to Western Europe, where he practiced his trade for the next twenty years. His chance to return to Poland came when the Krakow University (Krakow was then part of Austria-Hungary: auth.) invited him to occupy its chair of ophthalmology. However, the Austrian government did not approve the appointment because of his political views. Eventually, the Russian government gave him permission to join the Warsaw Ophthalmologic Institute in 1852. He remained there until the reopening of the Warsaw University when he joined its ophthalmologic faculty. He resigned in 1871 in protest of the edict to make Russian the language of instruction at the university. Yet, in spite of his anti-Russian stance, Szokalski was awarded the Order of St. Vladimir on the fiftieth year of his medical practice (1889). Szokalski published most of his work in local Polish and Russian medical journals.

Perhaps the most prolific contributor from Russia to the German ophthalmological literature was Joseph Talko (1838–1906), a graduate of the Kiev University (1861) and a military physician for most of his professional life. After being stationed in the various areas of the Russian Empire, he was appointed chief oculist for the Warsaw military district. While there, he founded the Lublin Medical Society. Talko's articles on the clinical practice of ophthalmology were restricted to case reports and were thus of limited value scientifically. His main contribution was his faithful documentation, in the German literature, of events in Russia pertaining to ophthalmology, and this represents a valuable source for the medical historian.

Talko provided a remarkably complete statistical study of eye diseases in the Russian army, especially in his own Polish military district (231). Of the 871,000 enlisted men and 36,000 officers in 1890 in the Russian army, 30,400 were treated for eye disorders with 30,000 full recoveries. The highest incidence of eye disease was in the Kiev military district, the lowest in the Don Cossack cavalry. The most prevalent disease was "catarrhal conjunctivitis" (48.8%) and trachoma (36.5%). In another note, Talko wrote about the eyeglasses worn by the Russian patriarchs (heads of the Russian Orthodox Church) between the years 1589 and 1700 (232). All were round and convex and were poorly polished. They were constructed in the pince-nez style. Talko could not establish if they were produced in Russia or abroad.

In one of the more amusing communications, Talko reported on a Saratov police case where a murderer was identified supposedly by an "optogram" taken from the retina of the victim's eye (233). The validity of such an identification was disputed by several scientists including Talko himself, and by Richard Thiele, a well-known photographer. The latter wrote that the police probably used a spot on the cornea rather than the retina. It was not said whether or not such evidence was admitted in court, or what became of the accused. In the same communication (233), Talko reported on the use of fish skin where human skin grafts were not successful. This method was pioneered by one Katsaurov, an ophthalmologist with the Yaroslav General Hospital. He used the skin of an eel pout *(Lota vulgaris)* to repair damaged eyelids. Apparently, the attached fish skin acted as support for the growth of epidermis cells, which eventually covered the entire fish skin transplant. This method was also used by the surgery department of that hospital to cover the stumps of amputated limbs. Talko also provided an interesting observation on the ophthalmologic aspects of Tsar Alexander II's fatal injuries in 1881 as a result of the explosion of an anarchist's bomb (234). Though the official cause of death was given as massive hemorrhage from the tsar's shattered limbs, Talko doubted this diagnosis because of a report by a Dr. Dvoryashin, who had attended the tsar following his injuries. He apparently noted a clonic cramp in the left eye of the tsar, which did not disappear even after the heroic measures used to save the tsar's life resulted in raising his blood pressure, improved his heartbeat, and permitted the monarch to remain conscious. Talko felt that the cause of death might have been cerebral hemorrhage (stroke?) since the clonic cramp observed usually indicates damage to the fourth and sixth cranial nerves.

The grandfather of Russian ophthalmologic surgery was probably Georg Jaesche (1815–1876) who, contrary to many medical historians, was not an ophthalmologist but was instead a general surgeon. Jaesche was born in Dorpat and obtained his medical education in his hometown. He then occupied various positions in several Russian provincial hospitals, his last job being as a chief physician in the Nizhnii Novgorod General Hospital. Georg Jaesche's brother, Emanuel, became an ophthalmologist and frequent contributor to medical literature, e.g., his method for alleviating obstruction of the tear duct (235). We are indebted to Emanuel Jaesche for calling our attention to his brother's invention of what we today call the Arlt-Jaesche operation (236). This procedure corrects entropium and distichia, or inversion of the eyelid and inversion of the eyelashes, respectively. Jaesche published his operative method in 1844 in a local Russian medical journal, and his article thus received no attention in the Western world. The operation consisted of three steps: the ciliary edge was separated from the eyelid; an oval piece was cut from the front of the eyelid; and the ciliary edge (with the eyelashes) was folded into the front and sewn over the area that was exposed in step 2 above. The whole procedure took 3–5 min. to accomplish.

The various eye diseases were studied by a number of investigators in Russia. Glaucoma was then, as it is now, a challenging problem. Not much was accomplished at that time in elucidating its etiology, and most research work on glaucoma was being done on producing a diagnosis and symptomatic treatment of the disease. The man who did much to improve the methods for determination of intraocular pressure was L. G. Bellarminov (1859–1930), a graduate of the Military-Medical Academy (1886) and later a professor of ophthalmology at the academy. One of his contributions was the construction of a monometric apparatus with a recorder and a canula to be inserted into the eye (237). With this apparatus he found in animals that intraocular pressure was generally parallel to blood pressure, though in many cases there was a lag period between changes of blood pressure and those of intraocular pressure. Thus, the stimulation of the sympathetics elicited a rise in the intraocular pressure, which was later followed by a rise in blood pressure. There was no connection between intraocular pressure and eye movements. The development of glaucoma following cataract surgeries was described by Graefe in 1869, but this complication apparently escaped the attention of most ophthalmologists. The man who "rediscovered" this problem was K. Rumshevich, a Kiev University professor who described several such cases (238). He treated such induced glaucoma cases by iridectomy. Rumshevich was also a pioneer in the investigation of malignant tumors of the eye. For example, he described two cases of corneal sarcoma, whose origin was elsewhere in the body (239).

As was mentioned previously, the method of choice for the treatment of glaucoma in the nineteenth century was iridectomy, and this operation was introduced into Russia by Froebelius. It was substantially improved by S. Logechnikov (1838–1911), a Moscow University professor and a sometime student of Arlt and Graefe (240). The operation, officially called *iridosclerotomy*, was described by Logechnikov in a Russian ophthalmological journal in 1889, whereas in 1894, much to Logechnikov's consternation (241), Kniess from Heidelberg claimed to have developed this procedure. In the West, Logechnikov is probably best known for his discovery of a symptom characteristic of diffuse scleroderma (242). He described a seventeen-year-old girl suffering from this disease whose eyes were completely immobilized, i.e., no movement up or down, or sideways could be accomplished. The eyes were otherwise completely normal, especially their accommodation functions. The patient eventually recovered, and the eye movements returned to normal therewith. Logechnikov felt that this symptom was due to a disturbance in the structure of eye muscles rather than a disorder of the nervous system.

The trachomatose process was studied in some detail by Emmanuel Mandelstamm (1839–1912), an ophthalmologist in Southern Russia where the disease was very prevalent. Mandelstamm (243) was born in Kovno in a Jewish merchant's family and studied medicine in Dorpat and Kharkov Universities. In 1868, he defended his doctoral dissertation in St. Petersburg. From 1875 to 1880, Mandelstamm was the acting head of the Ophthalmology Department of Kiev University substituting for the ailing Ivanov (see below). Upon

Ivanov's death, the faculty elected Mandelstamm to become chair of ophthalmology, but the university administration did not approve the appointment, apparently because Mandelstamm was Jewish. His appointment unfortunately came for review during the turbulent times following the assassination of Tsar Alexander II, which was blamed, at least in part, on a Jewish conspiracy. Mandelstamm thereupon resigned from the university and opened a private eye clinic, where he practiced ophthalmology for a number of years. In examining a number of trachomatose patients (some 38% of Mantelstamm's clinic patients had this disease), Mandelstamm concluded that the basic trachomatose process appeared in several forms, which theretofore had been considered to represent separate diseases (e.g., follicular catarrh, acute trachoma, chronic trachoma, etc.) (244). He felt that trachoma was a nonspecific inflammatory process caused by accumulation of leukocytes in the adenoids, and that methods for treating inflammations in general were applicable for treatment of trachoma. Mandelstamm also published a number of case reports, mostly in *Graefe's Archiv.*

One of the foremost Russian eye pathologists of his time was Alexander Ivanov (1836–1880) (245). He was a graduate of Moscow University (1859) and, after a study abroad, he was appointed chair of the Ophthalmology Department at Kiev University (1869). Throughout his entire life Ivanov was ill with tuberculosis, and attempted to keep the disease in check by frequent trips to Southern Europe. The last four years of his life (1876–1880) were, in fact, spent in Mentone, in Southern France. According to Russian law, a professor could not absent himself from his post for more than 4 months without losing his job. However, an exception was made in Ivanov's case by the tsar himself, who permitted Ivanov to keep his chair in Kiev until his death. Ivanov's reputation as an eye pathologist and pathological anatomist was so great that investigators from all over Europe sent him specimens for pathological examination. As a result, he was able to make a number of original observations on the nature of such processes as inflammation of the retina and optic nerve, vitreous body, and the development of Blessig's (also called Ivanov's) cysts. Detachment of the vitreous body was most frequently observed in eyes damaged by a foreign body (246). In one typical case, the foreign body had brought about damage to the cornea, pus formation in both the retina and choroid, and degeneration of nerve cells in the retina. The latter were replaced by connective tissue. The pus cells had originated from the blood vessels of the retina, and Ivanov proposed (correctly) that these were actually circulating leukocytes. The vitreous body was detached from the retina except for a spot near the origin of the optic nerve. It was also infiltrated by pus cells. The space between the retina and vitreous body was filled by a gelatinous exudate, which also contained pus cells. In several cases, the vitreous body contained abscesses in addition to the other pathological changes. Detached vitreous body was also observed by Ivanov in absence of any accidental injury, such as following cataract operations, in myopia, and tumors of the eye.

Ivanov provided a detailed description of Blessig's cysts, which he called edema of the retina (246). This disorder did not favor any specific age group, and was not localized in any specific topographical area of the retina. The cysts were present largely in its outer layers and were especially large in the outer and inner nuclear layers. The size of such cysts sometimes reached that of the pea. The smaller cysts were filled with a liquid rich in protein, the larger ones—with a gelatinous material. Ivanov thought that the development of retinal cysts was due to disorders in the local capillary and venous circulation.

Physiology of the eye and its deviations from the norm were studied in Russia by a number of workers, most prominent of whom were Adrian Krickov (1849–1908) and M. V. Voinov (d. 1875) from

Moscow, and Vladimir Dobrovolsky (1838–1904) of St. Petersburg. The latter inherited Junge's chair of ophthalmology when Junge retired in 1893, and Dobrovolsky's post was, in turn, taken over by Bellarminov when he retired.

Dobrovolsky's interests included the mechanism of accommodation, the etiology of myopia and astigmatism, and the physiology of the retina. He believed that there were two types of myopia, the inborn type where the eye axis was lengthened as a result of a congenital disorder, and the acquired type where the lengthened axis was due to the atrophy of the optic nerve. He was a strong proponent of the theory that taught that the sole mediator of accommodation was the ciliary muscle (247). From his study of astigmatism in a number of patients, Dobrovolsky concluded that the disorder was accompanied in most cases by amblyopia (where a patient sees the object for the first minute or two, and which then becomes blurred), and that the proper diagnosis of astigmatism could not be accomplished without the use of topical atropine (248). Dobrovolsky was able to bring about artificial astigmatism in himself and a number of his colleagues by fitting various types of eyeglasses and doing ophthalmologic examinations on his subjects. He found that in every case of artificially-produced astigmatism, there was a distance where the subject could clearly and sharply see writing and other objects. This phenomenon, Dobrovolsky reasoned, could be produced only if the focal point of the light rays entering the eye was exactly at position of the retina. When the object was not at this optical distance from the eye, the focal point of the light rays entering the eye was either behind or in front of the retina, the result being blurring of vision. Dobrovolsky felt that astigmatism was due to the inability of the eye to accommodate to the various object distances, and proposed that this was due to an inability of the lens to change its shape. This was, in turn, due to an abnormality in the function of the ciliary muscle. He then presented a number of cases where proven changes in the physiology of ciliary muscle were accompanied by astigmatism.

Dobrovolsky's work on the physiology of the retina (249) was a follow-up on Voinov's theory of color perception (see below). He investigated in great detail the perception of colors (red, green; blue and yellow) by the various areas of the retina, finding that the intensity of color perceptions vary unequally throughout the surface of the retina. For example, Dobrovolsky found that the best color perception was in the center of the retina, and that it declined toward periphery of that structure. The center of retina was least sensitive to red, whereas the perception of blue and yellow colors was strongest in its center. Much of Voinov's work on this subject originated from his tenure at Helmholz's laboratory in Germany. He was concerned with the various aspects of optometry, such as the determination of the strength of eyeglasses to be prescribed (250), the causes of binocular vision, and the physiology of color blindness (251). However, his most important contribution appears to be his paper on color perception, which was first communicated in January 1874 at a medical society meeting in Moscow and was published in 1875 (252). His proposal was a significant step beyond the then dominant Young-Helmholz theory, and one wonders why Voinov's contribution is so seldom if ever mentioned in the pertinent literature. Perhaps one part of this error of omission is due to the fact that in May 1874, Hernig, a German ophthalmologist, published a similar theory, so that his name is usually associated with it. Voinov proposed that there were two types of light receptors in retina: the color receptors and the receptors that distinguish the shades of dark and light. The peripheral retinal areas contained the latter receptors only, whereas the center of the retina contained both. He also proposed the existence of four types of color receptors: the red, yellow, green, and blue. All of these were present at the center of the retina, but moving toward its periphery, the yellow and red receptors become less plentiful, and finally there existed an area in the retina where only the blue and yellow receptors were present. The red and green receptors were most sensitive to injury. If all the receptors were stimulated at the same time, the subject perceived white, according to Voinov. White was also perceived when complementary receptors (red and green, and blue and yellow) were stimulated simultaneously. If noncomplementary receptors were stimulated, different colors were perceived: purple when blue and red were stimulated, and orange when red and yellow were involved. Voinov ascribed the property of dark-light perception to the rods of retina, which, he noted, were much more plentiful in night animals than in day animals. The perception of colors he ascribed to cones, noting their great prevalence in day animals. His theory was also supported by the fact that rods were predominant in the periphery of the retina, where color perception was minimal. Voinov's theory differed from that of Young and Helmholz in that Voinov proposed the existence of two types of light receptors, and that the number of color receptors was proposed to be four instead of three.

Russia has produced a number of pioneers in the field of otology, though few accounts of its development in Russia or biographies of its practitioners can be found in the West. One of the most accomplished otologists in Russia was Robert Wreden of St. Petersburg, who was active in the 1860s and 1870s. His practice included the doctoring of the tsar's family, yet he treated the peasants and common folk with equal diligence and professionalism. Wreden established and operated the St. Petersburg Ear Hospital, whose first patient was, strangely enough, an individual suffering from otomycosis, which Wreden is commonly credited with discovering. Eventually, Wreden saw five other cases of otomycosis (253) and found that the condition was characterized by a rather lengthy period of deafness, pain, and ringing in the ear. The tympanic membrane was covered with a white film that was difficult to remove unless the ear was soaked frequently in warm water, or a solution of sodium bicarbonate. When such white plugs were eventually removed, all proved to be masses of Aspergillus microorganisms. He was never able to observe the presence of Penicillium microorganisms in the ear, even though Aspergillus and Penicillium spores were known to be closely associated with each other in the air. He felt that the best treatment for the disease was to clear out the ear first, then irrigate with a solution of tannin, lead acetate, or corrosive sublimate. Full cure usually followed the therapy.

Wreden wrote extensively about middle ear infections in children (254). He made a survey of 80 deceased children, ages 12 to 14 months, and found middle ear infections or other abnormalities in 84% of them. In 16% he found acute *otitis media,* in 21% there was a chronic middle ear infection, and 45% had purulent *otitis media.* He ascribed the high degree of pathology in the middle ear to the fact that most of the children examined had died of some respiratory problem, and he was therefore able to stress the connection between ear and respiratory diseases. In addition, he noted a very high correlation between ear disease and brain affliction. Such brain involvement was usually characterized by an inflammation of the *pia mater*, though he frequently found cysts, edema, and meningitis in the brain. Wreden was first to describe internal and external diphtherial otitis media, and extensively described gangrenous otitis media. Especially pathetic was the case of a little girl whose father came to Wreden with a story that the ear of his daughter had fallen off. Wreden went to the man's house, and found that the entire temporal bone with the ear attached to it had come off, so that the child's brain was visible. The girl lived for another 10 hours thereafter. Gangrenous otitis media, Wreden found, was usually a complication of measles, and was also common in children with hereditary syphilis.

Wreden was also a successful ear surgeon and surgical innovator. He noted that in most cases, perforation of the tympanic membrane to improve hearing did not result in a permanent cure, since the opening had a tendency to close itself. His modification of this procedure involved the excision of a large piece of the tympanic membrane, and in addition, removal of the malleus handle. He designed his own instruments for the performance of this operation and demonstrated to other otological luminaries of his time, F. E. Weber and Voltolini, his technique. Both were apparently impressed, and Voltolini eventually published a modification of Wreden's procedure (255)

Wreden's concept of the sense of hearing involved the Stapedius as the principal component of this sequence of events (256). According to this theory (*Binnenmuskeltheorie*), hearing was initiated by the contraction of the Stapedius muscle, which forced the plate of the stapes into the vestibule, where the increased pressure was then responsible for stimulation of the appropriate nerve endings. Woldemar Poorten of Riga argued against Wreden's theory, proposing that the contraction of Tensor tympani was, instead, the initiator of hearing.

One of the more distinguished Russian ear anatomists was Alexander Prussak (1839–1897), a graduate of the Military-Medical Academy and later a professor of otology at his *alma mater.* He investigated the blood supply to the tympanic membrane (257) and found that its blood vessels were situated in both the circular and radiating patterns. The tympanic membrane was especially rich in veins, and the arteries were more often than not seen to join the veins directly without going through a capillary network. Many of the veins were found to drain into the Sinus cavernosus, so that the blood flow was extremely rapid with minimal blood pressure. Prussak was particularly interested in the malleus-tympanic membrane interaction (258). He determined that the portion of the malleus in contact with the tympanic membrane was of collagenous character, and was continuous with the circular fibers of the inner layer of the tympanic membrane. He found no synovial fluid and no synovial sack in that area as had been claimed by Troeltasch. The fibers joining the tympanic membrane with the malleus are known as Prussak's fibers. He also described a space between the cartilaginous portion of the malleus and the Pars flaccida. This space is today known as Prussak's pouch or sack.

Among the Moscow ear specialists, perhaps the best known person was Stanislaus v. Stein (b. 1855). He is best known in the West for his detailed study of labyrinthine diseases, and what we today call the von Stein's tests for labyrinthine disease (259). The best known of these is the situation where the patient is unable to stand on one leg with his eyes closed. Another interesting symptom is the hopping maneuver: the patient is blindfolded and told to hop forwarding a given direction. If the patient has a labyrinthine disease, he would hop for a few steps then would continue to hop in one place, all the while believing that he was moving forward. Other phenomena that v. Stein investigated were the behavior of his patients on inclined surfaces, their ability to turn around, and their ability to move backward and sideways. He published an extensive monograph on his observations. Von Stein's contributions also included an interesting discovery in regard to the transmission of sound waves by facial structures (260). He confirmed previous findings that the ear perceives not only the air vibrations, but also bone vibrations. However, he was able to show that the local nervous system had to be intact for this to occur. He, for example, had a patient whose trigeminal and glossopharyngeal nerves were anesthetized because of a head cold. When a tuning fork was placed against the areas served by the affected nerve endings, the patient heard nothing. On the other hand, if the fork was placed against a healthy area, the tone was heard.

And finally, the title of "the most distinguished Russian ear physiologist" should probably go to Elie de Cyon. His contributions are described in detail in the next chapter.

Surgery. Russian surgery may be conveniently divided into the pre- and post-Pirogov periods. The former, spanning some 150 years prior to the 1850–1860s, was dominated by the military surgeons and had few if any civilian counterparts. Even in the latter part of the nineteenth and the early twentieth centuries, the influence of the military surgeon was felt throughout the Russian medical profession, since the most prominent civilian surgeons had begun their careers in the armed forces. Russian military surgery apparently reached its zenith during the Russo-Japanese war when it was an object of study by many foreign medical officers. Of special interest was the Russian method of performing abdominal surgery on the wounded as soon as was possible after the injury was sustained. For this purpose, there were a number of medical railroad trains operating in the nearest possible proximity of the battle line, which were equipped with all the standard surgical and medical facilities. Today, this function is of course performed by the helicopters that transport the wounded to stationary hospitals for immediate surgery. Of special prominence in connection with the Russian hospital trains was a female surgeon, Princess Vera Ignatievna Gedroits, who performed complicated surgeries practically within the range of Japanese rifles (261).

Surgery began its expansion in Russia during the reign of Catherine the Great, though most surgeons of that time were foreigners. For instance, Mohrenheim (see above) was called from Vienna to teach surgery and medicine at a newly-established hospital medical school in St. Petersburg, as was Christoph E. H. Knackstedt (1749–1799) of Braunschweig, Germany. Of the native Russian surgeons of that early period one can mention Peter Dubovitsky (1815–1867) and Nicholas Arendt (1780–1859). The former was a graduate of Moscow University, and was one of the first surgery professors at Kazan University (1837). In 1841, Dubovitsky went to the Military-Medical Academy where he served as president from 1851 to 1867. He published his own medical journal between 1843 and 1848. Nicholas Arendt (262) is said to have been the most prominent surgeon in Russia before Pirogov. He was primarily a military doctor, but was also personal physician to Tsar Nicholas I. He participated in the Prussian campaign (1806–1807), the Swedish campaign (1808–1809), and the Napoleonic war (1812–1814). He was especially well-known for his method of tying blood vessels.

Before the advent of antiseptic techniques in the latter half of the nineteenth century, surgical procedures in general were largely limited to amputations of extremities, the removal of topical tumors, and the removal of stones from the genital-urinary tract. The problem of bladder stones was especially acute in central Russia, so that in the 1860s as many as 20% of all surgical beds in area hospitals were occupied by patients with such stones (263). Most stones that were examined were of mixed character: a uric acid nucleus, followed by a layer of calcium oxalate, and lastly by a layer of phosphates. There was also a large incidence of purely phosphate stones (263). Uric acid stones were rare in Russia in contrast to their incidence in Germany, France, and England. Oxalate stones were most common in children and young adults, whereas phosphate and urate stones were seen mostly in older persons. It was speculated that the high incidence of stones observed was due to largely vegetable diets among the populace and frequent intake of acidic beverages such as the fermented bread drink called *kvas.* The removal of stones in Russia was done by the lithotomy method, since lithotripsy apparently resulted in higher mortality rates. The Moscow University surgical clinic reported a mortality rate of 12% among 4486 lithotomies performed, compared to a rate of 14% in England and France. Ilya Buyalsky (1789–1864) is said to have been one of the most skillful stone surgeons in Russia of that time.

Many innovations on the techniques of amputations were made by Russian surgeons, the best known being the osteoplastic amputation of the foot described by Nicholas I. Pirogov (1810–1881) in 1854 (264). It is today known as the Pirogov-Syme operation. Pirogov's most important contribution to medicine was, however, the popularization of surgery in Russia and its elevation to a high level of competency. His name was honored by the Russian medical profession that termed its annual medical conventions "The Pirogov Medical Congresses"; by the tsarist government that bestowed upon him the highest civilian and military honors; and by the Soviet regime that portrayed him as one of the greatest personalities of the medical profession. Pirogov was born in Moscow and graduated from Moscow University as a physician at the age of seventeen. After some postgraduate work at Dorpat, Pirogov went abroad to study with the Weber brothers in Leipzig, Rokitansky in Vienna, and Schoenlein in Zurich. He returned to Russia in 1835 and a year later, he was appointed to the surgery faculty at Dorpat, thus being one of the first Russian-nationality professors at that German-dominated university. In 1840, Pirogov moved to St. Petersburg to occupy the chair of surgery at the Military-Medical Academy. Upon his suggestion, a surgical clinic was established under the chair of surgery.

Pirogov's pedagogic activities were frequently interrupted by his participation in various military campaigns of that era, such as the Caucasian campaign of 1847, where he first used ether as an anesthetic, and the Crimean War of 1855–1856. During the latter campaign, Pirogov organized a nurses' corps to help with the wounded and fought, up to the emperor himself, for an improvement of conditions in military hospitals. In 1856, he resigned his position in St. Petersburg and was appointed to oversee development of the educational system in Southern Russia. In 1861, he was appointed to study educational systems abroad for the purpose of recommending foreign institutions where Russian students could get the best possible postgraduate higher education. During the Russo-Turkish war in 1877–1878, Pirogov was busy taking care of wounded Russian soldiers in the war's European theater.

Pirogov made many contributions to medicine in addition to describing his osteoplastic operation technique. He was one of the first physicians in Europe to use ether as an anesthetic, though contrary to his Western colleagues, he preferred to administer ether rectally. He also instituted in 1852 the use of frozen body sections as an aid for the study of anatomy in Russian medical schools. He wrote extensively on the subject of military medicine on the basis of his experiences gathered in the wars in Caucasus and Crimea (265). His works differed from those of his British and French counterparts in that instead of merely presenting casualty statistics, he wrote mostly on methods that he believed to be helpful in diminishing mortality and morbidity rates among the wounded. He felt that in this regard, the proper organization of the medical aid system, a proper system of nursing care, and minimizing transportation of the wounded was far more important than the quantity and quality of attending physicians.

The tradition of excellence created by Pirogov at Dorpat was very ably continued by Georg Adelmann (1811–1888) who assumed Dorpat's chair of surgery upon the departure of Pirogov (266). Adelmann was born and educated in Germany and went to Dorpat in 1841 upon Chelius' recommendation. He stayed at Dorpat for thirty years, the customary length of time in the Russian civil service, then retired to Berlin. During this time, Adelmann trained an entire generation of Russian surgeons, among whom were such men as Szymanovsky, Grube, Bornhaupt, and Rayher. Adelmann's contributions to surgery were many and included a method for performing splenectomy and a method for the amputation of fingers. The latter procedure carries his name to this day.

One of the more prominent students of Adelmann's clinics was Ernst von Bergmann (1836–1907), who was born in Riga and educated at Dorpat. He was in the Prussian service from 1866 to 1870, and took part in the Franco-Prussian war. In 1871, he joined the faculty at Dorpat University, and in 1878 left Russia to accept Langenbeck's chair in Berlin. Bergmann's contributions to surgery were varied and numerous. He is credited with the development of steam sterilization of instruments in 1886, the discovery of antibacterial properties of mercurial salts, and with bridging the gap between the antiseptic surgery of the nineteenth century and the aseptic technique of the twentieth. Most of these far-reaching discoveries were made by Bergmann during his activities in Berlin, and their detailed discussion is well beyond the scope of this opus.

In spite of the great potential of the surgical approach to the treatment of disease, which was evident in the middle of the nineteenth century, the civilian medical establishment of Russia was reluctant to grant surgery a place equal to that of therapeutic medicine in Russia's hospitals and medical schools. Thus, by 1861, there was still no surgical clinic at Kiev University, and at one of Russia's most progressive medical schools, the Military-Medical Academy, a surgical clinic was not established until 1840 after the arrival of Pirogov. According to Walther (267), this was due to the fact that internists were the head physicians in all civilian medical establishments, and they apparently thought little of surgeons (Auth.: residual connection between medieval barbers and surgeons?). Nevertheless, in spite of such early obstacles, the post-Pirogov period of Russian surgery is characterized by innovation and discovery, though occasional statements such as "Little is known of the Russian surgeons who wrote only in Russian, for very few of their works have been considered worth translating..." (268) are found in Western medical literature. The works of Russian surgeons, whether printed in Russian, German, or French, were abstracted by the *Centralblatt fuer Chirurgie,* and were thus available for those unable to understand Russian. As a matter of fact, prominent Russian surgeons such as Velyaminov and Fedorov rarely published their papers in foreign-language journals.

Surgery, as stated above, was in the 1850s and 1860s mostly concerned with the amputation of extremities that had been damaged or become gangrenous, the removal of stones from the genital-urinary tract, and the excision of topical tumors. Abdominal surgery was not practiced because of the high mortality rate from peritonitis and pyemia. However, the development of huge ovarian cysts in some women prompted both the patient and the surgeon to undertake the risk in an attempt to remove such tumors that sometimes amounted to 50–100 pounds in weight. Abdominal surgery thus began with ovariectomies both in Western Europe and Russia. The first successful ovariectomy was apparently performed in the United States in the early 1800s, but attempts at this in Europe met with little success and were not performed until the 1850s and 1860s. Thus, by 1866, 904 ovariectomies had been performed in the world with a mortality rate of 41%. By the same time, Russia accounted for 28 ovariectomies with a 43% mortality rate (269). The first surgeon to perform a successful ovariectomy in Russia was Krassovsky (1862), who was already mentioned in a previous section. By 1866 he had performed 15 ovariectomies, of which 6 ended fatally. Others followed his example, including Szymanovsky, Grube, and Sklifossovsky (270). The latter removed a 31-pound cyst from one woman, and a 54-pound tumor from another. Both survived. The first bilateral ovariectomy in Russia was performed by Maslovsky, an assistant in Krassovsky's clinic (269). He removed a 34-pound cyst from the left ovary of his patient and a smaller one from her right one. The patient recovered. It may be noted that all of these operations were performed without the benefit of antisepsis or asepsis, and may thus be considered to represent a truly heroic period in the history of surgery.

Antiseptic technique was introduced into Russia by Paul P. Pelekhin (1842–1917) who published the first papers on the subject in Russia in 1868 and 1869, shortly after visiting Lister's clinic in England. He was also sent to the United States to learn the methods of military surgery, especially the methods of handling casualties in the Civil War. Pelekhin was on the faculty of the Military-Medical Academy, retiring from service in 1889. Pelekhin's efforts in the field of antisepsis were expanded by Nicholas Sklifossovsky (1836–1904), who in the 1880s designed Moscow University's surgical clinics for antiseptic surgery. Aseptic techniques were popularized in Russia by Sergey P. Fedorov, who in 1893 installed the first autoclave in Russia at the Moscow University. An ingenious method for purifying air in the operating room was devised in 1890 by a Dr. Sapeshko, a gynecologist in Prof. Rein's gynecological clinic at Kiev University. Sapeshko installed a water pipe system that sprayed a fine aerosol of water throughout the operating room before starting the surgery. This process thus precipitated all dust and airborne bacteria (271).

The methodology of anesthesia practiced in Russia paralleled that of Western Europe. As stated above, Pirogov was one of the first physicians in Europe to use ether for this purpose, though ether was later largely replaced by chloroform. An area of anesthesiology that was pioneered in Russia was the intravenous administration of anesthetics. The most popular compound used for this purpose in Russia was hedonal (ethyl carbamate or urethane), which was first proposed in 1895 by Schmiedeberg. Hedonal was first tried in dogs, then on human subjects by Kravkov, a professor of pharmacology in St. Petersburg (272), who preferred to use hedonal (3 g.) in conjunction with small doses of chloroform (10–15 g/hr). Fedorov

(1869–1936) used hedonal without chloroform (273, 274) and noted that there were minimal aftereffects following surgery, that the anesthetic did not bring about significant changes in respiration and pulse rates, and that no renal damage could be observed. Fedorov also noted that an intravenous anesthetic such as hedonal was especially useful in neck and head surgery, where the surgeon would not be hampered by a face mask. In addition to his work on aseptic surgery and anesthesia, Fedorov is also credited with establishing the specialty of urological surgery in Russia. He founded the Urological Society in St. Petersburg in 1907 and served as the vice president of the International Urological Congress in 1914. His work involved the development of techniques for the removal of stones from the urinary tract, electrocoagulation, and nephrectomy (275). He is said to have installed the first X-ray apparatus in Russia only one year after its discovery in 1895.

One of the first Russian surgeons to attain world prominence was Julius Szymanovsky (267). He was born in Riga in 1829 and is said to having been related to a Polish line of royalty and the dukes of Kurland. By the time of his birth, however, his family had become completely Germanized, so that Szymanovsky could not or would not speak correct Russian up until his death. This language difficulty created special problems for Szymanovsky during his activities as professor at Kiev University, though the students did not appear to have been particularly disturbed by all this. At any rate, in spite of his language problem, he was able to write an excellent text on plastic surgery ("Dermatoplastic") in the Russian language in 1865. His research papers were, however, published mostly in German. Szymanovsky received his medical education at Dorpat under Adelmann, graduating in 1856. By 1858 he was a professor of surgery at the Helsingfors University (Finland), and in 1861 he moved to Kiev in the same capacity. Szymanovsky's brilliant career was cut short by a malignant tumor, and in spite of surgery by Pirogov, he succumbed to it in 1868 at the age of thirty-eight.

Szymanovsky's contributions included the construction of new surgical instruments (e.g., a concave chisel to resect portions of the hip joint), the designing of a cast that immobilized the entire pelvic area, a technique for pelvic surgery (276), modification of Pirogov's osteoplastic foot amputation technique (277), and the advocacy to perform foot surgery on the basis of its anatomy and function as a locomotor device (278). He also developed a technique for the postoperative care of surgical wounds, where the stump left after an amputation was permitted to heal while immersed in cold water (279). This method minimized both the morbidity and mortality rates in the wards of Dorpat University Hospital, where the overall mortality rate of "only" 22.9 percent was reported following amputations. This rate was very good by the standards of the early 1860s. Szymanovsky explained the beneficial effects of his immersion therapy by the fact that air was excluded from contact with the wound, and that the pus forming in the wound would sink to the bottom of the immersion fluid.

Szymanovsky was an ardent advocate of the conservative approach to surgery. In this connection, he reported on two cases of cerebral hernia that were at first mistaken to be cysts (280). In the first case involving a nineteen-year-old girl, he began operating for removal of the "cyst" but upon incision of the skin, he found an artery not normally seen in that location. Proceeding more cautiously, he punctured the tumor and obtained a clear fluid that was similar to cerebrospinal fluid. He then realized that he had exposed the dura mater and immediately proceeded to close the wound. Contrary to the experience of other surgeons, his patient survived without developing meningitis, and was discharged from the hospital with instruction to wear a protective head covering. In another instance, a soldier came to the hospital with a similar "tumor." Other surgeons diagnosed a cyst and were ready to operate when Szymanovsky believed that he recognized the lesion as being another case of cerebral hernia. He did not allow the surgery and recommended that the soldier be discharged from the army. The patient, however, refused to do this and was discharged from the hospital. In both cases, the lesions were present on the side of the head—one on the left, the other on the right side of otherwise healthy individuals. The tumors were apparently there for as long as the patient could remember, and Szymanovsky concluded that they were of the congenital defect origin. He urged the surgeons to carefully investigate the history of each cyst, and if such tumors were found to be present since birth, chances were that the surgeon was dealing with a cerebral hernia rather than a cyst. After the appearance of the above report (280), a certain Dr. Mamorsky from Kiev published a paper questioning Szymanovsky's approach to cyst removal, since the soldier discharged by Szymanovsky had tried to have his tumor removed elsewhere and had thus consulted Dr. Mamorsky. The latter obliged and removed a congenital cyst from the soldier's head. Dr. Mamorsky advised that in case of doubt, the surgeon should probe at the base of the tumor until it either hit the bone or elicited cerebral symptoms, as would be the case if a cerebral hernia was involved. Szymanovsky's response to this approach was violent (281). He noted that Dr. Mamorsky had gambled and the gamble paid off. Had the tumor been a cerebral hernia, Dr. Mamorsky would have done an autopsy rather than writing a journal article. In cases like this, Szymanovsky advised, it was better for the patient not to take the risk of surgery and to learn to live with the tumor.

Szymanovsky's sentiments were strongly reiterated by Georg Jaesche, developer of the Jaesche-Arlt operation (see above). He reminded the medical profession (this was in the preantisepsis era) that even simple operations carried a high risk of mortality, such as his own experience with thirteen relatively uncomplicated procedures (e.g., corrections of foot defects), where six patients expired (282). Jaesche also reported on a case similar to that of Szymanovsky's, where a patient was making an excellent recovery after a brain surgery. But as soon as the patient got up for the first time after surgery, he collapsed and died immediately (283). He wrote that persons like Dr. Mamorsky, who unnecessarily take risks with patients' lives, may occasionally be successful, but sooner or later Providence will cut them down to their size through the suffering and needless deaths of their patients.

A very prominent Moscow University surgeon whom the Soviets had considered to have established a Russian surgical school was Alexander A. Bobrov (1850–1904). Numerous Russian surgeons were trained in his clinic and he developed a number of surgical procedures such as a hernia operation (Bobrov-Champagnier method), the removal of liver cysts, and the corrective technique for spina bifida (284). The latter method was first used by Bobrov on an eight-year-old child who had no bowel or bladder control functions. After replacing the spinal cord in the spinal column, Bobrov grafted a piece of the ilium to the spinal column to cover the opening. Eventually, acceptable bowel and bladder controls were developed in the patient, and the child was discharged as cured. Bobrov considered this method best suited for spina bifida in the sacral and sacrolumbar regions. With spina bifida in higher regions of the spinal column, Bobrov suggested using a piece of a rib as the grafting device.

One of the pioneer vascular surgeons in Russia was Johann Minkievicz (1826–1897), a Pole by birth and a Russian military physician for most of his professional life. Most of his research work was done on animals, and it is doubtful that he ever applied his proposals to human beings. He investigated various methods for tying veins with a view of devising best ways of treating varicose veins in human patients (285). He found that the least number of complications was obtained when a section of the vein was removed and the two stumps were ligated with a metallic thread. He noted that the function of the excised vein was readily taken over by other veins, i.e., there was a development of collateral circulation. The physiology of arteries was also investigated by Minkievicz with the purpose of finding ways to treat aneurisms (286). He made incomplete ligatures on arteries using metallic thread and found that the vessels were constricted to ½ to ⅓ of their original size above and below the ligature. He thus proposed that a partial metallic ligature be placed above and below the aneurism, which then should proceed to shrink. He was unable to find a suitable subject among his human patients to test his procedure.

Several Russian surgeons other than Pirogov and Szymanovsky made significant innovations in osteoplastic and orthopedic surgeries. One of these, Vladimir Vladimirov (1837–1903), was the developer of the tarsectomy procedure (287). Vladimirov was a graduate of the Kazan University (1860) and, after working in Germany and France for some two years, he returned to Russia as an assistant at the Kazan University. There he defended his doctoral thesis, the subject of which was a new osteoplastic operation on the foot, whereby the astragalus and calcaneus only were removed leaving the rest of the foot intact. This procedure is known by the name of "tarsectomy," and it was rediscovered five years later by Johann Mikulicz, a German surgeon. It is consequently also known as the Vladimirov-Mikulicz operation. Vladimirov tried twice to be appointed to the chair of surgery at Kazan University, but was unsuccessful and consequently devoted most of his professional life to Zemstvo service and private practice. Toward the end of his life falling upon hard times because of his miserly Zemstvo pension, Vladimirov made his home in a room of a cheap hotel. He died in the Penza Zemstvo Hospital of gangrene and was buried in the hospital's cemetery. The author of Vladimirov's obituary (287) bitterly compared the lives of the Russian surgeon and his German counterpart, Johann Mikulicz. The latter had a coveted position in a university in Breslau, a large private surgical clinic, and a spacious villa on expansive grounds. Vladimirov, on the other hand, being no less talented, could not afford anything better than a hotel room in his old age, and had to die in a small provincial hospital forgotten by everyone.

A surgeon who devoted most of his efforts to osteoplastic and orthopedic surgery was Ivan F. Sabaneyev (b. 1856), a graduate of Kiev University and professor of Odessa University. He emigrated to Turkey following the Bolshevik coup, and his later fate seems to be unknown. Among several procedures proposed by Sabaneyev, his osteoplastic resection of the knee is probably of the greatest interest (288). The operation was carried out mostly in patients with tuberculosis of the joints and involved the removal of patella and the bone endings of the femur and tibia, followed by the joining of the bone stumps. The muscular insertions were thus retained, and the wound was covered by the normal inner surface of the skin. Sabaneyev is best known in the West for his development of the gastrostomy technique, which was also described by Rudolf Frank of Vienna. It is today known as the Sabaneyev-Frank operation, and involves the construction of a fistula into the upper portion of the stomach for purpose of artificial feeding. Sabaneyev used this procedure on four patients in whom the esophagus had been removed because of cancer (289). The patients survived for four days to 2 months, and autopsy of the longer survivors showed that the stomach-skin junction had healed satisfactorily.

Other Russian orthopedic surgeons of some accomplishment included Robert R. Wreden (b. 1867; not to be confused with Robert Wreden, the otologist mentioned above) and Sergey Saltykov (b. 1874). Saltykov was a graduate of the Kharkov University and worked abroad until the First World War. He then returned to Russia, only to leave after the Bolshevik coup. He finally settled in Zagreb,

Yugoslavia, as a professor of pathology. Saltykov's work involved mostly bone transplantation and in one of his papers (290), he mentioned a Dr. Radzimovsky of Kiev University who, in 1881, was apparently one of the first to recognize immunologically mediated rejection of bone transplants. Up to that time, a bone transplant either "took" or did not "take," with no further thought being given to the process.

One of the most versatile Russian surgeons was Nicholas M. Volkovich (1858–1928), who was first a student and then professor at Kiev University. He was responsible for establishing in 1908 the Kiev Surgical Society. Volkovich's work involved several seemingly unrelated areas of surgery, such as plastic surgery of the nose, the repair of urogenital tract in both the male and female, the removal of hemorrhoids, for which purpose he constructed a special set of instruments (291), and orthopedic surgery. In the latter area he was especially concerned with the handling of tuberculous joints, where he advocated a radical resection approach based on Sabaneyev's and Vladimirov's experiences (292). As much as possible, the joint was not to be disturbed, was sawed off at the top and bottom, and was then removed as one piece. The bone endings were then brought together, so that no empty space remained in the appendage. He handled twenty-eight patients in this manner. Volkovich also described a new symptom that he had observed in chronic appendicitis (293). He found that the tone and elasticity of the abdominal muscles on the right side of the patient were markedly decreased in such cases. He confirmed his observations using a tonometer designed by Exner and Tandler.

The most influential Russian surgeon of the years just preceding the Bolshevik coup was Nicholas A. Velyaminov (1855–1920), who was often asked to minister the tsar and his family. He was a graduate of Moscow University, served as a military surgeon between 1878 and 1884, and was appointed to the faculty of the Military-Medical

Academy in 1894. He served as its president between 1910 and 1912. During the Russo-Japanese and First World Wars, Velyaminov directed in a major capacity the medical services of Russian armed forces. He was the founder of the St. Petersburg Surgical Society and of the first major Russian surgical journal, the *Khirurgicheskii Vestnik* (1895). It later merged with another Russian surgical publication, the *Letopis Russkoi Khirurgii.* In 1910, the combined journal was named *Velyaminov's Surgical Archives;* it maintained publication until 1917. Throughout the thirty-two-year period of the journal's existence, Velyaminov served as its senior editor.

Velyaminov's contributions to medicine were concerned with the disorders of the joints and the significance of endocrinology in surgical practice. In 1910, he published an extensive treatise on diseases of joints and their classification. He was especially interested in tuberculous joints and surgical treatment of tuberculosis in general. He was responsible for the establishment of an institution for tuberculous children on the Baltic Sea in the town of Wenden (Ventspils) in the Kurland province of Imperial Russia (in Latvia today). In 1908 Velyaminov published a paper describing a form of polyarthritis, which he saw in patients who had recovered from infectious diseases such as typhoid and scarlet fevers (294). The disease was characterized by a stiffness of all joints, muscular atrophy, damage to the spinal cord, and mediastinal goiter that was diagnosed via X-rays. He believed that the causative agent of the polyarthritis was a toxin produced by the thyroid gland. He treated his patients with a thyroid extract, and a group of such patients experienced full recoveries. In the other group where no dramatic results were seen after the treatment, full cure was affected by thyroidectomy. This apparent paradox was explained by Velyaminov as follows (295): the goitrous thyroid gland produces toxic substances which either enhance the action of the thyroid hormone, or negate its effects. The former instance results in the typical Basedow disease symptoms, whereas in the latter, a myxedemalike disease is produced that Velyaminov called *disthyresis.* Some patients could thus be helped by administration of a thyroid extract, whereas others—by thyroidectomy.

Velyaminov was the recipient of many honors both in his native land and abroad. For instance, he was an honorary member of the British Royal College of Surgeons. The Soviet regime considered Velyaminov to have been one of the most distinguished Russian surgeons, and he occupies an honored place in most texts on the history of medicine printed in the Soviet Union. He is said to having starved to death during one of the famines that occurred in Russia after the 1917 revolution.

Miscellaneous Russian Medical Innovators. There were a number of Russian physicians whose activities cannot be reasonably classified under one above topic or another. Among such influential physicians was V. A. Manassein (1841–1901), an internist, superb diagnostician, physiologist, and social activist. His major research papers were concerned with the physiology of fever (296, 297), where he described inducing fevers in rabbits by ichor injection, then extracting the muscle of both normal and pyretic animals with various solvents. He found much more nitrogen in both the water and alcohol extracts of the feverish animals than in those of normal animals. In addition, the ratio of water-extractable to alcohol-extractable residue was 1:1.2 in normal animals, whereas it was 1:1.8 in pyretic rabbits. He concluded that fevers affected the metabolism of the muscle. However, he also noted that even more pronounced differences were seen in muscle compositions in starved and healthy animals, and since feverish animals do not eat as well as normal ones, the differences observed may have been due to the resultant nutritional states. Manassein is also said to have discovered the antimicrobial properties of Penicillium glaucum, and to have published with Polotebnev a couple of papers on this subject in the *Military-Medical Journal* (1871) and in *Meditsinskiy Vesmik* (1871).

Manassein apparently led an exciting life as a student. He was expelled from both the Kazan and Moscow Universities for his antigovernment activities. He also wrote numerous letters to the editors of Russian newspapers protesting the excessive amount of German influence on the Russian academic life. Yet he then proceeded to apply for admission to the Dorpat University, which, as stated above, was totally dominated by the Baltic German elite. Not surprisingly he did not get admitted. He was finally admitted to the Military-Medical Academy and graduated from it in 1867. He then worked for Botkin for a number of years, then went to Germany to work with Hoppe-Seyler, and upon his return to Russia was appointed to the faculty of Military-Medical Academy. In his medical practice, Manassein was an ardent promoter of physical therapy and nutritional therapy of diseases.

Though Polotebnev was mentioned as the founder of the systematic discipline of dermatology in Russia, the Russian dermatologist best known in the West is probably Peter Nikolsky. He was born in 1858 in Tambov and obtained his medical education at Kiev University. From 1884 to 1887 he served there as an assistant in the dermatology clinic and in 1910 he was appointed to the chair of dermatology at Warsaw University. In 1915, he moved with the University to Rostov-on-Don. Nikolsky is of course responsible for the discovery of what we today call the Nikolsky sign, which is used for the diagnosis of pemphigus (298). It is a disease characterized by blistering of the skin and is invariably fatal. Nikolsky pointed out that the blistering itself was not a reliable diagnosis of the disease, since similar symptoms are found in other conditions, e.g., erythema. Moreover, histological findings are also not conclusive. Instead, Nikolsky proposed that the diagnosis of pemphigus be based on the fact that the skin of such patients loses its upper layer if it is rubbed lightly. He had apparently observed this symptom as early as in 1894, and had noted that it was present at early stages of the disease, even before the appearance of blistering. His patients were also suffering from methemoglobinemia, though he did not speculate on the connection between this blood condition and the skin symptoms. Nikolsky thought that pemphigus was due to a hereditary weakening of the cohesion forces responsible for the maintenance of normal skin structures.

One of the more prominent laryngologists in Russia was E. M. Stepanov of St. Petersburg, who also served as an editor of the German publication *Monatschrift fuer Ohrenheilkunde*. His work is exemplified by a paper describing hemorrhagic laryngitis (299), which he felt was a novel disease. He distinguished three stages of the disorder: the initial catarrhal stage, the hemorrhagic stage, and the last catarrhal stage. The disease generally lasted up to 6 weeks and generally resulted in the patient's full recovery. Laryngoscopic examination showed blood clots distributed throughout the larynx, and cauterization did not stop the hemorrhage.

A pioneer in the field of rheumatic diseases was I. P. Popov (1833–1892), the chief physician in the Nikolayev Naval Hospital (300). In 1886 he was able to culture cocci from the blood of a rheumatic patient that when injected into rabbits, brought about symptoms of acute rheumatism of the joints, pericarditis, and endocarditis. The bacteria could then be recovered from the rabbit bloods and, upon culturing, gave colonies identical to those obtained from the sick patient's blood. Though this work represents an important milestone in the understanding of the etiology of rheumatic fever, it does not appear that Popov's work was ever known abroad, and remained buried in a local scientific publication.

Anatoly Bezkorovainy, J. D., Ph. D.

REFERENCES

GENERAL

Biographisches Lexikon der hervorragenden Aerzte aller Zeiten und Voelker. Urban und Schwarzenberger, Publ., Berlin, 1929–1935

Biographisches Lexikon der hervorragenden Aerzte der letzten fuenfzig Jahre. Urban und Schwarzenberger, Publ., Berlin, 1932–1933.

A. N. Bakulev, Ed. *Bol'shaya Meditsinskaya Entsiklopediya*. State Publ. House, Moscow, 1963.

R. Kuznetsov, Ed. *Lyudi Russkoi Nauki*. State Publ. House, Moscow, 1963.

CITED REFERENCES

1. F. H. Garrison. *Russian Medicine Under the Old Regime*. Bull. N. Y. Acad. Med. 7:693, 1931.
2. Horsley Gantt. *Russian Medicine*. Paul B. Haeber, Publ., New York, 1937.
3. *Medicine, Past and Present in Russia*. The Lancet 1897 (II):343.
4. C. E. A. Winslow. *Public Health Administration in Russia in 1917*. Weekly Public Health Rep. (Washington) 32:2191, 1917.
5. S. Ramer. *Who was the Russian Feldsher?* Bull. Hist. Med. 50:213, 1976.
6. A. Suvorin. *Russkiy Kalendar' na 1885 g.* Suvorin Press, St. Petersburg, 1884, p. 156.
7. Editorial, Public Health Rep. (Washington) 27:23, 1912.
8. Editorial, The Lancet 1915 (I):1054.
9. Editorial. *National Insurance Against Sickness in Russia*. The Lancet 1912 (I):1071.
10. P. F. Krug. *The Debate Over the Delivery of Health Care in Rural Russia: the Moscow Zemstvo 1864–1878*. Bull. Hist. Med. 50:226, 1976.
11. Ph. Biedert. *Die Kindernaehrung in Saeulingsalter und die Pflege von Mutter und Kind*. Verl. V. Ferd. Enke, Stuttgart, 1905.

12. Edith Rimpel. *Ivan Ivanovic Molleson.* Osteuropa Institut, West Berlin, 1968.
13. R. E. McGrew. *The First Russian Cholera Epidemic: Themes and Opportunities.* Bull. Hist. Med. 36:220, 1962.
14. N. Howard-Jones. *Cholera Therapy in the Nineteenth Century.* J. Hist. Med. & Allied Sci. 27:373, 1972.
15. Editorial. Weekly Public Health Rep. (Washington) 29:163, 1914.
16. Editorial. *Plague in Russia.* Brit. Med. J. 1907 (I):1209.
17. A. O. Kawal. *Danilo Samoilowitz: an Eighteenth Century Ukrainian Epidemiologist and his role in the Moscow Plague (1770–1772).* J. Hist. Med. & Allied Sci. 27:434, 1972.
18. Editorial. *Plague in Russia.* Weekly Public Health Rep. (Washington) 29:163, 1914.
19. K. Dewhurst. *Anton Chekhov (1860–1904). Pioneer in Social Medicine.* J. Hist. Med. 10, 1955.
20. N. A. Semashko. *Friedrich Erisman.* Bull. Hist. Med. 20:1, 1946.
21. N. M. Frieden. *Physicians in Pre-Revolutionary Russia: Professionals or Servants of the State?* Bull. Hist. Med. 49:20, 1975.
22. G. P. Sacharoff. *Rudolf Virchow und die Russische Medizin.* Virchow's Archiv 235:329, 1921.
23. S. Botkin. *Ueber die Wirkung der Saltze auf die circulierenden rothen Blutkoerperchen.* Virchow's Archiv 15:173, 1858.
24. E. Botkin. *Zur Morphologie des Blutes und der Lymphe.* Virchow's Archiv 145:369, 1896.
25. Professor Gregory Antonovitch Zakharin. The Lancet 1898 (I):193.
26. G. Sacharjin. *Zur Blutlehre.* Virchow's Archiv 21:337, 1861.
27. Erich Ebstein. *Ueber Milchkuren in aelterer und neuerer Zeit.* Med. Klin. 4:1464, 1908.
28. A. Jarotzky. *Ueber die diaetetische Behandlung des runden Magengeschwuers.* St. Petersb. Med. Wochenschr. 36:13, 1911.
29. A. Jarotzky. *Zur diaetetischen Behandlung des Geschwuers des Magens und des Duodenums waehrend seiner akuten Periode.* Acta Med. Scand. Suppl. 35:3, 1930.
30. A. I. Jarotzky. *Ueber die Veraenderungen in der Groesse und im Bau der Pankreaszellen bei einigen Arten der Inanition.* Virchow's Archiv 156:409, 1899.
31. N. S. Korotkov. Izv. Voyenno-Med. Akad. 11;365, 1905. Quoted by George Pickering in *High Blood Pressure*, Grune and Stratton, New York, 1968.
32. N. S. Korotkov. *Opyt opredeleniya sily arterial'nykh kollateraley,* P. P. Saikin, Publ. St. Petersburg, 1910. Quoted by *Index-Catalogue of the Library of the Surgeon General's Office, U. S. Army* (3rd series), 7:312, 1928.

33. W. C. Krylow. *Ueber die fettige Degeneration der Herzmuskulatur.* Virchow's Archiv 44:477, 1868.

34. A. Ignatovski. *Wirkung tierischen Eiweisses auf die Aorta und parenchymatosen Organe des Kaninchens.* Virchow's Archiv 198:248, 1909.

35. N. Anitschkov and S. Chalatov. *Ueber experimentelle Cholesterinsteatose und ihre Bedeutung fuer die Entstehung einiger pathologischer Prozesse.* Centralbl. f. allg. Pathol. u. Pathol. Anat. 24:1,1913.

36. N. Anitschkov. *Die pathologischen Veraenderungen innerer Organe bei experimenteller Chelesterinesterverfettung.* Deutsch. Med. Wochenschr. 39:741, 1913.

37. A. N. Klimov, L. P. Rodionova, and L. G. Petrova-Maslakova. *Experimental Atherosclerosis Induced by Repeated Intravenous Administration of Hypercholesterolemic serum.* Cor. Vasa 8:225, 1966. Quoted by Daniel Steinberg's *Cholesterol Wars,* Elsevier Press, 2007, pp. 23 & 26.

38. Daniel Steinberg, *Ibid.* p. 20.

39. *E.* S. London. *Zur Lehre von den Bacquerelstrahlen und ihren physiologisch-pathologischen Bedeutungen.* Berl. Klin. Wochenschr. 40:523, 1903.

40. S. W. Goldberg and E. S, London. *Zur Frage der Beziehungen zwischen Bacquerelstrahlen und Hautaffectionen.* Dermatol. Z. 10:457, 1903.

41. A. J. Abrikosoff. *Ueber einen Fall von multiplen Myom mit diffusen Verbreitung im Knochenmark.* Virchow's Archiv 173:335, 1903.

42. A. I. Abrikosoff. *Ein Fall von multiplem Rhabdomyom des Herzens und gleichzeitiger herdfoermiger Kongenitaler Sklerose des Gehirns.* Beitrag z. pathol. Anat. u. z. allgem. Pathol. 45:376, 1909.

43. A. Abrikosoff. *Ueber Myome, ausgehend von der quergestreiften willkuerlichen Muskulatur.* Virchow's Archiv 260:215, 1926.

44. Mstislawus Nowinsky. *Zur Frage ueber die Impfung der krebsigen Geschwuelste.* Centralbl. f. d. med. Wissenschaften 14:790, 1876.

45. K. Buinevitsch. *Meine Theorie der Harnbildung.* Zentralbl. f. inn. Med. 40:410, 1928.

46. Dr. Winogradoff. *Ueber kuenstlichen und natuerlichen Diabetes mellitus.* Virchow's Archiv 24, 600, 1862.

47. Dr. Winogradoff. *Beitrage zur Lehre vom Diabetes mellitus.* Virchow's Archiv 27:533, 1863.

48. L. W. Ssobolew. *Zur normalen und pathologischen Morphologie der inneren Secretion der Bauchspeisdruese.* Virchow's Archiv 168:91, 1902.

49. J. J. Schirokogoroff. *Die Mitochondrien in den erwachsenen Nervenzellen des Zentralnervensystems.* Anat. Anz. 43:522, 1913.

50. J. J. Schirokogoroff. *Die sklerotische Erkrankung der Arterien nach Adrenalininjectionen.* Virchow's Archiv 191:482, 1908.

51. J. J. Schirokogoroff. *Primaeres Sarcom des Pancreas.* Virchow's Archiv 193:395, 1908.

52. Webb Haymaker, Ed. *The Founders of Neurology.* Charles C. Thomas Press, Springfield, 1953.

53. L. R. Grote. *Die Medizin der Gegenwart in Selbstdarstellungen. Wladimir Bechterew.* Verlag v. Felix Meiner, Leipzig, 1927, p. 1.

54. Wl. v. Bechterew. *Von der Verwachsung oder Steifigkeit der Wirbelsaeule.* Deutsche Z. Nervenheilk. 11:327, 1897.

55. W. v. Bechterew. *Ueber ankylosirende Entzuendung der Wirbelsaeule und der grossen Extremitatengelenke.* Deutsche Z. Nervenheilk. 15:37, 1899

56. W. v. Bechterew. *Neue Beobachtungen und pathologisch-anatomische Untersuchungen ueber Steifigkeit der Wirbelsaeule. Ibid.,* p. 45.

57. W. v. Bechterew. *Ueber Epilepsia choreica.* Deutsche Z. Nervenheilk. 12:266, 1898.

58. W. v. Bechterew. *Hemotonia apoplectica.* Deutsche Z. Nervenheilk. 15:437, 1899.

59. W. v. Bechterew. *Ueber die Localisation der Geschmackcentra in der Gehirnrinde.* Arch. Anat. u. Physiol. (Suppl. Physiol. Abt.), p. 145, 1900.

60. W. v. Bechterew. *Ueber das corticale Sehcentrum.* Monatschr. f. Psychiatrie u. Neurologie 10:432, 1901.

61. V. A. Betz. *Anatomischer Nachweis zweier Gehirncentra.* Centralbl. f. d. Med. Wissenschaften 12:578 & 595, 1874.

62. W. Betz. *Ueber die feinere Struktur der Gehirnrinde des Menschen.* Centralbl. f. d. Med. Wissenschaften 19:193, 209, and 231, 1881.

63. A. S. Dogiel. *Die sensiblen* Nervenendungen *im Herzen und in den Blutgefaessen der Saeugetiere.* Arch. f. d. mikrosk. Anat. u. Entwicklungsgeschichte 52:44, 1898.

64. A. S. Dogiel. *Ueber den Bau der Ganglien in den Geflechten des Darmes und der Gallenblase des Menschen und der Saeugethiere.* Arch. f. Anat. u. Entwicklungsgeschichte, Anat. Abt. 1899, p. 130.

65. L. Darkschewitsch. *Einige Bemerkungen ueber den Fasernverlauf in der hinteren Commissur des Gehirns.* Neurol. Centralbl. 5:99, 1886.

66. L. Darkschewitsch. *Versuche ueber die Durchschneidung der hinteren Gehirnkommissur beim Kaninchen.* Arch. f. d. gesammte Physiol. 38:120, 1886.

67. L. Darkschewitsch. *Affection der Gelenke und Muskeln bei cerebrallen Hemiplegien.* Arch f. d. Psychiatrie u. Nervenheilk. 24:534, 1892.

68. Wladimir Muratow. *Zur Casuistik der acuten Hirnkrankheiten des Kindesalters.* Neur. Centralbl. 14:817, 1895.

69. Wladimir Muratow. *Sekundaere Degenerationen nach Durchschneidung des Balkens.* Neurol. Centralbl. 12:714, 1893

70. Wladimir Muratow. *Beitrag zur Pathologie der Zwangsbewegungen bei zerebralen Herderkrankungen,* Monatschr. f. Psychiatrie u. Neurologie 23:510,1908.

71. Ios. Kupressov. *K fiziologii zhoma mochevago puzyrya.* Thesis, St. Petersburg, 1870. Also: J. Kupressow. *Zur Physiologie des Blasenschliessmuskels.* Pflueger's Archiv 5:291, 1872.

72. W. Kernig. *Ueber die Beugenkontraktur im Kniegelenk bei Meningitis.* Z. f. klin. Med. 64:19, 1907.

73. W. Kernig. *Ueber objektiv nachweisbare Veraenderungen am Herzen, namentlich auch ueber Pericarditis nach Anfallen von Angina pectoris.* Berl. klin. Wochenschr. 42:10, 1905.

74. G. Rossolimo. *Le reflexe profond du gros orteil.* Rev. Neurolog. 10:723, 1902.

75. G. J. Rossolimo. *Ueber Dysphagia amyotactica.* Neurol. Centralbl. 20:146, 1901.

76. A. Y. Koschewnikov. *Eine besondere Form von corticaler Epilepsie.* Neurol. Centralbl. 14:47, 1895.

77. S. S. Korsakov. *Ueber eine besondere Form psychischer Stoerung combiniert mit multipler Neuritis.* Arch. f. Psychiatrie u. Nervenheilk. 21:669, 1889–1890.

78. L. Minor. *Centrale Haematomyelie.* Arch. f. d. Psychiatrie u. Nervenheilk. 24:693, 1892.

79. L. Minor. *Klinische Beobachtungen ueber centrale Haematomyelie.* Arch. f. Psychiatrie u. Nervenheilk. 28:256, 1895–1896.

80. L. Minor. *Ein neuer Thermaestesiometer mit Mischvorrichtung.* Neurol. Centralbl. 30:1037, 1911.

81. V. K. Rot. *Meralgia Paraesthetica .* Med. Obozreniye 43:678, 1895.

82. Adolf Hasenclever. *Ungedrueckte Briefe Justus Christian von Loders an den Nationaloekonomen Ludwig Heinrich Jakob aus den Jahren 1810–1813.* Arch. Gesch. Med. 11:300, 1919.

83. Editorial. *Ludwig Stieda.* Anat. Anz. 52:131, 1919–1920.

84. Ludwig Stieda. *Ueber Halsrippen.* Virchow's Archiv 36:425, 1866.

85. L. Stieda. *Der Talus und das Os trigonum Bardelebens beim Menschen.* Anat. Anz. 4:305, 1889.

86. Wenzel Gruber. *Nachtraege zu den Bildungshemmungen der Mesenterien und zu der Hernia interna mesogastrica ueberhaupt; und Abhandlung eines Falles mit einem Mesenterium commune fuer den Duenn-Dickarm einer betraechtlichen Hernia interna mesogastrica dextra und einer enorm grossen Hernia scrotalis dextra besonders.* Virchow's Archiv 44:215, 1868.

87. W. Gruber. *Ueber einen Fall nicht incarcerierter, aber mit Incarceration des Ilium durch das Omentum complicierter Hernia interna mesogastrica.* Zeitschr. f. prakt. Heilk. 9:325, 1863.

88. N. Melnikow-Raswedenkow. *Eine neue Conservirungsmethode anatomischer Praeparate.* Beitr. z. pathol. Anat. u. allgem. Pathol. 21:172, 1897.

89. P. Lesshaft. *Der anatomische Unterricht der Gegenwart.* Anat. Anz. 12:395, 1896.

90. P. Lesshaft. *Ueber die Bedeutung der Bauchpresse fuer die Erhaltung der Baucheingeweide in ihrer Lage.* Anat. Anz. 3:823, 1888.

91. P. Lesshaft. *Die Architektur des Beckens.* Anat. Hefte 3:171, 1894.

92. Seraphima Schachowa. *Untersuchungen ueber die Nieren.* Thesis, Bern,1876 (National Library of Medicine, Bethesda, MD).

93. A. Schklarewsky. *Ueber das Blut und die Suspensionsfluessigkeiten.* Pflueger's Archiv 1:603, 1868.

94. A. Schklarewsky. *Zur Extravasation der weissen Blutkoerperchen.* Pflueger's Archiv 1:657, 1868.

95. N. Bubnoff. *Ueber die Organisation des Thrombus.* Virchow's Archiv 44:462, 1868.

96. M. Nikiforow. *Mikroskopische-technische Notizen. 3. Eine einfache Methode zur Fixation von Deckglaspraeparaten, namentlich solcher von Blut.* Zeitschr. f. wissenschaftl. Mikrosk. 5:337, 1888.

97. M. N. Nikiforow. *Ueber Stoerungen der Blutzirkulation bei Fett- und Parenchymzellenembolie.* Folia Haematol. 1:323, 1904.

98. M. Nikiforoff. *Untersuchungen ueber den Bau und die Enwicklungsgeschichte der Granulationsgewebes.* Beitr. z. pathol. Anat. u. allgem. Pathol. 8:400, 1890.

99. D. Romanowsky. *Zur Frage der Parasitologie und Therapie der Malaria.* St. Petersb. Med. Wochenschr. 16:297, 1891.

100. W. Bloom. *Alexander A. Maximov.* Zeitschr. f. Zellforschung u. mikrosk. Anat. 8:801, 1928–1929.

101. A. Maximow. *Ueber die Struktur und Entkernung der rothen Blutkoerperchen der Saeugenthiere und ueber die Herkunft der Blutplaettchen.* Arch, f. Anat. u. Entwicklungsgeschichte, Anat. abt. 1899, p. 33.

102. A. Maximow. *Zur Lehre von der Parenchymzellen-Embolie der Lungenarterie.* Virchow's Archiv 151:297, 1898.

103. A. Maximow. *Der Lymphozyt als gemeinsame Stammzelle der verschiedenen Blutelemente in der embryonalen Entwicklung und im postfetalen Leben der Saeugentiere.* Folia Haematol. 7:125, 1909.

104. A. Maximow. *Ueber krebsaehnliche Verwandlung der Milchdruese in Gewebskulturen.* Virchow's Archiv 256:813, 1925.

105. S. A, Waksman. *The brilliant and tragic life of W. H. W. Haffkine, bacteriologist.* Rutgers Univ. Press, New Brunswick, 1964.

106. W. M. Haffkine. *A lecture on vaccination against cholera.* Brit. Med. J. 2:1541, 1895.

107. Ernest Hart. *A summary of Dr. Waldemar Haffkine's work against cholera.* J. Amer. Med. Assn. 27:1204, 1896.
108. Bella Nemanova. *Two encounters: Moscow-New York* (in Russian). Novoye Russkoye Slovo, May 6, 1978.
109. Editorial. *Progress in vaccines.* Chem. Eng. News 48:19, 1970 (No. 45).
110. W. M. Haffkine. *A discourse on preventive inoculation.* The Lancet 1:1694, 1899.
111. Leon Korwacki. *Ueber die Schutzimpfung gegen Cholera von Standpunkte der specifischen Veraenderungen.* Zeitschr. f. Hyg. U. Infektionskr. 54:39, 1906.
112. D. K. Zabolotny. *Ein Ueberblick ueber die Entwicklung der Choleraepidemie in Russland im Jahre 1907–1908 und ueber die Anticholera Massregeln.* Centralbl. f. Bakt., Ref. 45:705, 1910.
113. S. Fedoroff. *Zur Blutserumtherapie der Cholera asiatica.* Zeitschr. f. Hyg. u. Infektionskr. 15:423, 1893.
114. Dr. Issaeff. *Untersuchungen ueber die kuenstliche Immunitaet gegen Cholera.* Zeitschr. f. Hyg. u. Infektionskr.. 16:287, 1894.
115. R. Pfeiffer and V. Issaeff. *Ueber die specifische Bedeutung der Choleraimmunitaet.* Zeitschr. f. Hyg. u. Infektionskr. 17:355, 1894.
116. S. J. Zlatogoroff. *Aus den Beobachtungen der Cholera-epidemie von 1907 im Saratowschen Gouvernemant. Centralbl. f. Bakt., Ref. 41:797, 1908.*
117. S. J. Zlatogoroff. *Zur Frage der Diagnostik der Choleravibrionen.* Centralbl. f. Bact., 1. Abt. 48:684, 1909.
118. S. J. Zlatogoroff. *Ueber die Aufenthaltsdauer der Choleravibrionen im Darmkanal des Kranken und ueber die Veraenderlichkeit ihrer biologischen Eigenschaften.* Centralbl. f. Bakt., 1. Abt. 58:14, 1911.
119. S. J. Zlatogoroff. *Zur Morphologie und Biologie des Mikroben derBubonenpest und des Pseudotuberkulosebacillus der Nagetiere.Centralbl. f. Bakt., Abt. 37:345, 513, & 654, 1904.*
120. D. Zabolotny and Dr. Maslekowetz. *Beobachtungen ueber Beweglichkeit und Agglutination der Spirochaete pallida.* Centralbl. f. Bakt.., 1. Abt., 44:532, 1907.
121. M. N. Sakharoff. *Spirachaeta Anserina et la Septicemie des Oies.* Ann. de l'Inst. Past. 5:564, 1891.
122. H. Zorn. *Die Febris recurrens. Nach Beobachtungen auf der maennlichen Abteilung des Obuchoff'schen Hospitales.* Centralbl. f. d. med. Wissenschaften 3:729, 1865.
123. S. Botkin. *Vorlaeufige Mitteilungen ueber die Epidemie der Febris recurrens in St. Petersburg.* Centralbl. f. d. med. Wissenschaften 3:62, 1865.
124. O. Motschutkoffsky. *Experimentelle Studien ueber die Impfbarkeit typhoeser Fieber.* Centralbl. f. d. med. Wissenschaften 14:193, 1876.

125. O. Motschutkoffsky. *Beobachtungen ueber den Rueckfalltyphus.* Deutsch. Arch. f. klin. Med. 30:165, 1882.

126. O. Motschutkoffsky. *The treatment of certain diseases of the spinal chord by means of suspension.* Brain, A Journal of Neurology 12:326, 1890.

127. O. O. Motschutkoffsky. *Ein Apparat zur Pruefung der Schmerzempfindung der Haut-Algesiometer.* Neur. Zentralbl. 14:146, 1895.

128. G, Gabritschewsky. *Therapie des Febris recurrens.* Zeitschr. f. klin. Med. 56:43, 1905.

129. G. Gabritschewsky. *Ueber einige Streitfragen in der Pathologie der Spirochaeteninfektionen.* Centralbl. f. Bakt., 1. Abt. 26:294 & 486, 1899.

130. G. Gabritschewsky. *Ueber Streptokokkenvaccine und deren Verwendung bei der Druse der Pferde und dem Scharlach des Menschen.* Centralbl. f. Bakt., Abt. 1, 41:719 & 844, 1906.

131. N. Langowoy. *Beobachtungen ueber die Wirkung der Scharlach-Streptokokkenvaccine.* Centralbl. f. Bakt., 1. Abt. 42:362 & 463, 1906.

132. G. Gabritschewsky. *Die Versuche einer rationellen Malariabekaempfung in Russland.* Zeitschr. f, Hyg. u. Infektionskr. 54:227, 1906.

133. G. Gabritschewsky. *Ueber prophylaktische Massnahmen im Kampfe gegen die Dphterie.* Centralbl. f. Bakt.., 1. Abt. 26:490, 1899.

134. G. Gabritschewsky. *Zur Prophylaxe der Diphtherie.* Centralbl. f. Bakt., Ref. 36:45, 1901.

135. Harry Wain. *A history of Preventive Medicine.* Charles C. Thomas, Publ., Springfield, 1970, p. 291.

136. H. Harold Scott. *A history of tropical medicine.* Williams & Wilkins, Baltimore, 1939, p. 1074.

137. K. J. Wrublewski. *Win Trypanosoma des Wisent von Bielowesch.* Centralbl. f. Bakt., 1. Abt. 48:162, 1909.

138. B. Danilewsky. *Zur Frage ueber die Identitaet der pathogenen Blutparasiten des Menschen und der Haematozoen der gesunden Thiere.* Centralbl. f. d. med. Wissenschaften 24:737 & 753, 1886.

139. C. A. Hoare. *Early discoveries regarding the parasite of oriental sore (with an English translation of the memoir by P. F. Borovsky "On sart sore 1898).* Transact. Roy. Soc. Trop. Med. Hyg. 32:67, 1938–1939.

140. N. Berestnew. *Actinomycosis.* Zeitschr. f. Hyg. u. Infektionskr. 29:94, 1899.

141. A. Petroff. *Zur Inpfbarkeit der Tuberkulose.* Virchow's Archiv 44:129, 1868.

142. S. Metalnikoff. *Die Tuberkulose bei der Bienenmotte (Galeria melonella),* Centralbl. f. Bakt.., 1. Abt. 41:54 & 188, 1906.

143. S. Metalnikoff. *Ein Beitrag zu der Frage ueber die Immunitaet gegen die Infektion mit Tuberkulose.* Centralbl. f. Bakt. 1. Abt. 41:391, 1906.

144. W. J. Kedrowski. *Ueber die Kultur der Lepraerreger*. Zeitschr, f, Hyg. u. *Infektionskr*. 37:52, 1901.

145. N. Klodnitzky. *Neue Methode der bakteriologischen Blutuntersuchung*. Centralbl. f. Bakt., Ref.. 41:561, 1908.

146. N. Klodnitzky. *Die Methodik der bakteriologischen Blutuntersuchung bei Infektionskrankheiten*. Centralbl. f. Bakt., 1. Abt. 58:376, *1911*.

147. Dr. Meierowitsch. *Zur Aetiologie des Erysipels*. Centralb. f. Bakt. 3:406, 1886.

148. A. Pawlowsky. *Ueber die Mikroorganismen des Erysipels*. Centralbl. f. Bakt. 3:754, 1888.

149. L. Rosenthal. *Zur Aetiologie der Dysenterie*. Deutsche med. Wochenschr. 29:97, 1903.

150. L. Rosenthal. *Das Dysenterietoxin (auf natuerlichen Wege gewonnen)*. Deutsche. Med. Wochenschr. 30:235, 1904.

151. L. Rosenthal. *Ein neues Dysenterieheilserum und seine Anwendung bei der Dysenterie*. Deutsche med. Wochenschr. 30:691, 1904.

152. N. P. Vasilyev. *Infektsionnaya zheltukha*. Stasulevich Publ. Co., St. Petersburg, 1888 (in Russian) (Natl. Library of Medicine, Bethesda, MD).

153. Wladislaw Hubert. *Der gegenwaertige Stand der Frage ueber die Kindersterblichkeit in Russland und deren Bekaempfung*. Arch. f. Kinderheilk. 57:351, 1912.

154. C. Rauchfuss. *Nil Filatov (obituary)*. Jahrbuch f. Kinderheilkunde 55:517, 1902.

155. Abt's Pediatrics, vol. 6, 1925, p. 421,

156. Nil Filatov. *Ueber die Diagnose der Initialperiode der Fieberkrankheiten bei Kindern*. Arch. f. Kinderheilk. 8:180, 1887.

157. Nil Filatov. *Influenza bei Kindern*. Arch. f. Kinderheilk. 5:357, 1884.

158. O. Heubner. *Karl Rauchfuss (obituary)*. Jahrbuch f. Kinderheilkunde. 83:80, 1916.

159. C. Rauchfuss. *Zur Casuistik der Gefaessverschliesungen*. Virchow's Archiv 18:537, 1860.

160. N. Etlinger. *Casuistik der angeborenen Herzfehlern*. Zeitschr. f. Kinderheilk. 6:117, 1885.

161. C. Rauchfuss. *Ueber die paravertebrale Daempfung auf der gesunden Brustseite bei Pleuraerguessen*. Deutsche Arch. f. kilin. Med. 89:186, 1906–1907.

162. S. Lewaschew. *Zur operativen Behandlung der exsudativen Pleuritiden*. Deutsche Med. Wochenschr. 16:1229, 1890.

163. A. A. Kissel. *Zur Behandlung der eitrigen Pleuritiden bei Kindern nach der Methode Prof. Lewaschew's*. Jahrb. f. Kinderheilk. 50:51, 1899.

164. A. Kissel. *Ein diagnostisch schwieriger Fall von Herzklappenerkrankung bei einem 12-jaerigen Knaben*. Jahrb. f. Kinderheilk. 53:717, 1901.

165. A. N. Schkarin. *Eitrige Pleuritiden bei Saeuglingen.* Bakteriologie. Jahrb. f. Kinderheilk. 51:650, 1900.

166. D. Ssokolow. *Zur Klinik der exsudativen Pleuritis bei Kindern.* Jahrb. f. Kinderheilk. 67:70, 1908.

167. D. Ssokolow. *Der differentielle Pneumograph und seine Anwendung bei Kindern.* Jahrb. f. Kinderheilk. 75:265, 1912.

168. D. Sokolow. *Ueber Druesenerkrankungen bei Kindern.* Arch. f. Kinderheilk. 58:103, 1912.

169. W. Tschernow. *Ueber akute chronische Bronchiektasie bei Kindern.* Jahrb. f. Kinderheilk. 69:64, 1909.

170. W. E. Tschernow. *Zur Diagnose und Behandlung der sporadischen und epidemischen Cerebrospinalmeningitis.* Jahrb. f. Kinderheilk. 67:161, 1908.

171. J. W. Troitzky. *Ueber die Funktionsstoerungen des Herzens bei Scharlach.* Arch. f. Kunderheilk. 45:393, 1907.

172. G. E. Wladimiroff. *Ueber die Himbeerzunge der Kinder.* Archiv f. Kinderheilk. 57:127, 1912.

173. G. E. Wladimiroff. *Die Hospitalmasern und Sterbcasuistik nach Masern.* Archiv f. Kinderheilk. 58:346, 1912.

174. A. D. Sotow. *Drei seltene Faelle von Complicationen bei Masern.* Jahrb. f. Kinderheilk. 50:1,1899.

175. W. Moltschanoff. *Zur Frage ueber die Rolle der Nebennieren in der Pathologie und Therapie der Diphtherie und anderen Infektionskrankheiten.* Jahrb. f. Kinderheilk. 76:200 (E-H), 1912.

176. W. N. Klimenko. *Die Aetologie des Keuchhustens.* Centralbl. f. Bakt., 1. Abt. 48:64, 1909.

177. W. P. Schukowsky and A. A. Baron. *Congenitale Atresie der Speiseroehre mit Trachealfistel.* Archiv f. Kinderheilk. 58:191, 1912.

178. W. P. Schukowsky. *Congenitales Sarkom der Nebenniere bei einem achttaetigen Kinde.* Jahrb. f. Kinderheilk. 69:213, 1909.

179. W. P. Schukowsky and A. A. Baron. *Hirngeschwuelste im Kindesalter. Tumor cerebri bei einem 5-jaehrigen Maedchen mit Amaunose wegen Sehnervenatrophie.* Archiv f. Kinderheilk. 58:307, 1912.

180. J. Winocouroff. *Zur Casuistik der multiplen boesartigen Geschwuelste im Kindesalter.* Arch. f. Kinderheilk., 52:33, 1910.

181. H. Kowarski. *Einiges ueber Rachitis.* Jahrb. f. Kinderheilk. 67:171, 1908.

182. A. Schkarin and W. Kurajeff. *Beitraege zur Frage ueber die Wirkung von Solbaedern auf den kindlichen Organismus.* Zeitschr. f. Kinderheilk. 7:413, 1913.

183. J. A. Schabad. *Zur Bedeutung des Kalkes in der Pathologie der Rachitis.* Archiv f. Kinderheilk. 52:47 and 68, 1910.

184. J. A. Schabad. *Der Kalkgehalt der Frauenmilch. Zur Frage der ungenuegenden Kalkzufuhr als Ursache der Rachitis.* Jahrb. f. Kinderheilk. 74:511, 1911.

185. H. Kowarski. *Sechs Faelle von Idiotia amaurotica progressive familiaris infantilis.* Jahrb. f. Kinderheilk. 76:58, 1912.

186. D. v. Ott. *Zum Neubau des Kaiserlichen Klinischen Institutes fuer Geburtshuelfe und Gynaekologie.* Monatsschr. f. Geburtsh. u. Gynaekol. 11:809, 1900.

187. C. J. v. Siebold. *Versuch einer Geschichte der Geburtshuelfe.* Verl. v. Franz Pietzker. Tuebingen, 1902.

188. D. v. Ott. *Eduard Anton Krassowsky.* Monatsschr. f. Geburtsh. u. Gynaekol. 7:690, 1898.

189. A. Martin. Kronid von Slavianski. Monatsschr. f. Geburtsh. u Gynaekol. 8:527, 1898.

190. Review of "Festschrift aus Anlass der 25-jaehrigen Lehrtaetigkeit des Professor D. v. Ott" (St. Petersburg, 1904). Monatsschr. f. Geburtsh. u. Gynaekol. 25:140, 1907.

191. A. E. Feoktisow. *Einige Worte ueber die Ursachen und den Zweck des Menstrualprocesses.* Arch. f. Gynaekol. 27:379, 1886.

192. D. v. Ott. *Gesetz der Periodicitaet der physiologischen Funktionen im weiblichen Organismus.* Arch. f. Gynaekol. 39:130, 1891.

193. D. v. Ott. *Ueber den Stoffwechsel zwischen Frucht und Mutter.* Arch. f. Gynaekol. 27:129, 1886.

194. D. v. Ott. *Ueber die Folgen der Kochsaltztransfusion fuer den Organismus.* Arch. f. Gynaekol. 20:334, 1882.

195. D. v. Ott. *Kaiserschnitt nach der Methode von Porro mit guenstigem Erfolge fuer Mutter und Frucht.* Arch. f. Gynaekol. 37:88, 1890.

196. D. v. Ott. *Zur operativen Behandlung der durch Zerstoerung der Urethra complicierten Vesicovaginalfistern.* Monatsschr. f. Geburtsh. u. Gynaekol. 2:231, 1895.

197. N. Wolkowitsch. *Eine Methode der Wiederstellung der weiblichen Harnroehre mit gleichzeitiger Beseitigung einer Blasen-Scheidenfistel.* Monatsschr. f. Geburtsh. U. Gynaekol. 20:1253, 1904.

198. A. v. Gubaroff. *Ueber einige Bedingungen zur Beurtheilung der Vorzuege der einzelnen Operationsmethoden bei Uterusmyomen.* Monatsschr. f. Geburtsh. u. Gynaekol. 8:30 and 157, 1898.

199. W. Sneguireff. *83 Faelle von Myomo-hysterectomie abdominalis totalis nach modificierten Doyen'schen Verfahren.* Monatsschr. f. Geburtsh. u. Gynaekol. 4:525, 1896.

200. N. Altuchoff and W. Sneguireff. *Eine neue Methode der Unterbindung der Arteriae uterinae par laparatomiam.* Monatsschr. f. Geburtsh. u. Gynaekol. 3:453, 1896.

201. W. F. Sneguireff. *Endometritis dolorosa.* Arch. f. Gynaekol. 59:277, 1899.

202. J. Lazarewitsch. *Ueber die Unterinjektion und 3 Faelle kuenstlichen Fruegeburt.* Schmidt's Jahrb. d. ges. Med. 119:58, 1863.

203. J. Lazarewitsch. No title given. Schmidt's Jahrb. d. ges. Med. 142:244, 1869.

204. A. v. Gubaroff. *Ueber einen Fall von Placenta praevia.* Monatsschr. f. Geburtsh. u. Gynaekol. 7:25, 1898.

205. N. Pobedinsky. *Zur Frage ueber die Behandlung der durch Myome des Uterus complicierten Schwangerschaft und Geburt.* Monatsschr. f. Geburtsh. u. Gynaekol. 12:292, 1900.

206. W. Stroganoff. *58 Faelle von Eklampsie ohne Todesfall von dieser Erkrankung.* Monatsschr. f. Geburtsh. u. Gynaekol. 12:422, 1900.

207. W. W. Stroganoff. *Bakteriologische Untersuchungen des Genitalkanals beim Weibe in verschiedenen Perioden ihres Lebens.* Monatsschr. f. Geburtsh. u. Gynaekol. 2:365, 1895.

208. W. Stroganoff. *Ueber die Anwendung der russischen Teemaschine (Samovar) als Sterilisator bei aseptischen Operationen.* Monatsschr. f. Geburtsh. u. Gynaekol. 1:503, 1895.

209. J. Grammatikati. *Meine Methode der Therapie der entzuendlichen Zustaende des Uterus und seiner Adnexa.* Monatsschr. f. Geburtsh. u. Gynaekol. 8:171, 1898.

210. A. Sitsinsky. *Die Behandlung der septischen Wochenbetterkrankungen in der Gebaermutterhoehle.* Monatsschr. f. Geburtsh. u. Gynaekol. 20:640, 1904.

211. E. M. Kurdinowsky. *Der Geburtsact, am isoliertem Uterus beobachtet. Adrenalin als ein Gebaermuttermittel.* Arch. f. Gynaekol. 73:425, 1904.

212. N. Fenomenow. *Die Anwendung des Adrenalins in der Gynaekologie.* Monatsschr. f. Geburtsh. u. Gynaekol. 20:1173, 1904

213. E. M. Kurdinowsky. *Weitere Studien zur Pharmakologie des Uterus und deren klinische Wuerdigung.* Arch. f. Gynaekol. 78:539, 1906.

214. K. Ulesko-Stroganowa. *Beitrag zur Kenntnis des epithelioiden Gewebes in dem Genitalapparate des Weibes.* Monatsschr. f. Geburtsh.u. Gynaekol. 25:16 and 160, 1907.

215. K. Ulesko-Stroganoff. *Zur Frage von dem feinsten Bau des Deciduagewebes, seiner Histogenese, Bedeutung und dem Orte seiner Entwicklung in Genitalapparat der Frau.* Arch. f. Gynaekol. 86:542, 1908.

216. K. Ulesko-Stroganowa. *Beitraege zur Lehre von mikroskopischen Bau der Placenta.* Monatsschr. f. Geburtsh. u. Gynaekol. 3:207, 1896.

217. J. Talko. Klin. Monatsbl. f. Augenheilk. 31:401, 1893.

218. A. J/. Skrebitzki. *Ueber Verbreitung und Intensitaet der Erblindungen in Russland und die Vertheilung der Blinden ueber die verschiedenen Gegenden des Reiches.* Klin. Monatsbl. f. Augenheilk. 24:107, 1886.

219. W. Dobrovolsky. *Zur Frage ueber die Verbreitung der Erblindungen in Russland.* Klin. Monatsbl. f. Augenheilk. 24:324, 1886.

220. Editorial. Klin. Monatsbl. f. Augenheilk. 24:155, 1886.

221. Jos. Talko. *Professor Dr.Wladimir Karawajew.* Klin. Monatsbl. f. Augenheilk. 30:327, 1892.

222. J. Talko. *Eduard Junge.* Klin. Monatsbl. f. Augenheilk. 36:413, 1898.

223. E. Junge. *Ophthalmologisch-mikroskopische Notizen.* Graefe's Archiv 5(2): 191, 1859.

224. E. Junge. *Beitraege zur pathologischen Anatomie in der "getiegerten Netzhaut."* Graefe's Archiv 5(1):49, 1859.

225. Gustav Braun. *Bemerkungen zur Lehre von Glaucom.* Graefe's Archiv 9(2):222, 1863.

226. W. Froebelius. *Zur Technik der Iridektomie bei Glaukom.* Graefe's Archiv 7(2):119, 1860.

227. Editorial. *Robert Blessig.* Klin. Monatsbl. f. Augenheilk. 16:240, 1878.

228. Dr. Blessig. *Ein Fall von Embolie der Arteria centralis retinae.* Graefe's Archiv 8(1):216, 1861.

229. Dr. Blessig. *Ueber die Veraenderungen der Niere nach Unterbindung der Nierenarterie.* Virchov's Archiv 16:120, 1859.

230. J. Talko. *Viktor Felix Szokalski.* Klin. Monatsbl. f. Augenheilk. 29:78, 1891.

231. J. Talko. *Ueber Augenerkrankungen in der Russischen Armee.* Klin. Monatsbl. f. Augenheilk. 31:144, 1893.

232. J. Talko. *Zur Geschichte der Brillen in Russland.* Klin. Monatsbl. f. Augenheilk. 31:217, 1893.

233. J. Talko. *Offene Correspondenz Jaroslav a. d. Wolga. Zur Optographie. Ophthalmoskopische Gasbilder. Transplantation von Fishhaut.* Klin. Monatsbl. f. Augenheilk. 30:356, 1892.

234. J. Talko. *Erscheinungen in den Augen des verwundeted Kaisers Alexander der Zweiten.* Klin. Monatsbl. f. Augenheilk. 19:168, 1881.

235. E. Jaesche. *Zur Behandlung der Traenenschlauch-Obstruktionen.* Graefe's Archiv 10(2):166, 1864.

236. E. Jaesche. *Jaesche's Operation fuer Entropium und Distichiasis.* Klin.. Monatsbl. f. Augenheilk. 11:97, 1873.

237. L. G. Bellarminoff. *Anwendung der graphischen Methode bei Untersuchung des intraocularem Druckes.* Pfluegers Archiv 39:449, 1886.

238. K. Rumschewitsch. *Casuistik des Glaukoms nach Staar-Operatiuonen.* Klin. Monatsbl. f. Augenheilk. 34:191, 1896.

239. K. Rumschewitsch. *Zur Casuistik der cornealen Neubildungen.* Klin. Monatsbl. f. Augenheilk. 31:50, 1893.
240. A. Natanson. *S. N. Logetschnikoff.* Centralbl. f. prakt. Augenheilk. 35:225, 1911.
241. S. Logetschnikoff. *Eine Notiz zur "neuen Behandlung des Glaucoms."* Klin. Monatsbl. f. Augenheilk. 32:96, 1894.
242. S. N. Logetschnikow. *Ein seltener Fall von schwerer diffusen Sklerodermie.* Zeitschr. f. Augenheilk. 10:355, 1903.
243. Dr. Brogi. *Max Emanuel Mandelstamm.* Zentralbl. f. prakt. Augenheilk. 36:162, 1912.
244. E. Mandelstamm. *Der trachomatoese Process.* Graefe's Archiv 20(1):52, 1883.
245. W. Dobrowolsky. *Alexander Iwanoff.* Klin. Monatsbl. f. Augenheilk. 9:218, 1881.
246. A. Iwanoff. *Beitraege zur normalen und pathologischen Anatomie des Augens.* Graefe's Archiv 15(2), 1869.
247. W. Dobrowolsky. *Beitraege zur Lehre von den Anomalien der Refraktion und Accomodation des Auges.* Klin. Monatsbl. f. Augenheilk. 6 (Beilagehefte 1–2), 1868.
248. W. Dobrowolsky. *Ueber verschiedene Veraenderungen des Astigmatismus unter dem Einflusse der Accomodation.* Graefe's Archiv 14(3), 57, 1868.
249. W. Dobrowolsky. *Ueber die Empfindlichkeit des normalen Auges gegen Farbentoene auf der Peripherie der Netzhaut.* Graefe's Archiv 32(1):9, 1886.
250. M. Woinow. *Zur Bestimmung der Sehschaerfe bei Ametropie.* Graefe's Archiv 15(2): 144, 1869.
251. M. Woinow. *Zur Diagnose der Farbenblindheit.* Graefe's Archiv 17(2):241, 1871.
252. M. Woinow. *Beitraege zur Farbenlehre.* Graefe's Archiv 21(1):223, 1875.
253. R. Wreden. *Sechs Faelle von Myringomykosis.* Arch. f. Ohrenheilk. 3:1, 1867.
254. R. Wreden. *Die Otitis media neonotarum.* Monatsschr. f. Ohrenheilk. 2:97, 113, 129, 149, and 165, 1868.
255. R. Wreden. *Sphyrotomie. Ein neues operatives Verfahren gegen gewisse Faelle von Taubheit und Ohrensause.* Monatsschr. f. Ohrenheilk. 1:22, 1867.
256. R. Wreden. St. Petersburg Med. Z., 2:317, 1871. Quoted by W. Poorten in "Anatomisch-physiologische Bemerkungen gegen die Wreden'sche Binnenmuskeltheorie." Monatsschr. f. Ohrenheilk. 6:54, 1872.
257. A. Prussak. *Zur Physiologie und Anatomie des Blutstromes in der Trommelhoehle.* Monatsschr. f. Ohrenheilk. 2:176, 1868.

258. A. Prussak. *Zur Anatomie des menschlichen Trommelfels.* Arch. f. Ohrenheilk. 3:255, 1867.

259. S. v. Stein. *Ueber Gleichgewichtsstoerungen bei Ohrenleiden.* Arch. f. Ohrenheilk. 39:312, 1895.

260. S. v. Stein. *Ein Beitrag zur Kopfknochenleitung.* Arch. f. Ohrenheilk. 28:201, 1889.

261. F. F. Cartwright. *The development of modern surgery.* Thomas Y. Crowell Publ. Co., New York, 1967.

262. Editorial. *Nicholas v. Arendt. Nekrolog.* Arch. f. klin. Chir. 1:332, 1861.

263. Dr. Klien. *Ueber die Steinkrankheit und ihre Behandlung in Russland.* Arch. f. klin. Chir. 6:78, 1865.

264. G. Halperin. *Nikolai Ivanovich Pirogov—surgeon, anatomist, educator.* Bull. Hist. Med. 30:347, 1956.

265. N. Pirogoff. *Grundzuege der allgemeinen Kriegs-Chirurgie nach Reminiscenzen aus den Kriegen in der Krim und im Kaukasus nund aus Hospitalpraxis.* Leipzig, 1864. Reviewed in Arch. f. klin. Chir. 8:55, 1867 (Jahresbericht 1863–1865).

266. E. v. Bergmann. *Georg Franz Blasius Adelmann (Nekrolog).* Arch. f. klin. Chir. 37:888, 1888.

267. A. Walther. *Julius v. Szymanowsky. Nekrolog.* Arch. f. klin. Chir. 9:370, 1868.

268. J. S. Billings. *The history and literature of surgery.* Argosy-Antiquarian Ltd., New York, 1970.

269. J. Maslowsky. *Extirpation beider Ovarien, mit einem kurzem Berichte ueber die Geschichte der Ovariotomie in Russland.* Arch. f. klin. Chir. 9:527, 1868.

270. Dr. Sklifossoffsky. *Zwei Faelle von Ovariotomie mit gluecklichem Ausgange.* Arch. f. klin. Chir. 9:24, 1868.

271. R. Schaeffer. *Experimentelle Beitraege zur Asepsis bei Laparatomien.* Monatsschr. f. Geburtsh. u. Gynaekol. 8:156, 1898.

272. N. P. Kravkov. *Ueber die Hedonal-Chloroform Narkose.* Arch. f. exper. Pathol. U. Pharmakol. Suppl. 1908, p. 317.

273. S. P. Fedoroff and A. P. Jeremitsch. *Ueber allgemeine Hedonalnarkose.* Centralbl. f. Chir. 37:316, 1910.

274. S. P. Fedoroff. *Die intravenoese Hedonalnarkose.* Centralbl. f. Chir. 37:675, 1910.

275. S. Fedoroff. *Ueber Nephroktomien mit Anlegung von Klemmpincetten ae demure.* Centralbl. f. Chir. 23:617, 1896.

276. Prof. Szymanowsky. *Ueber Resektion den Hueftgelenkes.* Arch. f. klin. Chir. 6:787, 1865.

277. Prof. Szymanowsky. *Modifikation der Pirogoff'schen Osteoplastik.* Schmidt's Jahrb. 109:212, 1861.

278. Prof. Szymanowsky. *Kritik der partiellen Fussamputationen.* Arch. f. klin. Chir. 1:366, 1860.

279. J. Szymanowsky. *Beitrag zur Amputation nebst Erfahrungen ueber die Immersion und Irrigation.* Schmidt's Jahrb. 110:323, 1861.

280. Prof. Szymanowsky. *Hernia lateralis (!) cerebri.* Arch. f. klin. Chir. 6:560, 1865.

281. Prof. Szymanowsky. *Die tellerfoermige Grube bei Belggeschwuelsten, am Schaedel.* Arch. f. klin. Chir. 6:777, 1865.

282. G. Jaesche. *Ein Beitrag zur Casuistik der Resektionen.* Arch. f. Chir. 8:162, 1867.

283. Georg Jaesche. *Auch eine operierte Geschwuelst am Schaedel.* Arch. f. klin. Chir., 8:183, 1867.

284. A. A. Bobroff. *Ein neues osteoplastisches Verfahren bei Spina bifida.* Centralbl. f. Chir. 19:465, 1892.

285. J. Minkiewicz. *Vergleichende Studien ueber alle gegen Varices empfohlenen Operationsverfahren.* Virchow's Archiv 25:193, 1862.

286. J. Minkiewicz. *Ueber die Behandlung einiger Aneurismen mittelst der unvollstaendigen metallenen Ligatur "a fil perdu."* Virchow's Archiv 63:201 and 474, 1875.

287. N. K. *Vl. Dm. Vladimirov (Nekrolog).* Khirurgiya 15:256, 1904 (in Russian, National Medical Library, Bethesda, MD).

288. J. Ssabanejew. *Die Amputatio femoris inter-condyloidea osteoplastica.* Centralbl. f. Chir. 17:413, 1890.

289. J. F. Ssabanejew. *Ueber die Anlegung einer rohrenfoermigen Magenfistel bei Verengerungen der Speiseroehre.* Centralbl. f. Chir. 20:862, 1893.

290. S. Saltykow. *Ueber Replantation lebender Knochen.* Beitr. z. pathol. Anat. u. allgem. Pathol. 45:440, 1909.

291. N. M. Wolkowitsch. *Normalferfahren der operativen Beseitigung der Haemorrhoidalknoten.* Centralbl. f. Chir. 22:630, 1895.

292. N. Wolkowitsch. *Zur Frage der operativen Behandlung der Tuberkulose der grossen Gelenke der Extremitaeten und speciell der Resektion derselben.* Centralbl. f. Chir. 31:1443, 1904.

293. N. Wolkowitsch. *Ueber eine bisher nicht beobachtete Erscheinung bei sich wiederholenden Appendicitisanfaellen.* Centralbl. f. Chir. 38:757, 1911.

294. N. A. Weljaminow. *Polyarthritis chronica progressive thyrotoxica.* Centralbl. f. Chir. 35:979, 1908.

295. N. A. Weljaminow. *Die Erkrankungen der Schilddruese und ihre chirurgische Behandlung.* Centralbl. f. Chir. 37:1590, 1910.

296. W. Manassein. *Chemische Beitraege zur Fieberlehre.* Virchow's Archiv 56:220, 1872.

297. W. Manassein. *Zur Lehre von den Temperaturherabsetzenden Mitteln.* Pflueger's Archiv 4:287, 1871.

298. Prof. P. V. Nikolsky. *Gruppa pemphigus.* Vrachebnaya Gazeta 9:269, 1902 (in Russian, National Library of Medicine, Bethesda, MD).
299. E. M. Stepanow. *Ueber Laryngitis haemorrhagica.* Monatsschr. f. Ohrenheilk. 18:1, 1884.
300. M. Palkin. *Anfaenge der experimentellen Erforschung des Rheumatismus in Russland.* Muench. Med. Wochensschr. 111:1601, 1969.

Tables and Illustrations

Table 1. Medical facilities and personnel in Imperial Russia

Year	Doctors	Feldshers	Midwives	Pharmacists	Dentists	Hospitals	Beds
1881	14,500	7600	3800	2000	-	3000	74000
1894	15,740	17,000	-	-	-	-	-
1903	17,658	-	-	-	-	-	-
1907	18,315	20,279	-	-	-	-	-
1909	19,900	25,000	13,000	11,400	5700	7050	193000
1912	24,000	28,500	14,000	13,400	7238	8100	220000

Table 2. Growth of Zemstvo medicine in Russia

	1870	1880	1890	1900	1910
Number of districts	530	925	1440	2010	2686
District radius (km)	36	26	21	18	15
Number of inhabitants per district (x1000)	95	58	42	33	28
Hospital beds per 10,000 persons	1.5	2.5	3.4	4.0	4.8

Table 3. Death rates in Imperial Russia and other nations (deaths/1000)

Year	Germany	Hungary	England	Italy	Austria	Spain	France	Russia
1885	27.2	32.9	19.3	27.3	29.4	-	25	35.6
1890	24.4	31.9	18.8	25.6	28.8	-	25	33.6
1895	23.3	31.8	18.7	23.6	27.9	-	23	30.7
1900	21.3	28.7	18.4	22.9	25.6	-	23	33.8
1904	-	-	15.6	-	-	-	22	29.1
1909	-	-	15.0	-	-	-	21	28.9
1912	14.8	22.9	14.7	19.3	18.8	22.2	19.6	25.9
1914	-	-	14.0	-	-	-	-	22.4
1920	15.1	-	12.4	18.8	-	-	18.0	33.2*

_____
* After the Bolshevik coup

Table 4. Birth rates in Imperial Russia and other nations (births/1000)

Year	Germany	Bavaria	Austria	England	Italy	France	Russia
1880	40	39	38	34	37	22	49
1890	-	36	38	31	37	19	47
1900	-	-	-	29	-	18	52
1910	-	-	-	25	-	16	44
1912	27	-	25	24	32	19	43

Table 5. Infant mortality rates in Imperial Russia and other nations (deaths/1000 births)

Year	Germany	Bavaria	Austria	England	Italy	France	Russia
1879	235	316	257	153	214	222	296
1893	-	279	249	146	190	217	268
1910	192	-	207	130	142	127	266

MILITARY-MEDICAL ACADEMY
IN ST. PETERSBURG IN THE NINETEENTH CENTURY.
PAVLOV TAUGHT PHYSIOLOGY HERE.

A cartoon in a Russian publication during the reign of Catherine II promoting smallpox vaccination among the Russian population. The vaccinated kids do not want to play with the unvaccinated ones, and the latter are asking their father why he did not allow them to be vaccinated. Source: an unknown American medical history journal.

Sergey P. Botkin (1832–1889) and his wife. Botkin is by many considered to be the "father of Russian medicine." Among other things, he is credited with establishing the infectious nature of hepatitis. It is called Botkin's Disease in Russia.

Anton Chekhov (1860–1904), dramatist and Zemstvo doctor

216

N. I. Pirogov (1810–1881) (left) and N. Sklifosovsky (1836–1904) (right), pioneers in Russian surgery. Pirogov was first to use ether as anesthetic in the Crimean War.

G. H. Minkh (1836–1896), bacteriologist,
expert on the leprosy disease.

N. P. Kravkov (1865–1924) pharmacologist
and pioneer in surgican anesthesia techniques

Daniel K. Zabolotny (1866–1929),
bacteriologist, developer of vaccines.

G. N. Gabrichevsky (1860–1907),
bacteriologist and developer of the scarlet fever vaccine.

Anatomy laboratories of the Kazan University Medical Faculty
in the nineteenth century

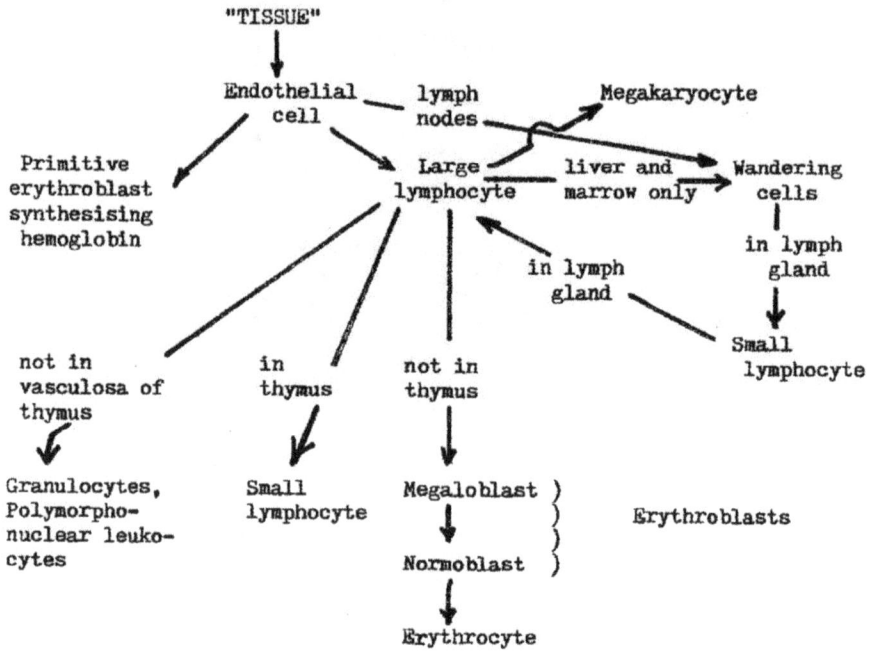

MAXIMOV'S CONCEPT OF HEMOPOIESIS IN EMBRYONIC,
LIVER, AND BONE MARROW TISSUES

Chapter III

BIOLOGICAL SCIENCES

Introduction

Life sciences were especially well represented in Imperial Russia, and the high quality of Russian biology is even more evident from the fact that the two Nobel Prizes awarded Russian scientists of the pre-Soviet era went to biologists. The recipients were Ivan Pavlov (1904) and Ilie Mechnikov (1905). This is not to say that other Russian scientists were not of Nobel Prize caliber, as, for instance, was Dimitry Mendeleyev, who in 1869 published his invention, the periodic table of elements. But his work was done well before the introduction of Nobel Prizes into our world; as well, there were notable anti-Mendeleyev machinations by the Royal Swedish Academy, which determines who wins and who doesn't win the Nobel (1). It should be also noted that the various fields of basic sciences, unlike those in the medical arts, were, and are to some extent, dominated by individuals of world stature. For example, the complex field of digestion enzymes was initiated and pursued by Pavlov and his disciples. On the other hand, the etiology of arteriosclerosis was investigated by a number of researchers over a period of years beginning with Ignatovsky and ending with Klimov. So, Pavlov won the Prize, and no one won anything for their work on the role of cholesterol in heart attacks. For such reasons, this chapter is concerned with individual personalities much more so than the preceding chapter discussing Russian medicine.

Physiology

Russian physiology was to a great extent directed toward the elucidation of the function of the nervous system in the living organism. In fact, it may be even stated that the basic premise of Russian physiology was that the nervous system was the main if not the only foundation of the phenomenon of life in the vertebrate organism. The original protagonist of this view was Ivan Mikhailovich Sechenov, who is considered by many to be the "father of Russian physiology." He was born in the village of Teply Stan in eastern Russia on August 1, 1829. From 1843 to 1848 he attended the Military Engineering Academy, then served as a lieutenant in the army until 1850. He then entered Moscow University to study medicine and graduated in 1856. As was the custom in those days, the government sent the new graduate abroad to study with Carl Ludwig in Leipzig, Emil du Bois-Raymond in Berlin, and Claude Bernard in Paris. In 1860, Sechenov was appointed to the chair of physiology at the Military-Medical Academy in St. Petersburg. In 1870 he resigned his position because the faculty voted not to appoint his protégé, Elie Mechnikov, to the chair of zoology. He then joined the faculty of Odessa University, but returned to St. Petersburg in 1876 to take charge of the chair of physiology at St. Petersburg University, a position for which he was recommended by Elie de Cyon. In 1888, he joined the faculty of Moscow University where he remained until his retirement in 1904. He died in the following year.

Sechenov's greatest experimental contribution was made in 1861 while on leave of absence in Claude Bernard's laboratory. Sechenov was then able to repeat and amplify Ernst Weber's work on central inhibition, where it was shown that the severance of spinal chord from cerebral hemispheres intensified spinal reflexes. Using a frog, Sechenov sectioned its cerebrum and noted the intensity of spinal reflexes after each section. There was an intensification of spinal reflexes, and the intensity was diminished upon stimulating the peripheral end of the section. In Russia, Sechenov was interested in the mode of carbon dioxide binding and transport in blood. He attempted to construct various model systems therefor by measuring absorption of carbon dioxide by various salt solutions. He found that the salts of weak acids were best in absorbing carbon dioxide (especially sodium acetate), and that salts of strong acids were not as effective. Among the latter, nitrates were the best, followed by chlorides and sulfates. He found that alkali was consumed by these model systems upon the absorption of carbon dioxide. Based upon these hardly unexpected findings, Sechenov and his student Shaternikov were able to design an apparatus for the measurement of carbon dioxide in expired air.

Sechenov's experimental contributions were thus rather limited. His main legacy remains his theoretical works and his contribution to education. He translated a number of physiology texts and pamphlets from German into Russian, and wrote several books himself. Among his better-known writings were the *Physiology of the Nervous System*, *Physiology of Nerve Centers*, *Reflexes of the Brain*, and *Who Should Study Psychology and How* (2). In the latter two publications, Sechenov argued that all actions of man were nothing more than reflexes or responses to external stimuli, and concluded that man himself was merely a rather complicated machine. He attacked the psychologists for their subjective view of the human organism and their allegedly metaphysical leanings. He proposed that physiologists rather than psychologists should study psychological phenomena. Sechenov's ideas on the reflex nature of human behavior were not new to the scientific world. As early as in 1845, Thomas Laycock published similar views, though Sechenov was apparently unaware of Laycock's work. Furthermore, it is said that the doctrine of associative psychology may have had an effect on the formation of Sechenov's ideas without him realizing this possibility (3).

Sechenov's assertion that, in view of the reflex nature of the brain, the existence of the soul is absurd and illogical elicited a strong reaction from among the Russian humanists, notably Dostoyevsky. Moreover, the government interpreted his writings as advocating the view that man was not to be held responsible for his actions, though Sechenov never said this directly. Since his writings appeared at a time characterized by various revolutionary movements in Russia, Sechenov did not enjoy any special favors from among the governing circles, and his book *Reflexes of the Brain* was withheld for some time from publication by the tsar's censor. Sechenov's autobiography indicates that he was not an opponent of the Russian monarchy, yet, at the same time, he was quite impatient with the tsar's bureaucracy and its stranglehold on higher education in Russia. He appears to have favored the establishment of private universities in Russia (all Russian universities were, as they were in Soviet Union, supported and controlled by the government), and the popularization of science among the lay public (4).

Of the numerous investigators influenced by Sechenov's ideas, the most outstanding ones of the tsarist period are probably Pavlov and Vedensky. Pavlov's work on reflexes may perhaps be characterized as psycho-physiological in nature, whereas that of Vedensky is essentially classically physiological and pharmacological. Nikolai E. Vedensky (also spelled Vvedensky) (1852–1922) was born in the Vologda province of Russia in the family of a village priest. His secondary education was obtained in a theological seminary, and his higher education at the St. Petersburg University. He then spent three years in jail for revolutionary agitation, yet in spite of all this he was appointed to the faculty of St. Petersburg University in 1889. He was elected to the St. Petersburg Academy of Sciences on 1909. According to Soviet sources, Vedensky was a staunch environmentalist (as opposed to geneticist?) who supposedly believed that given proper environmental stimuli, morphological and behavioral changes would be observed in human beings in two to three generations.

Vedensky's research interests largely involved the mechanisms of nerve conduction, especially the conduction of inhibitory impulses. He was a pioneer in the development of the telephone as a tool in the study of nerve conduction. He strongly opposed a concept prevalent at his time to the effect that there were special nerve fibers to conduct inhibitory impulses, and felt that a single nerve fiber could conduct both an inhibitory and stimulatory impulse under proper conditions (5).

The anesthetized nerve, in Vedensky's judgment, was an excellent model for the study of excitation and inhibition, and he consequently spent much time investigating the properties of narcotized nerves (6). He recognized four stages of nerve narcosis:

1. The stage of transformation, where all stimuli were inhibited. Yet, a weak stimulus could produce a weak muscle contraction in a nerve-muscle preparation, whereas a strong stimulus would produce a relatively strong response in the muscle.

2. The paradoxic stage, where strong stimuli are not conducted, but weak ones are.

3. The inhibitory stage, where the stimulation of the anesthetized portion of the nerve results in a contraction of the muscle, whereas stimulation of the unanesthetized portion of the nerve produces an inhibition of the contraction (i. e., a reversal of the effect of stimulation).

4. The deep narcosis stage, where absolutely no nerve conduction takes place.

All four stages were reversible. Vedensky thus concluded that the same nerve fiber could act as a conductor of both the inhibitory and excitatory impulses. The effects of anesthetics could be duplicated by various chemicals and sublethal temperatures. He believed that stage 3 above was in many ways similar to a situation where a nerve transmits inhibitory rather than excitatory messages. He termed such a state "parabiosis." It was also found that a potential difference existed between the nerve section in the parabiotic state and the normal segment of the same nerve. If the potential difference was lowered, the parabiotic state returned to either the paradoxical or transitional stage.

Vedensky found that if sufficient current was applied to the nerve-muscle preparation to cause tetanization of the muscle, the state of tetanization could be maintained almost indefinitely by continuously applying the stimulus. He thus discovered that a nerve does not become "exhausted" with continuous stimulation. If a very high current was passed through a nerve, the latter lost both its conductance and irritability. In this sense, the nerve was in a state similar to that of a deep narcosis (stage 4 above). The excitability of the nerve fiber was found to be greater than that of its end-plate. The end-plate responded to a weak stimulus in an excitatory manner, whereas strong stimuli caused it to respond in an inhibitory manner. Vedensky thought that the anesthetized nerve in its paradoxical stage, as well as in its transitional stage, becomes like the end-plate in its properties. Nerve excitability and conductance were separable. In the anesthetized nerve in the paradoxical stage, the ability to contract the muscle responded to middle-sized stimuli, whereas the conductivity remained with a weak current stimulus only.

Vedensky showed in a rather elegant experiment (7) that continuous stimulation of sensory nerves depressed the sensory centers in the brain. He noted that the tibial and fibular nerves may be expected to perform approximately similar function, and would thus be expected to send their stimuli to the same center in the brain. He then proceeded to stimulate the fibular nerve, and then determined the ability of the tibial nerve to bring forth a tetanization of the semitendenous muscle. He found that the constant stimulation of the fibular nerve decreased the ability of the tibial nerve to bring a forceful tetanus in a muscle. He likened this purely *in vitro* physiological experiment to the state of hysteria in some emotionally disturbed persons.

The other disciple of Sechenov's ideas mentioned above was Ivan Petrovich Pavlov, perhaps the best-known Russian scientist in the West. Much has been already written about him in the West, as well as in the Soviet Union. This volume will, for this reason, be limited to some highlights of Pavlov's illustrious career, the sources being Babkin's biographical work (8) and Pavlov's own books (9–11). Pavlov was born on September 15, 1849 in Ryazan. The son of a poor village priest, he knew both deprivation and hard work. His secondary education was obtained in the Ryazan theological seminary, but instead of following his father's footsteps, Pavlov enrolled in St. Petersburg University and graduated with a degree in natural sciences in 1875. He then entered the Military-Medical Academy, and graduated with a medical degree in 1879. His physiology professor there was the illustrious Elie de Cyon (see below). Between the years of 1879 and 1883, Pavlov worked in the research laboratories of S. P. Botkin's clinic (see Chapter II), and in 1883 he successfully defended a thesis for the degree of doctor of medicine. He was then sent abroad for further study, and worked in Ludwig's laboratory among other places. In 1890, Pavlov was appointed to the chair of pharmacology at the Military-Medical Academy and, at the same time, to the directorship of the physiology laboratory of the newly-formed Institute for Experimental Medicine. In 1895, Pavlov was appointed to the chair of physiology at the Military-Medical Academy, where he carried on the de Cyon tradition of teaching physiology from the experimental point of view. Pavlov remained at the academy until well after the Bolshevik coup of 1917; however, in time, he spent more and more of his time at the institute. He ran his laboratories there with the highest degree of efficiency, always insisting on the most stringent controls possible under the circumstances. Each project was a logical outgrowth of some previous investigation, yet he always made certain that all projects had some degree of overlap. This was designed to assure repeatability of all experimental results generated in his laboratory. Pavlov died in 1936 without ever retiring from his work.

There was apparently very little that Pavlov cared about other than his science. In contrast to many of his colleagues, he had no use for politics both during the Russian Imperial era or the Bolshevik state. His occasional public statements in regard to politics were made only when they served to promote the interests of science. Babkin, Pavlov's biographer (8), divides Pavlov's scientific activities into three periods: 1. The work done on the innervation of the heart (1878–1888), which was done in Botkin's laboratories and abroad, 2. The work on digestion (1889–1904), for which Pavlov won his Nobel Prize in 1904, and 3. The work on conditioned reflexes, which Pavlov pursued until his death.

Pavlov divided the nervous control of heart action into four categories: 1. Inhibition of the number of contractions, 2. Decrease in the force of contractions, 3. Increase in the rate of contractions, and 4. Increase in the volume of the heart. The first two effects were found to be controlled by the vagus nerves, the third was brought about by de Cyon's accelerator nerves (see below), and the fourth effect was controlled by the *augmenter nerve* discovered by Pavlov. Stimulation of this nerve caused an increase in the blood pressure without increasing the heart rate. Pavlov noted that stimulation of the accelerator nerve caused a dissociation of the auricular and ventricular beating, whereas stimulation of the augmenter nerve restored the balance. He thought that the augmenter nerve worked directly on the cardiac muscle by controlling its metabolism.

Pavlov's investigations on the digestive processes (9) began with the development of pertinent methodology. He noted that the previous investigations on the digestive organs were complicated by the fact that food got mixed with digestive tract secretions. To avoid this problem, he developed what is today known as the Pavlov pouch (Pavlov called it "the little stomach") for study of gastric secretions. This involved the isolation of a small section of the stomach with the innervation and blood supply intact, and an opening to the outside. To study pancreatic secretion, Pavlov modified the Haidenhain fistula by cutting out a piece of the intestine containing the pancreatic duct orifice, and suturing it unto the skin with an opening to the outside. Pavlov also developed a method called sham-feeding. This involved the cutting of the esophagus and diverting its upper end to the outside. Thus, when a dog with this arrangement was given food by mouth, nothing got to the stomach. The dog does not realize this and keeps eating, thus permitting an uncomplicated study of gastric and pancreatic secretions. This operation also permits the placing of food into the stomach of the dog directly without going through the mouth. In his work on the digestive processes, Pavlov preferred not to use vivisection. He instead performed his operations under sterile conditions to prepare the animal for sham-feeding, or to establish the Pavlov pouch or the Pavlov-Haidenhain fistula, permitting the animal to recover completely, then performing his experiments. The following enzymatic activities were determined in the various digestive juices studied: 1. Proteolytic activity, assayed by the rate of liquefaction of a column of coagulated egg albumen, 2. Amylase activity, assayed by digesting starch, and 3. Lipase activity, assayed by titration.

Pavlov's group found that the largest quantity of gastric juice was elaborated about one hour after food ingestion, whereas the largest amount of pancreatic juice was secreted some three hours after eating. Gastric juice exhibited the highest proteolytic activity about one hour after feeding, then dropped and showed another peak after some five hours. Enzymatic activities of the pancreatic juice showed maximum activities some two hours after feeding, and then fell steadily. The enzymatic activity and acidity of the gastric juice did not vary with the volume of the gastric juice elaborated; instead, the necessary quantity of the enzymes and acid was supplied by variation of the gastric juice volume, all other things being equal. Proteolytic activity of the gastric juice varied with the type of food ingested. Greatest proteolytic activity was observed with bread, followed by meat, and lastly by milk. It was shown that the protein present in the bread was actually responsible for eliciting the high proteolytic activity of gastric juice. The pancreas produced most amylase with bread diets, and most lipase with milk diets. Pavlov thus postulated that the composition of the digestive juices adapts itself to the type of food ingested by the animal.

The investigation of nerve control of gastric and pancreatic secretions was focused on the vagus nerves. Pavlov's group showed that if the dog's right vagus nerve was cut between its laryngeal and cardiac branches, and the left vagus was cut in the neck, no gastric juice was elaborated, even though such animals survived for months in apparently good health. Stimulation of the peripheral end of the vagus caused copious secretions in the stomach, which were inhibited by atropine. Stimulation of the vagus also caused a secretion of the pancreatic juice; however, there was a lag period of about one minute. Sham-feeding of dogs equipped with the Pavlov pouch and intact vagi produced a flow of gastric juice, but if the food was introduced into the stomach directly without the dog seeing it, there was a very scant elaboration of the gastric juice. Pavlov postulated that the flow of gastric juice during eating is largely due to psychic phenomena and is mediated through the vagus nerves. He called such secretions *appetite juice*. The possibility of chemical stimulation of gastric juice secretion was also investigated. Pavlov found that water and meat extracts stimulated the flow of gastric juice. Bicarbonate, pure protein, protein digest (peptone), starch, fat, and sugars did not elicit the secretion. Milk and gelatin were weak stimulants. Pavlov thus proposed that the initial surge of gastric juice was due to psychic stimulation, whereas later secretions were maintained *via* chemical stimulation.

Pancreatic juice was also secreted, though to a very small extent by psychic stimulation. The largest amount of pancreatic juice was produced upon the entrance of the acidic chyme into the duodenum. Pavlov recognized that it was the acidity of the chyme and not its composition that elicited the flow of pancreatic juice, and believed that it was nerve-mediated. Bayliss and Starling later showed that secretin, a hormone, rather than nerve impulses, was responsible for this effect. The movement of chyme into the duodenum was recognized by Pavlov not to be continuous. He found that the acidic character of the chyme elicited a reflex reaction in the pyloric sphincter, which closed the intestine to further entry of chyme. Upon its neutralization, the pyloric sphincter was able to relax and to permit the entry of another portion of the chyme.

Pavlov's group demonstrated that intestinal wall secretions (succus entericus) had an augmenting effect on the proteolytic activity of the pancreatic juice. Pavlov correctly proposed that the intestinal wall elaborated an "enzyme of enzymes," which he named *enterokinase,* and which is concerned with activation of the largely inactive pancreatic zymogens. Pavlov was concerned with the relatively acidic character of gastric juice and the alkaline character of the pancreatic secretion. He felt that the excess alkali remaining in blood after the secretion of the gastric juice into the stomach was taken up by the pancreas and excreted into the duodenum as part of the pancreatic juice. The properties of bile were also looked at by Pavlov's group. The placing of fat, meat extract, and digested protein elicited the flow of bile. Water, acids, undigested protein, and starch were ineffective. Bile, it was believed, contained zymogen activators. Lipase was especially augmented by bile.

Pavlov's work on the secretion of saliva was able to distinguish between parotid and submaxillary secretions. The former responded largely to dry foods or sand, and it was composed largely of water and salts. Submaxillary saliva was secreted in response to all foods, and it was rich in amylase and mucin. Saliva flowed readily in dogs upon their sight of food, whereas mechanical stimulation of the buccal cavity produced no saliva. Pavlov concluded that the secretion of saliva in response to food was also in large part psychic, and this finding served as an assay method for his later work on conditioned reflexes.

Pavlov's investigations on conditioned reflexes (10) were apparently influenced by Sechenov's theories, though Pavlov apparently never met Sechenov in person. In contrast to the latter, Pavlov, at least before the Bolshevik coup, never had occasion to make generalizations on the nature of man or the existence of the soul on the basis of his experiments. He was, however, in full agreement with Sechenov in regard to the desirability for a physiological approach to psychology, and frequently scolded the psychologists for their subjective rather than objective approach to the study of human behavior. Pavlov occasionally expressed the belief, especially under Soviet pressure, that some conditioned reflexes could be inherited, though in the absence of experimental evidence therefor, he did not stress the point. Pavlov's group was extremely scrupulous to use only healthy dogs in their work, and to include as many controls as possible in their experiments. Pavlov's results were often contrary to those obtained in Bekhterev's laboratory (Bekhterev's work is discussed in chapter II), and it seems that Pavlov took special pleasure in publicly demonstrating how the failure of some Bekhterev students to use proper controls led them to erroneous conclusions. Pavlov's assay system in his study of conditioned reflexes was the flow of saliva, obtained from cannulated salivary gland ducts of dogs in response to sight or smell of food.

Two types of responses by the salivary glands were recognized by Pavlov: 1. The involuntary reflex, and 2. The acquired or conditioned reflex. The unconditioned reflexes were characterized by their consistency: if sand was introduced in the mouth of a dog, saliva rich in water and salts was produced time and time again; if pebbles were placed in the mouth, no saliva was ever elaborated. If edible food was given, saliva rich in mucus was always produced. Pavlov reasoned that the sensory nerves (centripetal nerves) conveyed the information as to the type of material in the mouth to the salivary centers in the brain, and the efferent (centrifugal) nerves then conveyed the appropriate information to the salivary glands as to which type of saliva needs to be secreted. It was also found that if sand or food were shown to the dog without placing them into the mouth, watery saliva and mucus-rich saliva respectively were secreted. The latter type of response Pavlov called *conditioned reflexes* or *psychic reactions.* The conditioned reflex was much more complicated than the nonconditioned one. It had to be cultivated (it was noninstinctive). It was, in addition, very unstable, and soon disappeared if the animal was not constantly reconditioned. Whereas the unconditioned reflex depended on the "essential" property of the stimulant (e. g., dryness or acidity), the conditioned reflex was dependent on nonessential properties such as color, odor, or appearance. Pavlov concluded that the salivary centers were connected to the sensory (visual, olfactory) pathways in the brain, but the connections were very weak and easily broken.

Attempts to trace the pathways of conditioned reflexes were not very successful. When the lingual nerves were cut, the unconditioned reflexes were drastically inhibited if not abolished, whereas conditioned reflexes remained functional. No special areas of the brain could be shown to possess the property of controlling conditioned reflexes, though the extirpation of the cerebral hemispheres abolished the ability of animals to form conditioned reflexes. Pavlov postulated the existence of so-called afferent centers or analyzers in the cerebral hemispheres of the animal, which are supposed to receive the afferent impulses of the conditioned reflexes. The analyzers are, in turn, connected to the efferent (motor) centers, which could be identical to the unconditioned reflex centers, which then convey the efferent impulses to the *peripheral* targets. Removal of the frontal lobes of the cerebral hemispheres apparently did not destroy any analyzers; however, hearing and sight analyzers were partially abolished by extirpating the posterior lobes.

Conditioned reflexes could be abolished in the animal by several procedures. Thus, a constant elicitation of the conditioned reflex without its occasional reinforcement by the unconditioned reflex soon resulted in its disappearance. A conditioned reflex could also be specifically inhibited by another stimulus. Thus, if a conditioned reflex based on light was elicited several times in the presence of an auditory tone, the reflex soon disappeared. The animal, however, still responded to the light only. Pavlov concluded that the tone inhibited the reflex elicited by the light. To the light and tone, Pavlov's group added a third stimulus: the metronome. It was then found that whereas light alone elicited no response whatever, all three stimuli elicited a weak response. It was then proposed that the metronome acted to inhibit the inhibitor (tone), or to disinhibit the response to light in the presence of the tone. In time, the metronome became a weak stimulus of itself, bringing forth the response conditioned by the light.

During the Soviet era, Pavlov vigorously advocated the application of his work on conditioned reflexes to psychiatry (11). He envisioned physiology as being the foundation of the overall superscience dealing with human biology including psychology. A large dark gap existed between fundamental physiology and practical psychology, as Pavlov saw it. He regularly attended the psychiatric wards of Leningrad hospitals to learn more about the diseases involved, and suggest possible courses of treatment. It is not known whether or not any patients benefited from his advice.

In discussing Pavlov's contributions to the digestive tract physiology, it would be appropriate to mention one of his predecessors at the Military-Medical Academy, one Nikolai Vladimirovich Ekk (Eck), the developer of the so-called *Eck fistula,* which was utilized by Pavlov and many subsequent investigators of the digestive phenomena. Ekk (1847–1908) was a medical man, and worked in Professor Tarkhanov's (see below) laboratory when he was called into the army during the Russo-Turkish war in 1877. He wrote his now classical paper (12) while participating in the siege of Plevna, a Turkish fortress. Ekk had noted that the severance of the portal vein and its anastomosis with the inferior vena cava did not result in any adverse effects against the animal, which could be ascribed to the altered liver blood supply. However, of the eight dogs that he had operated on, one died less than a day after the operation, six lived two to six days, and one lived for two-and-a-half months only to run away from the lab because of an animal keeper's negligence. The deaths were apparently due to infections. It was not until Pavlov introduced aseptic operating techniques into physiological investigations on animals that Ekk's operation was successful in the majority of cases. Ekk had developed his operation as a possible way to treat ascetic abcesses due to hepatic obstruction. It is still in use as a means to treat portal hypertension in human subjects, though the principal beneficiaries of Ekk's discovery have been gastrointestinal and liver physiologists.

Russia of the tsarist era produced a number of distinguished cardiovascular physiologists. The most successful of these was undoubtedly Elie de Cyon, and it is possible to consider him the second most important Russian physiologist, just behind the famous Pavlov. De Cyon was born is Samara (Kuibyshev, in the Soviet era) on March 25, 1843. He received his university education in Berlin (Prussia), and in 1866 we already find him working as a research associate in Ludwig's laboratory in Leipzig, Germany. In the following year he worked with Claude Bernard, and returned to Russia in 1870 to accept a faculty position at St. Petersburg University. In 1872, de Cyon was appointed chair of physiology at the Military-Medical Academy, which had been vacated by Sechenov. De Cyon completely reorganized the method of teaching physiology from a largely theoretical subject as taught by Sechenov to a practical course with many demonstrations and student participation in the laboratory. De Cyon was knighted by Tsar Alexander II in recognition of his services and his accomplishments. De Cyon was apparently very proud of this, for thereafter he always used the titular prefix *de* or *von* with his name, depending on whether he published in French or German.

De Cyon's political views were very conservative, and his demands upon the students were quite rigorous. For these two reasons, he was not popular with the student bodies, so that on many occasions, policemen had to be stationed in his lecture hall to control disruptive activities by students. Finally, student harassment resulted in de Cyon's resignation from the academy and emigration to France. He occupied himself there by writing books and participating in political activities, since no university would hire him to join its faculty. Later, having been appointed an agent of the Russian minister of finance, de Cyon helped to procure French loans for Russia, and was thus instrumental in establishing the Franco-Russian alliance. He died in 1912 without returning home to Russia (13).

While working in Ludwig's laboratory in Leipzig in 1866, de Cyon was able to develop an apparatus for maintaining circulation through a frog's heart. This permitted him to do numerous experiments on the effects of various inhibitors and stimulants on the frog's heart action (14). For instance, the relationship between the heart rate and intensity of contractions was investigated by de Cyon by varying the temperature of the fluid with which the frog's heart was perfused. He found that the intensity of the contractions remained at maximum from about 6 to $20^{\circ}C$, then decreased steadily until no contractions were noted at 37 to $40^{\circ}C$. The rate of heartbeat, on the other hand, showed a very slow rise between 6 and $20^{\circ}C$, then rose sharply to a maximum near 30°, then dropped to 0 at $37–40^{\circ}$. From these experiments, de Cyon postulated a law that states that the heart's work capacity is constant and independent of temperature, and independent of the frequency of contractions.

De Cyon was a strong proponent of the nerve theory of heart action control as opposed to the then popular "myogen" theory. With his brother, de Cyon discovered the sympathetic accelerator nerves that arise in the lower cervical and upper thoracic ganglia and lead to the cardiac plexus. These nerves were observed in the rabbit, dog, and horse. Their function was found to be contrary to that of vagi: stimulation of the accelerator nerves increased the frequency of heart contractions, but decreased their intensity. The Cyon brothers also discovered that splanchnic nerves were powerful vasoconstrictors. Their resection caused a drop in the blood pressure of the animal, whereas their stimulation caused a rise in blood pressure. If the splanchnics were cut, the brothers reasoned, then any fluctuations in blood pressure could be ascribed to the action of the heart. The Cyon brothers then stimulated the spinal chord of animals with severed splanchnic nerves and observed that the heart rate increased without an increase in blood pressure. They ascribed this effect to the accelerator nerves.

While working in Ludwig's laboratory, de Cyon also discovered what he called the *depressor nerve.* It later became known as the *aortic presso-receptor nerve.* De Cyon found that this nerve originated in the neck region of the vagus and the superior laryngeal nerves, then ran parallel to the sympathetic trunk, anastomosed with the inferior cervical ganglion, and then separated into numerous branches that innervated the heart. If the nerve was cut and the peripheral end stimulated, no effect was seen. If the central end of the cut nerve was stimulated, the heart rate decreased and the blood pressure dropped. De Cyon thus thought that this nerve had a sensory function. Cutting the splanchnics abolished completely the blood pressure-lowering action of the depressor nerve. De Cyon thus thought that this nerve was connected to the vagus nerve and the sympathetic centers of the brain, and served to protect the heart against overwork; as soon as the heartbeat would become too rapid, the depressor nerve would be stimulated and would transmit its message to the splanchnic and vagal centers of the brain. The splanchnics would be inhibited, thus promoting vasodilation and a drop in the blood pressure, whereas vagal signals would slow down the rate of heartbeat.

While working in Claude Bernards's laboratory, de Cyon had occasion to perfuse frogs' hearts with fluids containing oxygen, carbon dioxide, and nitrogen. The heart required oxygen to continue beating, but the other two gasses caused it to stop. De Cyon ascribed the effects of both temperature and gasses upon heart function on the basis of nerve stimulation and inhibition. He felt that a slow rise in temperature caused stimulation of accelerator nerves and depression of the depressor nerve, and *vice versa* with decreasing the temperature. A sudden rise of temperature from 0 to 40°C caused tetanus of the heart, and this was explained on the basis of a complete inhibition of the depressor nerve and a stimulation of the accelerator nerves. De Cyon also proposed that the depressor nerve was responsible for the transmission of emotions from the brain to the heart.

De Cyon was also much interested in space and time perception and their relationship to the semicircular canals of the ear (15). He damaged the latter and was able to divide the resultant effects into three categories: 1. Loss of equilibrium; 2. Forceful movement (immediately after the operation); and 3. Dropping the head to the floor with erratic movements (3–4 days after the operation). He then proposed that the semicircular canals were responsible for perception in three dimensions, each of the three canals being responsible for one dimension. He noted that the lamprey with its two semicircular canals was able to move only in two dimensions, whereas the movements of the Japanese dancing mice (Tanzmaeusen) with their single semicircular canals were able to move only in one dimension (a circular pattern). Of the two branches of the acoustic nerve, the *nervus cochlearis* was presumably responsible for the hearing sensation, whereas the *nervus vestibularis* was involved in space perception abilities. The latter was also found to be connected with the eye, thus providing a connection of visual perception with the equilibrium centers in the semicircular canals. De Cyon was also involved in research of the thyroid gland (he discovered innervation of the thyroid gland), and on the function of *nervi errigentes.*

De Cyon was a prolific writer, being responsible for some fourteen books in the Russian, German, and French languages. His greatest accomplishment in this field was the *Methodik der physiologischen Experimente und Vivisektionen* (1876), a manual of physiological experimental methodology. It served as a standard teaching tool for many years thereafter. It can be safely stated that de Cyon was one of the great physiologists of our time, yet his name is seldom if ever mentioned in his native land or abroad. The reasons for this are obscure, though one does get the impression that had his political views been different, his memory would have been better respected.

The rate of blood flow was investigated in various experimental animals by Ivan M. Dogel' (Dogiel in the German literature) (1830–1916). For this purpose he constructed the now classical *Stromuhr,* a monometer-like device illustrated at the end of this chapter. Actually, Dogel' described two versions of his Stromuhr, one with a built-in stream alternator (*Stromwender*), and the other with a rotational mechanism (*Kugeldrehung*) (16). Dogel' was a graduate of the Military-Medical Academy, then trained with Helmholtz, Kirchhof, Bunsen, and Ludwig. In Russia, he was first a professor in the St. Petersburg University, but in 1869 he moved to Kazan University, where he remained until his death. Beside his work on blood flow rates, Dogel' also investigated the histology and structure of lymph glands, physiology of the eye, the structure of heart ganglia, and pharmacology of chloroform, ozone, arsenic, and various poisons.

An investigator who made one of the early attempts to measure blood volumes in animals and human beings was Ivan R. Tarkhanov. He was born in Tiflis, Caucasus, in 1846, graduated from the Military-Medical Academy in 1869, and served as a professor of physiology there from 1877 to 1895. He then became a professor at the St. Petersburg University and died in 1908. His method for blood volume measurement in the human being (17) involved the measurement of hemoglobin before and after loss of body water by the subject. Blood volume, x, was then calculated by

$$x = \frac{p\,a'}{a' - a}$$

where p is the amount of water lost and a and a' are hemoglobin concentrations in the subject's blood before and after water loss. He offered evidence to show that rapid loss of body water, whatever the route, occurs at the expense of blood volume, and that the tissues exchange water with blood quite slowly. To bring about the water loss in human subjects, Tarkhanov recommended the use of Russian steam bath, where people have been known to lose 300 to 900 grams of water. However, instead of sending his subjects to the public baths, Tarkhanov constructed a hot box in his laboratory, which was heated to 45–50°C, and the subject kept there for 15–30 min. The subjects, upon completing the hot box session, were told to urinate and spit out any saliva, and their sweat was carefully wiped off with a clean towel. The subject's weight loss was then corrected for the solids present in sweat, urine, and saliva, and the amount of expired carbon dioxide and the amount of water lost from blood was then calculated from the revised weight loss and density of water at 37°C. Tarkhanov presented the results of thirteen blood volume determinations in various individuals. For example, a laboratory diener weighing 66,900 grams with a blood hemoglobin value of 0.096 g/ml was kept in the hot box at 50°C for 30 minutes. When he emerged from his confinement, he weighed 66,320 grams and his hemoglobin value was 0.109 g/ml. Solids in the sweat, urine, and saliva amounted to 4 grams, so total water loss was 578 ml. Blood volume was calculated at 4850 ml, or 5130 g if one used 1.059 g/ml as the density of blood. The weight of the blood was thus 7.7% of total body weight. The range was between 6.5 and 8.7% in the thirteen individuals examined. In two anemic individuals, the percentages were 6.0 and 4.6%.

Tarkhanov's lengthy paper apparently inspired one of his coworkers, a certain Dozent Tupoumov, to modify the sweating method by which Tarkhanov elicited the water loss, and to propose to accomplish the same thing by purging the patient with a potent laxative (18). The method involved first the collection of the subject's normal stool and determination of his normal hemoglobin, then purging the patient some 12 hours after his last evacuation, collecting the liquid stool as well as the urine, and determining the blood hemoglobin. After determining the solid residue content of his normal stool, the subject's blood volume was determined by using Tarkhanov's formula shown above. The editors of *Pflueger's Archiv,* where Tupoumov's paper was published, eventually got wind of the fact that no Dozent Tupoumov ever existed at the Military-Medical Academy, and that the article was published as a joke by one of Tarkhanov's students or associates (19). Incidentally, the Russian name Tupoumov translates into "the dull-minded one." The amazing thing that had happened with such a publication is that the stodgy German medical journal editors had no clue that the entire article was a ruse.

Tarkhanov is also well-known for his discovery and work on the psychogalvanic currents (20). He found that a strong current was produced on the surface of the skin if it was tickled with a feather or a similar object. These currents were also produced upon stimulating the ear with loud noises, the olfactory system with various odors such as acetic acid and ammonia, the eyes with light, and the tongue with sugar or acetic acid. Similar currents appeared if a person only imagined being stimulated. An especially strong current was produced if a person's hand was burned. It was also established that the skin areas rich in sweat glands (e. g., palms of hands) produced "negative" electricity, whereas areas poor in sweat glands produced "positive" electricity. The currents were produced minutes after the stimulus had been removed, and then faded gradually rather than abruptly. Continuous stimulation of skin produced progressively weaker current, until it completely disappeared. The current did, however, reappear as a result of mental effort. Exhausted individuals (physically or mentally) were not able to produce psychogalvanic currents. Tarkhanov felt that these currents were intimately related to sweat secretion.

Kazan University was, in addition to St. Petersburg, an important physiological research center in Imperial Russia. In fact, many medical and physiological researchers that eventually settled in the much desired city of St. Petersburg obtained their start at Kazan laboratories. Among the men who at one time were associated with Kazan University's Physiology Institute were Ivan Dogiel, Bekhterev, Lesshaft, Ovsyannikov, and Kovalevsky. The first three have already been mentioned in this chapter or chapter II above, and it now remains to discuss the contributions of the latter two. Phillip Ovsyannikov was born in 1827 and studied medicine at Dorpat University. He then joined the army, but was soon released to participate with von Baer (see below) in a study of Russian fishing industry. As a result of this, Ovsyannikov and von Baer became lifelong friends. In 1858, Ovsyannikov was appointed to a professorship at Kazan University, which he held until 1863 when he joined the Academy of Sciences in St. Petersburg upon von Baer's retirement. Ovsyannikov had numerous interests, one of which was nerve control of respiration (21). He was able to show that a strong stimulation of the central end of a severed vagus nerve stopped breathing in a deep expirational phase, whereas a weak stimulus had no effect.

Nicholas Kovalevsky (1840–1891) was probably the man who was most instrumental in establishing Kazan as a distinguished center for the study of physiology. He was born and studied medicine at Kazan, and after some studies with Bruecke, Ludwig, and Kolbe abroad, settled in Kazan as a professor of physiology. He issued a journal named *Reports of the Kazan Physiological Laboratories,* and published frequent reports in *Pfluegers Archiv* on papers presented at the local physiological society meetings. Kovalevsky's interests spanned a rather broad spectrum of subjects. At the beginning of his career, he was interested in the histology of the spleen (22), where he showed that the cells of the reticuloendothelial system were morphologically similar and functionally dissimilar to smooth muscle cells. His later interests included the effects of breathing on blood pressure and the rate of blood flow. As an example, in a paper coauthored by Ivan Dogiel (23) and using Dogiel's stromuhr, it was found that in dogs, interrupted respiration (dyspnea) resulted in an increased arterial blood pressure and a decreased rate of blood flow. Kovalevsky found that this had nothing to do with the vasomotor centers of the brain, and proposed that the affected arteries were actually contracting as a result of the dyspnea. Venous blood flow rate was, on the other hand, increased during the dyspnea episode.

Microbiology and Immunology

The stature of Elie I. Mechnikov is by far the greatest among Russian immunologists. He has been variously described as a victim of tsarist oppression (24), a virtual madman who most of the time had no idea of what he wanted or was trying to accomplish (25), or a methodical and calm scientist with a definite goal throughout his life (26). His personality in his youth was undoubtedly unstable. He attempted to commit suicide three times, always by some ineffective means: by taking an overdose of morphine that was promptly brought back by its emetic action; by running out in a cold winter day immediately after taking a hot bath in an attempt to catch fatal pneumonia; and by inoculating himself with a culture of relapsing fever microorganisms. Robinson (24) ascribes Mechnikov's emotional difficulties to a virtual plot engineered by the tsar's bureaucrats. Yet in his later years (27, p. 249), Mechnikov explained his youthful behavior on the basis of his own egocentricity and expectation that others should have worshipped his genius. He never mentioned any specific cases of persecution generated by the tsarist regime.

Mechnikov was born in 1845 in the Kharkov province in the family of a retired army officer and his Jewish wife. Having little confidence in Russian educational institutions, young Elie persuaded his parents to send him to the Wuerzburg University to study with Kolliker. The latter was, however, absent when Mechnikov arrived in Wuerzburg, and the young hopeful turned to a group of Russian students for moral support in his new and unfamiliar surroundings. He was not well received by by his countrymen because, some speculate, he was partly Jewish. However, neither Mechnikov himself nor his wife in her biography of him (26) have ever mentioned that his failures or successes in life occurred because he was partly Jewish. The disappointed Mechnikov did not remain in Germany for long and grudgingly entrolled in Kharkov University, which he finished in two years with a degree in zoology. The Russian government then granted him a stipend of 1600 rubles to further study abroad. He worked in various places in Germany and Italy, during which time he managed to accuse Leuckart, one of his mentors, of stealing his discovery on the life cycle of nematodes. In 1867, Mechnikov returned to Russia and was appointed to the faculty of Odessa University. Shortly thereafter, he got himself into a wild disagreement with an older professor, resigned, and went to teach zoology at St. Petersburg University. He was apparently quite unhappy there, and decided that marriage might improve things. He proceeded to marry a thirteen-year old daughter of an acquaintance. Unfortunately, she was ill with tuberculosis, and in part because of this, Mechnikov returned to the more southern climate of the Odessa University. His wife, nevertheless, died in 1873, and his first two attempts at suicide followed. He married his second wife and later biographer, Olga, in 1875, and a period of relative tranquility in his life followed. In 1881, his wife became ill with typhoid fever, and

Mechnikov himself developed cardiac symptoms. A period of depression followed, and a third attempt at suicide was made. In 1882, he resigned his position at the Odessa University and went to Messina, Italy, to study marine life. It was there that he discovered phagocytosis.

In 1885, Mechnikov was asked to become the director of a newly created bacteriological institute in Odessa, whose purpose was to develop methods for the protection of cattle and poultry against disease. Mechnikov accepted the offer, but being an idealist, he was not very enthusiastic at working on practical problems. He preferred to leave the operation of the institute to his subordinates, while he himself continued working on phagocytosis. The prestige of the new institute suffered a severe blow when Mechnikov and his associates tried to rid the fields of Little Russia (today's Ukraine) of mice by infecting them with the newly discovered chicken cholera microorganism. An ignorant public protested believing that this might bring about an outbreak of human Asiatic cholera epidemic, and the scheme had to be abandoned. In 1887, Mechnikov was invited to assume the directorship of a bacteriological institute in St. Petersburg, but he refused because of the severe weather conditions that existed there in winter time. His Odessa institute continued to be neglected by its director, and in 1888, his assistants Gamaleya and Bardach committed a more serious blunder: in an attempt to vaccinate sheep against anthrax, some 5000 animals perished. Thereupon, Mechnikov fled the wrath of the peasants and the Zemstvos that supported the vaccination program, and settled in Paris, France, where he joined the Pasteur Institute and remained there until his death in 1916. During his tenure in the Pasteur Institute, Mechnikov served as a preceptor to numerous Russian students, e. g., Besredka, Nefediev, Chistovich, Savchenko, Metalnikov, and others. In 1911, he was invited by the Russian government to study the endemic plague loci in eastern

Russia. He headed an expedition into that area and gathered information about not only the plague, but also about the epidemiology of tuberculosis among the native Kalmuks in that part of the country.

Mechnikov's political views have been subject of considerable number of discussions. Robinson (24) hints that Mechnikov dabbled in nihilism, though Olga Mechnikov states that he actually discouraged his students from pursuing political causes. He was an idealist, to be sure; his almost religious belief in evolution and his conviction that science would eventually solve all the problems of this world were well publicized. Referring specifically to Russia, Mechnikov stated that its problems would be solved by its intellectuals "apart from the government and in opposition to it." Yet a socialist or anarchist he was not in spite of his association with the exiled Russian revolutionaries Herzen and Bakunin. Mechnikov felt that collective ownership and collective means of production would restrict individualism too much (27, p. 223).

Mechnikov's formal university training was in the field of zoology, and his first scientific investigations were carried out in that field. Being a passionate believer in Darwin's theory of evolution, Mechnikov tried to establish the stages of gastrulation in lower organisms. He described the development of *Trachymedusae* and *Narcomedusae* (*Phylum Coelenterata*), *Tricladida* and *Pylictia* (*Phylum Platyhelmintes*), and the holothurians (*Phylum Echinodermata*). His greatest discovery came when he was vacationing in Messina, Italy, in 1882–1883, when he observed that the wandering cells of the transparent larvae of an echinoderm (starfish) devoured some thorns that he had introduced into these organisms. Later, while back in Odessa, he confirmed this phenomenon in a transparent crustacean, the *Daphnia.* He found that both the fungal spores and anthrax bacilli were devoured by the wandering (amoeboid) cells. He called this process *phagocytosis,* and the wandering cells involved, the *phagocytes.* He extended his findings to the mammalian organism and determined that the macrophages and microphages were involved in phagocytosis. He proposed that inflammation was a defensive reaction, and that white blood cells gathered in the inflamed tissue were used to phagocytose bacteria. His ideas were at first not readily accepted, and he spent much time in his later years defending and amplifying the theory of phagocytosis (27).

Phagocytes were, according to Mechnikov, the principal mechanism of immunity in the mammalian organism. He postulated that the macrophages contained a proteolytic enzyme, which he called *microcytase*. These enzymes served to digest the phagocytosed materials. In naturally immune animals, macrophages were primarily responsible for the immobilization of foreign red cells, whereas microphages were used to immobilize bacteria. In his study of red cell agglutination and lysis, Mechnikov recognized the presence of soluble "fixatives" and correctly argued that the fixative was necessary to cause the agglutination of red cells and bacteria. He felt that the cytases (complement) elaborated by the phagocytes were necessary to lyse these cells. Mechnikov also recognized the heterogeneity of these "fixatives," since the red cells of one species were not agglutinated by its own serum but were by that of another. He felt that the "fixatives" were produced by the phagocytes and excreted into the blood stream. Acquired immunity was, in Mechnikov's views, a stimulation of production of specific "fixatives." He also correctly viewed the antigen-antibody precipitate as being a combination of the fixative and a foreign substance.

Complementing Mechnikov's ideas of cellular immunity, there was the theory of humoral immunity promoted by Ehrlich and Bordet. Ehrlich discovered what we today call complement, a serum protein complex inactivated by heating at 50–60°C, and which is necessary for the lysis of red cells by a naturally immune serum. Mechnikov considered complement to be identical to his cytases, which were produced by the disintegration of phagocytes. Mechnikov and Ehrlich shared the 1905 Nobel Prize for their contributions to immunology.

Mechnikov's contribution to practical medicine came in 1903 when he and his collaborator Roux discovered that syphilis microorganisms remained localized at the site of the inoculation for some hours after it. This made it possible to develop a topical ointment, which, when applied to the infected area, would inhibit the spread of the microorganism and thus prevent infection. Such a substance proved to be calomel, and this drug came to be used extensively before the advent of antibiotics.

In his later years, Mechnikov became preoccupied with senility and death (27). He developed a theory of the ageing of man, which even at this time sounds plausible. He felt that tissues were being constantly altered by external poisons, or poisons produced by intestinal parasites. When the tissues were sufficiently weakened, they were attacked and destroyed by phagocytes. This was supposed to be especially true in case of atherosclerosis, which, Mechnikov thought, was brought about by phenolic compounds generated by intestinal bacteria. Even the graying of hair was, according to Mechnikov, the result of phagocytic activity. In summary, he believed that ageing was an autoimmune process. In attempting to determine the causes of longevity in some animals, Mechnikov noted that fish, reptiles, and some birds lived relatively long lives. In all such species, he noted that the large intestine was poorly developed. Thus, the long-lived bats have no large intestine at all. He further noted that the animals that store their wastes in the large intestine for a long time (e. g., horses and cattle) are not known for their long lifespans. He felt that the sole purpose of the large intestine was to store waste, and that such storage results in breeding of harmful microbes. The answer was, therefore, to sterilize the gut; however, his experiments with feeding antiseptics to animals did not succeed too well. Then noting that the contents of the large intestine were alkaline, Mechnikov suggested that the harmful bacteria could be driven out of the gut by acidifying it. To accomplish this, he suggested that people consume large quantities of sour milk containing an inoculum of *Lactobacillus bulgaricus,* a microorganism he found in yogurt samples of some Bulgarian natives, who were known for their long lifespans. Sincerely believing in the benefits of acidified alimentary canals, Mechnikov endorsed the products of a dairy manufacturer who marketed products to Mechnikov's specifications. Because of this, he was severely attacked by the French press in 1911 and 1912 for alleged conflict of interest and profiteering. These accusations brought much grief to the venerable professor, since he made not a penny on his endorsements and acted on what he believed would be in public interest.

One of the more distinguished students of Mechnikov was one Nikolai F. Gamaleya. He was born in Odessa in 1859, graduated from the local university in 1880, then obtained his medical degree at the Military-Medical Academy in 1883. In 1886, he was sent to the Pasteur Institute in Paris to learn vaccination techniques, then joined the ill-fated bacteriological institute in Odessa, which was headed by Mechnikov. But in 1892, Gamaleya was responsible for organizing the Microbiology Department of the Military-Medical Academy, where he remained until 1899. He then went back to Odessa to establish a bacteriological institute there. He published and edited the public health journal *Gigiena i Sanitariya* (Hygiene and Sanitation) from 1910 to 1913. Gamaleya was a tireless crusader for better public health conditions in Russia. The Soviet press described him as an enthusiastic follower of the Michurin school of biology (see below), and credited him with discovery of bacteriolysis, chemical vaccination, and elucidation of the role of lice in propagating typhus fever. Gamaleya died in 1949.

Of the better-documented accomplishments of Gamaleya is his discovery in 1887 of the organism that causes cattle plague (Rinderpest) (28). He showed that it was a small rod that the blood of the infected animals was able to transmit to the healthy animals, and that the infectious factor was filterable. The chicken cholera organism was isolated by Gamaleya in 1888 (29), and he named it *Vibrio metchnikovii*. These organisms were observed in the intestines and blood of infected chickens, and were transmittable to other chickens and birds, especially pigeons. The bacteria were killed by heating at $50^{o}C$ for 5 minutes, but not at 45° for 10 minutes. Vaccination of pigeons against Asiatic cholera provided them with immunity against chicken cholera.

One of the truly great microbiologists of Russia as well as in the entire world was Sergey M. Vinogradsky. He was born in Kiev in 1856, and after an unrewarding study of law at Kiev University and music at the St. Petersburg Conservatory, he settled for natural sciences and graduated from St. Petersburg University in 1881. He took further training at the Universities of Strassburg and Zuerich, and in 1890, he joined the newly created Institute of Experimental Medicine in St. Petersburg. In 1902, he became its director. The Bolshevik coup forced him to leave Russia and to join the faculty of Belgrade University. In 1922, he accepted an invitation to join the Pasteur Institute of Paris, where he remained until his death in 1953. He was a member of the British Royal Society and the French Academy.

Vinogradsky's first important contribution was in the field of sulfur bacteria (30), a project he carried out during his training period in Strassburg in 1886. He chose the filamentous *Beggiatoa* organism for his studies, even though he never succeeded in growing these organisms in pure culture. He was able to confirm Hoppe-Seyler's work showing that the sulfur bacteria did not reduce sulfur to hydrogen sulfide, and that elemental sulfur is deposited in these organisms as a result of oxidation of hydrogen sulfide. He also showed that the hydrogen sulfide arose in such cultures through the decomposition of sulfur previously deposited in the filaments of the *Beggiatoa* and the reduction of the gypsum present in the culture medium by contaminating microorganisms. The sulfur content of the filaments varied with the conditions of the culturing operations. By growing the organisms in springwater that contained no sulfate, no sulfur was observed in the filaments. On the other hand, if the *Beggiatoa* were grown in the hydrogen sulfide-containing atmosphere, sulfur granules immediately appeared in the filaments, and disappeared again upon transferring the organism into the springwater. The chemical reactions involved were defined as follows:

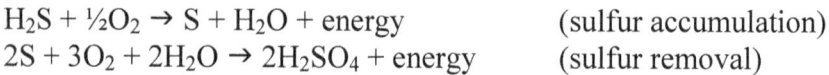

$$H_2S + \tfrac{1}{2}O_2 \rightarrow S + H_2O + \text{energy} \qquad \text{(sulfur accumulation)}$$
$$2S + 3O_2 + 2H_2O \rightarrow 2H_2SO_4 + \text{energy} \qquad \text{(sulfur removal)}$$

In his work on iron bacteria (31), Vinogradsky chose the species *Leptothrix ochracea*, previously described by Kuetzing. He noted that these organisms existed in iron-rich waters, and that the oxidation of ferrous to ferric iron does indeed take place inside the cell and the product, ferric oxide, is excreted and deposited outside the cell. The bacteria had an absolute requirement of iron for growth, though the amount of iron oxidized was unusually large compared to the scant amount of growth observed. Vinogradsky proposed that the energy required for growth of iron bacteria was derived from the oxidation of ferrous to ferric iron, and speculated that iron bacteria may have participated in the deposition of iron ores on this earth. Vinogradsky was thus first to elucidate the fundamental physiological mechanism in the iron autotrophs, and surprisingly, not much additional information has been gathered about these organisms since then.

Vinogradsky's greatest contribution to microbiology is said to be in the field of nitrogen fixation and nitrification of ammonia by soil organisms. Nitrogen fixation is a process whereby microorganisms convert atmospheric nitrogen into ammonia, the biologically utilizable form of nitrogen. Nitrification, on the other hand, is a process whereby the ammonia is converted to nitrite and then into nitrate. The latter is extremely important in sewage disposal technology. Vinogradsky began his work on nitrifying bacteria in Zuerich (30). From the Swiss soil he was able to isolate the organism *Nitrosomonas europaea,* which carried out the following reaction:

$$2NH_3 + 3O_2 \rightarrow HNO_2 + 2H_2O + energy$$

The organism was a small bacillus, which Vinogradsky could grow only in the complete absence of organic material. His culture consisted of purified ammonium sulfate (1 gram) and potassium phosphate (1 gram) diluted to 1000 ml with "lake water." To each culture vessel was added 0.5 to 1 gram of magnesium carbonate before sterilization. The microorganism could not be grown on gelatin plates. Vinogradsky later isolated other microorganisms that were able to carry out the nitrification reaction. These were *Nitrococcus* (1892), *Nitrospira briensis* (1931), *Nitrospira Antarctica,* which was isolated from Antarctic soil (1933), *Nitrocystis javanensis* (1892), and *Nitrocystis coccoides* (1933). The nitrite produced by the above organisms is oxidized to nitrate by the *Nitrobacter* organisms according to the following equation:

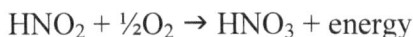

$$HNO_2 + \tfrac{1}{2}O_2 \rightarrow HNO_3 + energy$$

The first such organism to be isolated was *Nitrobacter winogrdskyi* discovered by Vinogrdasky in 1892. It was a short bacillus, and gave a Gram-negative reaction.

The first nitrogen-fixing organism, *Claustridium pasteurianum* (32), was discovered and grown in pure culture by Vinogradsky in 1895. It was a large rod with spores, which gave a positive Gram test. The organism fixed nitrogen under anaerobic conditions and produced butyric acid. Glucose was utilized in proportion to the nitrogen being fixed. The organism also grew well when peptone or ammonium sulfate was used as nitrogen sources. A typical nitrogen fixation experiment was carried out as follows: a flask containing 40 grams of glucose per liter and some inorganic salts was inoculated with the organism and was grown in the presence of nitrogen but no oxygen. After fermenting for 20 days, all sugar was depleted and a total of 53.6 mg nitrogen, 3.7 grams of acetic acid, and 14.2 grams of butyric acid were recovered from the medium. Hydrogen and carbon dioxide were also produced. When the culture was grown aerobically in the presence of ammonium sulfate or peptone, different butyric acid to acetic acid ratios were obtained. Thus, in the absence of ammonium sulfate under anaerobic conditions, the above ratio was 2.6, in the presence of trace amounts of ammonium sulfate the ratio was 4.0, and with large amounts of ammonium sulfate it was 6.0. With peptone, the butyric/acetic acid ratio was 23. In all cases, the acids accounted for 42–45% of the sugar utilized. *Claustridium pasteurianum* was isolated from the soil of the St. Petersburg area. Attempts to isolate the organism from Polish soil were not successful. Instead, Vinogradsky isolated a similar microorganism, which he called *Claustridium wolhynien,* from Poland's soil. It was identical to *Claustridium pasteurianum* with the exception that it did not produce butyric acid.

One of the more distinguished students of Vinogradsky was Vasilii L. Omeliansky (1867–1928), a graduate of St. Petersburg University and member of the St. Petersburg Academy of Sciences. He can be best characterized as a microbial biochemist. He studied the chemical decomposition of a number of substances by microorganisms, and, as a result, discovered several new species. His numerous papers on the nitrifying bacteria were basically an extension of Vinogradsky's work. One of such substances was cellulose. As a substrate for his studies he used Swedish filter paper, the purest example of cellulose available at his time. His bacterial sources were horse manure and mud. Fermentation of cellulose by the manure/mud inoculum resulted in the formation of methane and hydrogen (33). When the inoculum was heated to 75°C for 15 min., the ability to release methane was lost, though the ability to produce hydrogen remained. Omeliansky concluded that he was dealing with two microorganisms, one heat-sensitive, the other insensitive. The organisms grew in a medium containing ammonium sulfate, calcium phosphate, magnesium sulfate, and sodium chloride, and produced carbon dioxide in addition to methane and hydrogen. Eventually, the hydrogen-producing organism was isolated, and it is today known as *Claustridium omelianskii.* It is an anaerobe, it utilizes cellulose but not glucose for growth, and it produces carbon dioxide and hydrogen. It produces no methane. In 1916, Omeliansky reported the isolation of a methane-producing organism (34), which today is known as *Methanobacterium omelianskii.* This organism was found to utilize various primary alcohols for growth. Ethanol was an especially welcome source of carbon. Its concentration in the medium was 2–4%. After 2 weeks of fermenting, the organism produced gasses consisting of 11.5% carbon dioxide, 1.1% hydrogen, and 87.4% methane. The organism was long and thin, and it did not produce spores. Its sources were rabbit feces and soil.

Omeliansky's interests also included the degradation of organic acids by soil bacteria (35). He chose formic acid in the form of calcium formate as his model substance. Horse manure served as the inoculum. The reaction ran very well in the presence of peptone and deposited calcium carbonate as the reaction product. Omeliansky was able to isolate the organism causing this reaction, and called it *Bacterium formicum*. It is also known by the name of *Achromobacter formicum*. The organism was a Gram-negative rod and a facultative anaerobe. Using the pure culture, Omeliansky found that oxygen disappeared after the 7th day of fermentation, nitrogen on the 36th day (displaced by the gaseous products of fermentation), to be replaced by hydrogen and carbon dioxide (34 and 64% respectively by volume) at equilibrium. He thus proposed that the following transformation took place in the microorganism:

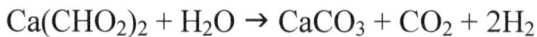

$$Ca(CHO_2)_2 + H_2O \rightarrow CaCO_3 + CO_2 + 2H_2$$

The organism utilized mannose for growth, producing carbon dioxide, hydrogen, ethyl alcohol, formic acid, acetic acid, and levorotatory lactic acid in quantities of 30.4, 1.2, 18.5, 0.7, 3.8, and 45.4% by weight, respectively. The fermentation of dulcite resulted in the production of hydrogen, carbon dioxide, acetic acid, formic acid, lactic acid, and succinic acid in the ratio of 1.0, 30.5, 11.2, 0.5, 25.8, and 31.0%, respectively.

Someone in Omeliansky's lab had once accidentally left a cylinder on a window sill filled with river water and some hay and gypsum at the bottom. After standing several months in the light, there was a burgundy-colored growth at the bottom of the cylinder. When an agar plate was inoculated with these bacteria, numerous colonies were obtained after a few days of incubation in light. Some colonies were colorless, others had varying shades of red pigment. Further culturing of these organisms was not successful. Under the microscope, Omeliansky observed a slightly curved bacillus with a flagellum of up to 6x the length of the organism. Characteristic rosettes were formed by the organism with the flagellae in the center and the body of the organism in the periphery. Omeliansky named his organism *Bacillus flagellatus* (36), and noted that in some respects, it was similar to an organism isolated from Chicago's water by Mabel Jones in 1907.

Virology can trace its beginnings in Russia to Dimitri I. Ivanovsky (1864–1920). He was a graduate of St. Petersburg University and a professor of botany there as well as at the University of Warsaw. Being a botanist and plant pathologist, Ivanovsky was interested in the etiology of the tobacco mosaic (Mosaikkrankheit) and the tobacco pox (Pockenkrankheit) diseases (37). Mayer considered the two diseases to represent the two stages of but a single disease; however, Ivanovsky maintained that these two conditions were two distinct entities that could exist at the same time in the same plant. He found that the leaf juice of the plant infected with the tobacco mosaic disease was infectious, and that infectivity could be abolished by boiling. Contrary to Mayer's finding, Ivanovsky established that the infectious factor was retarded by neither a double layer of filter paper, as the fungi would be, or by bacterial filters. He then concluded that the disease was caused by a very small microbe. This was the first pathologic condition that was shown to be caused by a microbial filterable agent. Such were the beginnings of virology. In addition to tobacco mosaic disease, Ivanovsky was responsible for a number of other contributions to the field of botany. When he was in Warsaw University (1901–1915), he apparently came under the influence of Mikhail Tsvett, the developer of chromatography (see below), and published a number of works on plant pigment chemistry (38). For instance, he was able to show that the absorption maxima of plant pigments depended on their physical state, such as the shifting of the absorption maximum of chlorophyll from 680 nm to 668 nm when the latter went from the solution to the colloidal state (39).

There were a host of microbiologists in Russia of perhaps a lesser stature than that of Mechnikov, Ivanovsky, or Vinogradsky, who nevertheless made far-reaching contributions to their fields of endeavor. Thus, a certain ophthalmologist, Katharina Kastalski (1864–1899), was first to isolate a pure culture of actinomycetes. The source was a tear duct of one of her patients. She also demonstrated that eye infections could be caused by seemingly nonpathological microorganisms. Another important microbiologist and teacher of microbiology and botany was Andrey S. Famintsin (1835–1921). He was a graduate of the St. Petersburg University and taught there from 1861 to 1889. He then founded the plant physiology laboratory at the St. Petersburg Academy of Sciences, and remained its director until his death. He worked on some aspects of photosynthesis, showing that natural and artificial lights were interchangeable in supporting plant growth. From his aquarium, Famintsin was able to isolate a new microorganism, which he named *Nevskia ramosa*. It was a large rod staining well with methylene blue and containing particles soluble in ethanol. Famintsin thought that these represented some sort of a lipid (40). Other investigators later thought that these might have been sulfur granules (41). The cells themselves were encased in a gelatinous coat and swam freely in the medium. The cell membrane was then very easily discernable. The bacteria apparently did not form spores.

During his activity in Paris, Mechnikov was an important contact for Russian immunologists and other life scientists. Among his students was one Ivan Savchenko (1862–1920), a graduate of the Kiev University and later a professor in Kazan University. In Russia, he carried on Mechnikov's work on cellular immunity and inflammation. Alexander Besredka (1870–1940) was a Russian by birth, but chose to spend most of his professional life in France. He was a graduate of the Odessa University and worked there with Mechnikov, Zelinsky, and A. Kovalevsky. After visiting France in 1892, he went to a medical school and graduated in 1897. He then remained at Pasteur Institute and did important work in the areas of anaphylaxis, virus diseases, and tuberculosis.

Zoology and Embryology

Russia was fortunate in having been the home base for the world's most accomplished embryologists: Kaspar Friedrich Wolff, who introduced the concept of differentiation of organs from primordial tissues; Heinrich Christian Pander, who discovered the three embryonic tissue layers; Karl Ernst von Baer, who discovered the mammalian egg; and Alexander Kovalevsky, who discovered the gastrula and the universality of germ layers.

Kaspar Friedrich Wolff was born in 1733 in Berlin, where he also studied medicine. After graduation, he participated in the Seven Years' War as a military physician, and then returned to Berlin to teach anatomy and physiology. The professorship that Wolff had expected to receive did not materialize, because, some say, there was opposition to Wolff's ideas by the clergy. Wolff then accepted an invitation to join the St. Petersburg Academy of Sciences, and left Germany for Russia in 1767. He remained in St. Petersburg until his death in 1794. Wolff's first monumental contribution was his *Theoria Generationis* (42), which was written in 1759 as a thesis for his doctorate in medicine. It was one of the first works to dispute the then-dominant "emboitment" (preformation) theory, whose champion was no other than the founder of histology, Marcello Malpighi (1628–1694). Wolff's approach was both philosophical, in the Aristotelean sense, and experimental, whereby the microscope was used to study both the plant and animal tissues. He noted that the various structures of the leaf were formed from seemingly homogeneous and undifferentiated tissue, and proposed that the same thing might be happening in chick embryo. He was then able to confirm this proposal microscopically. He also noted the fact that living tissues consist of "globules" (i. e., cells), and some authorities consider Wolff to be one of the originators of the cell theory of living matter for this reason. Wolff is also responsible for discovering the fact that specialized organs, such as the intestine and the spinal chord, are formed by an invagination or folding of primordial tissue. This work was published in 1768 in the St. Petersburg Academy proceedings, and did not receive wide attention until it was translated into German in 1812. Wolff is also the discoverer of the mesonephron, the embryonic kidney, and it is very often referred to as the Wolffian body.

Heinrich Christian Pander was born in Riga, Russia, in 1794 and studied medicine at Dorpat University and in Germany. In 1817, he published his work on the development of the chick embryo, where he described the three germ layers, which we today call the ectoderm, the endoderm, and the mesoderm. He also coined the term "blastoderm," which is used in connection with undifferentiated tissues. In 1820, Pander returned to Russia, was appointed to the Academy of Sciences (1823), and devoted the rest of his life to the study of geology and geography of Russia. He is credited with the discovery of the Silurian geological formations in Russia. Pander died in 1865.

Another embryologist whose contributions have perhaps unjustly been overshadowed by those of Wolff, Pander, and von Baer was Heinrich Rathke (1793–1860). He was born in Danzig, studied medicine at Goettingen and Berlin, and had his first faculty appointment at Dorpat University in Russia (1829–1835). He then went to Koenigsberg, where he occupied the position vacated by von Baer when the latter moved to Russia. By 1825, Rathke had recognized that the germ layers discovered by Pander were not restricted to vertebrates. This idea was later expanded by Kovalevsky. A number of embryonic structures were then discovered by Rathke, and many still carry his name. We thus have the Rathke's pouch, an invagination of the embryonic buccal cavity and of ectodermal origin, whence the anterior pituitary is formed. Associated with Rathke's pouch is Rathke's tumor, a congenital tumor arising from a faulty development of Rathke's pouch. Rathke's folds are the two mesodermic folds forming the rectum. Other structures are Rathke's columns, a pair of cartilaginous masses at the anterior end of the chorda dorsalis, and Rathke's duct, which is situated between the Muellerian duct and prostate.

The best-known embryologist associated with Russia was Karl Ernst von Baer. He was born in 1792 in the Estonian province of Russia, and was of Baltic German descent. He obtained his medical degree from the Dorpat University in 1814, then spent some time in Vienna, Wuerzburg, and Berlin studying botany, anatomy, and physiology. In 1816, he accepted a faculty position at the University of Koenigsberg in East Prussia, where in 1818 he submitted and successfully defended his doctoral thesis. Von Baer rose rapidly in the hierarchy of the university, being elected rector in 1825. His embryological investigations at Koenigsberg culminated with his monumental work entitled *De ovi mammalium et homini genesi. Epistola ad Academiam imperialem sciantiarum Petropolitanam,* which was sent simultaneously to the St. Petersburg Academy of Sciences and his publisher in Leipzig, Leopold Voss. The latter published von Baer's work in 1827, and its expanded version, *Commentar zu der Schrift: De ovi mammalium et hominis genesi. Epistola ad Academiam scient. Petropolitanam* was published in *Zeitschrift fuer organische Physik,* vol. 2, 1828.

The St. Petersburg Academy elected von Baer to full membership in 1829, when he spent a few months there working on the embryology of aquatic animals. In 1834, however, von Baer finally accepted a standing invitation from the Academy to join its staff, and moved back to his native land. He remained at the academy as an active scientist until his retirement in 1867. Von Baer died in 1876, some twelve days after submitting his last paper for publication. Beside the St. Petersburg Academy, von Baer was a member of the

Royal Society (1854) and the French Academy (1858). Von Baer's scientific interests were remarkably diverse (43). In addition to his classical work in embryology, which will be described below, von Baer contributed heavily to the fields of anthropology, physiology, and geography. He actually headed an expedition to the Arctic to explore Novaya Zemlya, and participated in expeditions to North Cape, the Sea of Azov, and the Caspian Sea. In his travels, von Baer discovered that in the northern hemisphere, the right banks of rivers become eroded in preference to the left banks. This fact is known as the Baer's Law. Von Baer investigated in detail the operations and methods of Russian fishing industry, and his suggestions contributed much to its development. He also authored many chapters in texts on physiology of that time, and published materials on the exploration, natural history, and populations of Russia.

Von Baer's best known discovery is the mammalian ovum, which he described in his *Epistola*. Up until that time, in analogy to the avian egg, the Graafian follicle of the ovary was considered to be the egg. Von Baer observed the true egg, at first visually, in the Graafian follicle of a pregnant dog, then confirmed his findings microscopically. The ova he saw were attached to the wall of the Graafian vesicle rather than being suspended in the vesicular fluid. Von Baer made an extensive comparative study of ovaries and eggs of a variety of animals. He pointed out that the Graafian follicles were encased by a capsule consisting of two layers lined on the inside with a granular membrane *(stratum granulosum)*. Similar structures were seen in the hen's egg. The follicular fluid of mammals was considered to be analogous to the yolk, and the ovum was analogous to the germinal layers of the eggs of lower animals. Mammals were thus considered by von Baer to give "eggs within eggs."

The mammalian eggs were found to be released from from the ovary surrounded by a layer of cells of Graafian follicle origin. Von

Baer then looked for and found eggs in dog oviducts and uteri, and showed them to be identical to those he saw in the ovary. Surrounding the ovum itself were two membranes, one relatively transparent, the other dotted with dark spots. The eggs floated free in the uterus for a while, then became attached to the uterine wall. Most attempts of detaching the eggs resulted in their lysis, but von Baer was able to recognize the two membranes remaining after such accidents. There is some controversy whether or not von Baer was able to observe a nucleus in the mammalian ovum. It is generally accepted that he did (it was called the Purkinje's vesicle at that time), yet failed to appreciate its function. Likewise, von Baer observed cell division in the fertilized ova without recognizing its true significance. He believed that the purpose of such division was to provide a better contact between the embryo and the uterine wall.

Von Baer's book entitled *Ueber Entwickelungsgeschichte der Thiere* (1828) consolidated all embryological knowledge of that era into a systematic science, and enumerated some fundamental principles that are valid even today: the formation of specialized organs takes place from a common (general) tissue, and the differentiation proceeds via formation of germ layers which then undergo histogenesis and morphogenesis. Until his death, von Baer did not accept the Darwinian theory of evolution. He felt that the basic assumption of this theory regarding the continuous nature of evolution was not valid. He pointed out that though the bones of extinct and modern species of animals were often unearthed in the same geological layer, no bones of intermediate species have been found. He felt that at some time during the existence of the earth, there was a force that created all the present and extinct animal species. He viewed evolution as being the development of the mind or spirit (Geist), and did not change his views even after the publication of Darwin's work. (44).

A Russian embryologist of more recent vintage was Alexander O. Kovalevsky (1840–1901). He was born in Duenaburg (Daugavpils) in today's Latvia, and graduated from St. Petersburg University in 1863 with a degree in natural sciences. He was a professor in the same university (1866–1868), Kazan University (1868–1869), Kiev University (1869–1874), Odessa University (1874–1891), and again at St. Petersburg University (1891–1894). Kovalevsky is considered to be the founder of comparative embryology. He carefully followed the development of embryos in a variety of lower animals (*Psolinus, Phoronis,* mollusks, and arthropods), as well as representatives of the chordate phylum (amphioxus, the tunicate, birds, and mammals). His detailed studies on ascidians (*Subphylum tunicata)* revealed the complexity of these animals and the presence of a notochord therein. He is thus credited by many with establishing the *Phylum chordata.* Kovalevsky's studies revealed that all animals developed from eggs that formed two germ layers *via* an invagination of the primary layer. These were later called the *endoderm and the ectoderm,* and the embryo at that stage was termed the *gastrula.* The third germ layer is then formed by an invagination of the endoderm. Kovelevsky's discovery of the universality of the germ layers (45) and the publication of the then novel theory of evolution by Darwin prompted the German embryologist E. Haeckel to propose in 1872 that all animals arose from a primitive life form that he called *gastraea.* This animal was supposed to have consisted of two cell layers similar to Kovalevsky's gastrula (note, however, that Kovalevsky did not coin the term *gastrula).* The latter vigorously opposed this theory, as did von Baer for perhaps different reasons. Nevertheless, because of its simplicity and logic, Haeckel's views (also called transcendentalism) were accepted by embryologists for many years thereafter.

Kovalevsky's work on *Phylum molluska* involved the study of development of the chiton (*Class amphineura)* and the dentalicum

(*Class scaphopoda*) with excellent illustrations that are used in some texts even today. Among the arthropods, Kovalevsky studied the development of musca (*Order diptera*), where he found that the lower level of the gastrula differentiated directly into mesoderm and endoderm (46). His other studies included the sea cucumber (*Phylum echinodermata, Class halothurioidea*). In 1867, he showed that the previously discovered *Actinotrocha branchiate* was actually the larval form of *Phoronis*, and showed that fertilization of its eggs occurred in its coelomic cavity.

A typical representative of classical zoology in Russia was Vladimir Zalensky (1847–1918). He was a graduate of Kharkov University (1867), and was on the faculty of Kazan and Odessa Universities. In 1897, he was appointed to the directorship of the zoological garden of St. Petersburg Academy of Sciences, and in 1901, director of the academy's biological station in Sevastopol. He was elected to the academy's membership in 1893. Zalensky's contributions were mostly in the field of lower animal development (46), especially the *Nemartines*. He described the gastrulation of *Prosorochmus vivaporus* and the development of this organism into a larva. He also followed the metamorphosis of the *Philidium*, maintaining that the longitudinal muscle layer of these organisms was of ectodermal rather than mesodermal origin, as was the belief at that time. He also investigated in great detail the development of this animal's digestive tract.

Botany

Russia produced a number of prominent botanists, many of whom also contributed significantly to other fields of the natural sciences. The most notable of the latter were Ivanovsky, the founder of virology, and Tswett, the developer of chromatography. A disproportionately large number of Russian botanists was interested in photosynthesis (39, 47), so that it is not surprising that one of the strongest areas of biology in the Soviet Union (which replaced Tsarist Russia in 1917) was the field of photosynthesis. One of its first Russian investigators was Alexander Nikolayevich Volkov (1849–1928). In his doctoral dissertation in 1866 (Dorpat University), Volkov reported that the amount of carbon dioxide utilized and oxygen produced by aquatic plants undergoing photosynthesis was proportional to the intensity of light available to the plant. He did this by a simple method of counting the number of oxygen bubbles rising to the surface per init time as a function of the intensity of light. He later worked with A. Meier in Heidelberg, and in 1875 was appointed to the faculty of Odessa University. In 1880, he resigned due to ill health and moved to Venice, where he became an accomplished painter. He wrote articles on art under the pseudonyms of Russov and Muromtsev, and his scientific legacy is limited to only four papers, including his now classical doctoral dissertation.

The relationship between the quantity of chlorophyll in a plant and the amount of light striking the plant was established by Vladimir N. Lyubimenko (1873–1937), a graduate of the St. Petersburg Forestry Institute (1898) and a staff member of the St. Petersburg Botanical Garden. He discovered in the period between 1905 and 1908 that shade plants (ombrophiles) had much more chlorophyll (0.4 to 0.8% by weight) than did plants commonly exposed to light (heliophiles, 0.1% chlorophyll by weight). The rate of photosynthesis in these types of plants was thus approximately equal. Under an intense light, the ombrophilic plants showed a retarded rate of photosynthesis. He later established that many plants could adjust to different light intensities. Physical aspects of photosynthesis were investigated in Russia mainly by Konstantin A. Purevich and Kliment A. Timiryazev, whom the Soviets had pronounced to be the "greatest" Russian botanist. He was born in St. Petersburg in 1843, graduated from St. Petersburg University with a degree in natural sciences, and then went abroad to Germany and France for further study. In 1869, he was appointed to the faculty of Moscow Agricultural Academy, and in 1877 became professor of plant anatomy at Moscow University. In part because Timiryazev was a militant opponent of the tsarist regime, he was forced to exercise his option to retire after thirty years of government (university) service in 1898. He died in 1920.

Timiryazev's main scientific contributions were in the field of photosynthesis. He proposed that in its process, chlorophyll is first decolorized, then the color gets regenerated. He was able to carry out this reaction *in vitro* by reducing and decolorizing chlorophyll chemically, then reoxidizing it and restoring the color by exposure to air. He called the colorless chlorophyll *protophyllin*. The apparent lack of relationship between chlorophyll and carbon dioxide fixation was explained by Timiryazev on the basis of "sensitization" of the pigment by the light, thus absorbing the solar energy and converting it to chemical energy to utilize the carbon dioxide. Until then, i. e., 1869, most botanists believed that the light was absorbed directly by the carbon dioxide that had been fixed by the plant, and that chlorophyll was merely a protective device for the plant. Timiryazev reported in 1903 that chlorophyll was situated on the surface of the leaf plastids, thus being readily accessible to light. He was an avid Darwinist, and his obsession with adaptation of species resulted in misinterpretation on his part of scientific data and a vitriolic controversy with the German physiologist-botanist Engelmann. The fact that maximum photosynthesis activity in plants takes place at the emission maximum of the sun (red region of the spectrum) was to Timiryazev the living proof of Darwin's theories. He refused to consider the fact that maximum energy was emitted by the sun in another region of the spectrum, and refused to believe that there existed a second photosynthesis maximum in the blue-violet region of such spectrum.

Purevich (1866–1916) was a graduate of and later a professor at the Kiev University. He determined by calorimetry that only 0.6 to 7.7% of the solar energy absorbed by the leaves was used for photosynthesis. He also worked on the respiration of plants showing that in *Aspergillus niger*, the respiratory coefficient was dependent on its nutritional state and temperature.

The chemical nature of chlorophyll and the chemical events accompanying the process of photosynthesis were studied by a number of Russian investigators. For example, the relationships between photosynthesis and carbohydrate metabolism were investigated by Vasilii V. Sapozhnikov (1861–1924), a student of Timiryazev and graduate of Moscow University. He found that there was a *de novo* synthesis of starch from sucrose, and that the process of photosynthesis would cease after the carbohydrate (starch) level would reach 17–25% of the dry weight in detached leaves of *Vitis lobrusca* and 23–29% in the leaves of *Vitis venifera*. He also found that, compared with the amount of carbon dioxide assimilated, remarkably little carbohydrate could be recovered from illuminated plant leaves undergoing vigorous photosynthesis. He concluded that the carbohydrate synthesized is rapidly converted to other substances in the plant. His results were later confirmed (in 1901) by his compatriot T. Krashennikov. Beside being a prominent botanist, Sapozhnikov was a traveler, explorer, and ecologist. A mountain in Siberia and a glacier in the Altai mountain range carry his name. He was apparently able to engage in his nonbotanical activities because of his professorial position at the Tomsk University in Siberia, which he occupied since 1893.

The chemistry of chlorophyll and its transformations during the process of photosynthesiswas first studied by Nikolai Avgustinovich Monteverde (1856–1929). He was of Spanish descent whose father had settled in Russia in 1823 upon the invitation of the Russian government for the purpose of designing public buildings. Monteverde, the son, graduated from St. Petersburg University in 1879 with a degree in natural sciences, and his thesis was executed with the help of Andrey Famintsin. He then worked with Ivan Borodin at the St. Petersburg Forestry Institute, where he defended his master's thesis on the subject of magnesium and calcium oxalates in plants. From 1892 to his death in 1929, Monteverde worked at the St. Petersburg Botanical Garden, first as a director of its museum and biological laboratories, then as chief of the section of medicinal plants. Of the 78 publications ascribed to Monteverde, only a handful was published in Western scientific journals. He is mostly remembered for his work on chlorophyll. In 1893, Monteverde isolated from plant leaves a compound he called *protochlorophyll*. He considered this to represent a precursor of chlorophyll. This view was modified when Monteverde, in collaboration with Lyubimenko, postulated a hypothesis in which protochlorophyll was, instead, a product of another precursor that it shared with chlorophyll. This common precursor was termed *chlorophyllinogen*:

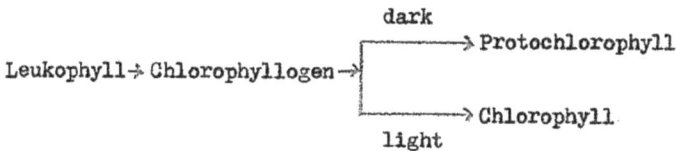

```
                              dark
                        ┌─────────→ Protochlorophyll
Leukophyll⇢Chlorophyllogen→│
                        └─────────→ Chlorophyll
                             light
```

In 1940, Hans Fischer finally showed that protochlorophyll was actually an oxidation product of chlorophyll.

The botanist who was able to separate the chlorophyll complex into its component parts was Mikhail Tswett (1872–1920). He was born in Italy of a Russian father and an Italian mother. He was educated in Geneva and Lausanne, Switzerland, and received his doctorate in botany in 1894. In 1896 he moved to St. Petersburg, where he taught plant anatomy and physiology He remained there until 1901 when he joined the faculty at the Warsaw University (Warsaw, Poland, was part of the Russian Empire at that time). In 1917, during World War I, as the German armies were advancing into Russian territory, he was evacuated with the university to Nizhnii Novgorod, Russia. In 1918, he joined the Yuriev University in Voronezh as a professor of botany and stayed there until his death in 1920 (48, 49). It is said that he died of tuberculosis. Tswett is best known among biochemists as the developer of the chromatographic technique (50). His purpose was of course not the development of a method for the biochemists' use, but was, instead, to separate plant pigments. Although it was at that time known that chlorophyll was not a homogeneous substance, no method for its resolution was yet available. Tswett extracted his pigments with benzene, petroleum ether, carbon tetrachloride, or carbon disulfide. He then placed the pigment extract on a sucrose or calcium carbonate column using carbon disulfide as the eluant, and such mixtures then resolved into five colored bands upon passage through the columns (51). The uppermost layer was yellow in color and was called beta-xanthophyll by Tswett. It absorbed light in the 462–475 nm range and at 430–445 nm in alcohol solution. It became blue when acidified. The second band was olive-green in color and was termed beta-chlorophyll. Its absorption maxima were at 450–465, 580–600, and 640–650 nm in petroleum ether. The third band was bluish-green and was termed alpha-chlorophyll by Tswett. It had two absorption maxima, one at 430–440 nm, and the other in the blue-violet range of the spectrum (unspecified). The next band represented a mixture of what Tswett called alpha' and alpha"-xanthophylls. These were separable on a column with benzene as eluant. The fifth and last band was alpha-xanthophyll with absorption maxima at 440–455 and 470–485 nm in petroleum ether solutions. The alpha-xanthophyll did not assume a blue color upon acidification. The carotenes were present in the void fraction of the chromatogram. In this way, Tswett was one of the first to show that carotenes and xanthophylls were different pigments.

Studying the pigments of algae, Tswett was able to show that their main pigment was xanthophyll, which was given the name of fucoxanthin by him. In algae he also discovered a third type of chlorophyll, which is today known as chlorophyll c. He was one of the first to study the fluorescence of plant pigments, and in 1911 he reported that chlorophyll a had a fluorescence band at 677.5 nm, and chlorophyll b at 655 nm. He was able to demonstrate that under the influence of light, chlorophyll changed to another compound of higher energy, which then emitted the excess energy that was then used for carbohydrate synthesis from carbon dioxide. He believed that the various methods then in use for the isolation of chlorophyll were too drastic to obtain chlorophyll in its native state. He thus showed that the crystalline chlorophyll previously isolated by Borodin was actually ethyl chlorophyll, an artifact of the preparation procedure. He published a paper summarizing all the known derivatives of chlorophyll, including all artificial varieties produced by various investigators (52). Tswett's findings and the validity of his chromatographic method were not accepted for several years, his main opponents being Willstaetter and Marchlewski, and a prolonged dialogue was waged by the latter and Tswett on the pages of *Biochemische Zeitschrift* and other journals.

Tswett was also a pioneer in the investigation of leucoanthocyanin chemistry, which Tswett termed "artificial anthocyanins" (53). He prepared these pigments by extracting apple peelings or pulp with 4–5 volumes of alcohol and one volume of HCl at room temperature for several days. When the extract was heated, it assumed at first a yellow, then an orange, and finally a brownish-red color. The HCl could be replaced by sulfuric acid but not by other acids, though the alcohol was replaceable by a number of other organic solvents. The pigment could be precipitated from the extract by an equal volume of water. Another method was to neutralize the acid, then to extract the solution with ether. The pigment was found in the ether phase and could be precipitated therefrom by repeated shaking with water. The absorption maximum of the pigment was at 530–570 nm before purification and 540–580 nm after. The absorption maximum changed to 580–610 nm in alkaline solution. The pigment was decolorized by sodium bisulfite and phenylhydrazine, hence Tswett concluded that the dye was an aldehyde or ketone.

Tswett was one of those rare individuals who was both an excellent chemist and biologist. He advocated the use of mild techniques for the isolation of biological materials, noting that the study of natural products was, at that time, performed mostly by organic chemists, who had no appreciation for the delicate nature of such compounds. He insisted that living systems were not merely mixtures of compounds, but carefully organized structures that could be easily disrupted by extraction procedures. He thus believed that chlorophyll was not present in plant leaves in the free form, but was instead bound to protein in a fashion similar to heme. It is this author's (A. B.'s) view that had Tswett not died in the chaotic days following the Bolshevik revolution in Russia, he would have won the Nobel for his invention of a technique (chromatography) which in later years "would revolutionize chemistry in the twentieth century" (49).

Russian contributions to algal photosynthesis and the nature of algal pigments can be represented by the work of Georgy Adamovich Nadson and Mikhail M. Gaidukov in addition to that of Tswett. Nadson was born in Kiev in 1867 and was still living in 1928 when he was proposed for membership in the Soviet Academy of Sciences. He graduated from the St. Petersburg University in 1889 with a degree in natural sciences, and in 1903 successfully defended his doctoral thesis.

He was the chief librarian of the St. Petersburg Botanical Garden and a professor of microbiology at the Women's Medical College in St. Petersburg. Nadson founded in 1914 the first Russian microbiological journal, the *Zhurnal Mikrobiologii*, and remained its editor until 1917. Nadson was a botanist and microbiologist, having worked on several aquatic plants, yeasts, bacteria, their physiology, reproduction, and structure of their pigments. He published over 100 papers, mostly in Russian journals. In 1912, Nadson described a new photosynthetic microorganism, which he called *Chlorobium limicola.* The organism required hydrogen sulfide for growth, converting it to sulfur. The sulfur was not stored in the organism, which was found in soil and stagnant water (41).

Gaidukov (1874–1928) graduated from Moscow University in 1898, then went to the St. Petersburg Botanical Garden to work with Monteverde. The summers of 1900–1902 he spent working with Engelmann in Berlin, and he apparently spent some time with Molish in Prague as well. In 1904, he successfully defended his magister thesis at the Kiev University, and joined its faculty for a year. In 1905, he moved to the Zeiss works in Jena, Germany, where he studied and developed microscopy until 1910, when he was appointed to the faculty of Moscow University. He remained there until 1913, when he once again moved to the St. Petersburg Botanical Garden. In 1919, he became professor of botany at the Ivano-Voznessensk Polytechnicum, and in 1924 he accepted an identical position at the newly established Minsk University. He remained there until his death.

Gaidukov provided much of the experimental evidence in support of Engelmann's theory of adaptation of algal pigments to the illuminating light via the adoption of a complementary color. Engelmann had postulated that deepwater algae were red because the light that filters down to them has a greenish-blue tinge, whereas surface algae are green because green is the complementary color to the sun's red emission band. Gaidukov was a staunch supporter of Engelmann in the latter's disputes with Timiryazev, Gaidukov's countryman, regarding the evolutionary implications of this phenomenon. Studying *Oscillaria sancta* and *Oscillaria caldariorum*, Gaidukov showed that these organisms grew best under the light whose color was complementary to the color of their pigments. However, under the light of an uncomplimentary color, the pigments of the organisms were able to assume coloration that was complementary to that of the light. This adaptation process took several weeks to several months to establish itself. Gaidukov also showed that this transformation could take place only in living cells (54).

In his later work (55), Gaidukov used blue-green alga *Phormidium tenue* and the red alga *Porphyra luciniata.* He illuminated both with a light much more intense than that used previously by him, and which, in addition, was broken down into its component parts *via* a spectroscope. Within 6–10 hours, the Phormidium exposed to the blue-green portion of the spectrum had assumed a brownish-yellow color, whereas the organism exposed to the red-yellow portion of the spectrum remained blue-green. The infrared and ultraviolet regions of the spectrum were lethal to the organism. Porphyra, on the other hand, became green under the red-yellow portion of the spectrum and remained red under blue-green portion of the spectrum. The fact that such profound changes on coloration had occurred within 10 hours of commencing the experiment showed that the transformation took place within the same cells, and that cell division was not necessary to bring about such change. Gaidukov was able to effect color changes of algal pigments by chemical means as well (56). He treated a number of algae with acids and bases, and showed that the red pigments were convertible to violet and green by alkali, and that violet algae could be rendered red by acids.

A botanist who worked on a variety of problems in the area of plant physiology was Ivan Borodin (1847–1930). His interests included photosynthesis, the chemistry of plant pigments, and nitrogen metabolism. Borodin received his magister degree in 1876, and was elected to the St. Petersburg Academy in 1902. He wrote a very successful textbook entitled *Plant Anatomy* in 1888, and established the Russian Botanical Society in St. Petersburg in 1916. In 1883, Borodin showed the heterogeneity of the yellow pigments of plants. He was able to crystallize two yellow pigments (carotene and xanthophyll?), one of which was soluble in benzene and insoluble in alcohol, whereas the other was insoluble in benzene and soluble in alcohol. It remained for Tswett in 1906 to effect a better separation of these pigments by the method of chromatography. In 1881, Borodin showed that respiration in plants was increased following an intense period of photosynthesis.

The question of asparagine in plants was an extremely puzzling phenomenon to botanists in Borodin's times. He found that asparagine was formed in leaves that were cut and placed in water in the dark, whereas it was absent in freshly cut leaves. He thought that asparagine originated from the breakdown of proteins and expected that asparagine would be reincorporated into proteins when photosynthesis would resume. The origin of asparagine was finally determined by Borodin's countryman, Dimitry N. Pryanishnikov. He disputed Borodin's conclusion regarding asparagine's origin from proteins by showing that it was not incorporated into proteins during photosynthesis even in the presence of ample supplies of carbohydrate. He instead proposed that proteins were synthesized from a series of amino acids, and that asparagine is a mechanism to detoxify ammonia that is produced in plants as a result of protein breakdown. Pryanishnikov was born in 1865 in Siberia. He graduated from Moscow University in 1887 and from the Moscow Agricultural

Academy in 1889. He became a professor at the Academy in 1895, and was elected to membership in the St. Petersburg Academy of Sciences in 1913. He died in 1948.

The state of plant pathology in Russia can be characterized by the contributions of either Ivanovsky mentioned above, or those of Mikhail S. Voronin (1838–1903). Voronin was a graduate of St. Petersburg University, and studied plant pathology in Heidelberg and Freiburg. In 1869, he became a lecturer at the St. Petersburg University and then at the Women's College in St. Petersburg. He became a member of the St. Petersburg Academy of Sciences in 1898. Most of Voronin's research was done at his home laboratory, which he supported from his own private funds. In 1866, Voronin was one of the first to observe the concentration of bacteria on the roots of legumes, but he did not realize the significance of his observations. He is chiefly remembered for his work on the pathology of the cabbage plant (57). In an effort to help the farmers of the St. Petersburg area, he undertook a study of the clubroot (hernia) of cabbage, a disease that brought great losses to them. In his study, he noted the young plants were frequently affected by a disease that caused them to bend at the root-stem junction and then to die. He isolated a causative fungus and called it *Chytridium brassicae.* The clubroot itself developed later, in more mature cabbage plants, and Voronin was able to demonstrate that the causative agent was a rather primitive fungus, which he called *Plasmadiaphora brassicae.* He classified this fungus as belonging to the myxomycete family. The life cycle of the parasite included the bursting of the affected root to release a large number of spores. These, in the moist ground, then turned into myxamebas, which entered the root hairs of healthy plants and there formed plasmodia, thus causing the root to swell and burst. Veronin suggested several remedial procedures to rid the truck gardens of the parasite. These included the removal and burning of roots after harvesting the cabbage, rotation of plants, and careful examination of seedlings before planting.

Pharmacology

Pharmacological research and training in Russia were concentrated at the Dorpat University. In fact, that university was the center of pharmacology in all of Europe, and was looked upon as the best source of pharmacology professors for European medical schools (58). This basic science of medicine was developed to its level of excellence by Rudolf Buchheim (1820–1879), who was born in Bautzen, Germany, and took his medical degree in Leipzig in 1845. He was summoned to teach pharmacology at Dorpat in 1846, and remained there for the next twenty years. During his tenure at Dorpat, Buchheim developed its pharmacology laboratory into a large institute with first-rate laboratories and numerous graduate students. In 1855, he was decorated by the Russian tsar with an order of St. Stanislaus for his accomplishments. In 1867, Buchheim migrated to Giessen in Germany, where he remained until his death. There, he spent much of his time in a vain attempt to stem the anti-pharmacology movement in Germany that was raging at that time. Pharmacology was believed to be an unnecessary course in the medical school curriculum, a subject that, the medical savants said, could be learned at bedside. With pharmacology thus relegated to a secondary place in European universities, Dorpat alone remained to carry the banner of pharmacology until that time when this anti-basic science trend was finally reversed.

Buchheim was succeeded at Dorpat by Oswald Schmiedeberg (1838–1921). He was a native Russian citizen, born in Laidsen of the Kurland province (today's Latvia), and received his medical degree from Dorpat in 1866. He remained there from 1867 to 1872, when he moved to the newly established Strassburg University. He remained there until the First World War. Schmiedeberg's scientific accomplishments are many and diverse. He discovered a form of digitalis *(Digitalinum verum),* did the classical work on the physiologic action of muscarine and nicotine, and introduced paraldehyde and urethane into medical practice. In the field of biochemistry, he discovered ferritin. With Naunyn, another Dorpater, Schmiedeberg established the still flourishing periodical *Archiv fuer experimentelle Pathalogie und Pharmacologie.* Schmiedeberg's successor at Dorpat, Rudolf Boehm, was the professor of pharmacology from 1872 to 1881, and later moved to occupy the chairs of pharmacology first at Marburg, then at Leipzig.

Rudolf Kobert was the pharmacology professor at Dorpat from 1886 to 1896. He established a pharmacological publication called *Arbeiten des Pharmacologischen Institutes zu Dorpat.* The purpose of the journal was to publish dissertations of the institute's students, and to demonstrate the high quality of work being performed there. The articles appearing in this journal, though cumbersome and ponderous, were incredibly detailed and indeed of high scientific quality. It may be seen from the very brief account given herein that Dorpat's pharmacology had a far-reaching influence and prestige among Europe's universities, and its graduates were often welcomed as chairs of European medical school pharmacology departments.

Cytology

Cytological work in Russia, as it probably was in other lands, was carried out by an assortment of botanists, histologists, and medical researchers. The originator of the cellular theory of living matter, Mathias Jakob Schleiden (1804–1881), was principally a botanist though he did, at one time, study medicine. He was a professor at

Dorpat University for a short period of time (1862–1863). Among the early Russian students of cell anatomy was I. Chistyakov (1843–1877), a Moscow University professor who was first to study mitosis in Russia, and first to describe "indirect" cell division in 1874. Ivan N. Gorozhankin (1848–1904), also of Moscow University, was first in 1883 to observe the union of gametes in a seed plant (the pine). And finally, Peremeshko's investigation on the regeneration of muscle fibers contributed to the understanding of the role of cell nuclei in the process of cell division (59).

The autonomy of cell nuclei was very ably demonstrated by Sergey M. Lukyanov (b. 1855). He looked at the cells of animals under various nutritional states including hunger, and noted that the cell nuclei retained their structures in spite of the drastic changes seen in the cell cytoplasm (60). He proposed that nuclei were semi-independent structures within the cells. Morpurgo complained that Lukyanov ignored his earlier work on the subject, to which Lukyanov replied that he did not consider Morpurgo's work to be reliable in view of the fact that Morpurgo had examined only "a few hundred" specimens, whereas he, Lukyanov, had looked at 18,000 (61). Lukyanov was a pathologist, and as such made contributions to that field both in the form of textbook writing and original research. He was one of the originators of the organ reserve theory, and investigated several types of damaged organs and their ability to carry out normal physiological functions. One of such projects was concerned with the ability of damaged liver to deliver bile (62). Guinea pigs were fitted with bile duct fistulas, and portions of their livers were excised. He found that the amount of bile elaborated by the damaged liver was considerably less than that of the controls, but the concentration of solids therein was not decreased, and was, in fact, somewhat elevated. He concluded that there was sufficient reserve to maintain the composition but not the volume of the bile elaborated by damaged guinea pig livers.

The term "Kurlov inclusion bodies" is commonly encountered in today's medical literature. These bodies were discovered by Mikhail G. Kurlov (b. 1859) in the white cells of spleenectomized guinea pigs (63). Kurlov recognized these as being different from other cytoplasmic structures because they stained with nuclear dyes. Today, the Kurlov inclusion bodies are recognized as storage sites for a macrophage-destroying substance that is probably a polysaccharide in nature (64). Kurlov himself was a graduate of the Military-Medical Academy in St. Petersburg and spent some time in Ehrlich's laboratory in Germany. He eventually became a professor of histology at Tomsk University.

The greatest Russian cytologist was probably Mikhail D. Lavdovsky (1846–1902). He was the son of a physician and a graduate of the Military-Medical Academy. He became a professor there in 1895, and remained in that position until his death. Lavdovsky's first interest was in the area of leukocyte morphology, where he was especially fascinated by "wandering cells" (65). He noted that both the amphibian and mammalian white cells would attach themselves to glass surfaces and would send out pseudopods. This phenomenon is today extremely important in the field of tissue culture. Lavdovsky was one of the first to note that the white cells of both the amphibian and mammalian bloods were not homogeneous populations. He believed that one type of these cells was nucleated, whereas the other was not.

Lavdovsky's greatest contribution was his elucidation of the function of the attraction sphere (Lavdovsky called it the astrosphere). Though he did not actually discover this structure, it is nevertheless often referred to as Lavdovsky's nucleid. Most authorities of his time felt that the centrioles originated from the nucleus, though nobody was successful in proving this. Lavdovsky studied this question in plant egg cells, since these have a very large nucleus. He first described in great detail the cell components, especially the appearance of the nucleolus with the centrosomes in it. He claimed that the centrosomes remained unchanged during mitosis, shifting only their position, sometimes disappearing completely from the nucleus and appearing in the cytoplasm, giving rise to the centrioles of the attraction sphere. He noted that the nucleolus gave rise to chromatin before mitosis began, and in doing so was disintegrated releasing the centrosomes. The latter then passed through the nuclear membrane into the cytoplasm. Lavdovsky was even able to see holes in the nuclear membrane during such process; in fact, the nuclear membrane appeared to have developed openings even before the appearance of chromatin in the nucleus. Such pore formation in the nuclear membrane was, in Lavdovsky's view, due to processes in the cytoplasm.

Lavdovsky was puzzled by the fact that the relatively small nucleolus could give rise to such a heavy structure as chromatin. He felt that the chromatin was actually sunthesized from nucleic acid present in the egg yolk in the form of granules, and which entered the nucleus through the above-mentioned pores. Once inside the nucleus, they appeared as what we today call the chromatin granules, which are seen in the nucleus during mitosis. These granules were then directly incorporated into the chromatin net without first being degraded chemically. Lavdovsky was one of the first to demonstrate unambiguously the achromatic spindle fibers that joined the chromosomes in the process of cell division. The spindle itself, according to Lavdovsky, was enclosed in an achromatic and transparent structure that he called *Markteil*. It appeared to originate in the achromatic portion of the nucleus (66).

Anatoly Bezkorovainy, J. D., Ph. D.

Biology and Politics in Russia

The preceding sketch of Russian biology shows that its practitioners were both competent and original in their work, and one might very well stop this narrative at that point. However, there is a very interesting aspect of Russian biology that developed in the late 1930s and ended with Stalin's (Soviet dictator's) death, namely the Lyssenkoist purges of classical geneticists. It would thus be of interest to briefly review this phenomenon and to inquire whether or not a precedent therefor could be found in the practice of the tsarist regime, or among the views of Russian biologists themselves. A more general question that can be posed is whether or not, beyond the middle ages, there was ever a situation in Europe when a secular government interfered with the collection and interpretation of scientific data to make it comply with a political doctrine. We can definitely state that there wasn't. Hitler's Germany came close, but never reached the extent to which the Soviet regime under Stalin was able to accomplish this goal, i. e., the political views of Karl Marx superseding scientific (genetic) data.

Unfortunately, Russia was probably the best candidate for Soviet views on genetics and evolution. There was a peculiar characteristic that was common to most if not all of the most prominent biologists of Russia of the pre-World War I era: their almost religious belief in Darwin's theory of evolution. The Russian scientific community was thus one of the first to bestow honors upon Darwin by electing him in 1870 to an honorary membership in the Moscow Society of Naturalists (67). To most Western scientists, on the other hand, the proposals of Darwin and his Russian disciples were strictly a scientific theory to be retained or rejected on the basis of evidence gathered. Some giants of Western biological thought were in fact in opposition to the assumptions and conclusions of Darwin (e. g., Rudolf Virchow and Du Bois-Raymond).

Joining and supporting the Russian Darwinists was the Russian school of environmentalists, exemplified by Sechenov, Pavlov, and Vedensky. To Sechenov, the existence of a central facilitation and inhibition systems meant that man was nothing but a mechanism responding to various stimuli, whereas to Pavlov, dogs salivating at the sight of meat indicated that in the future all emotional problems of man would be handled by altering or modifying his conditioned reflexes. These scientists successfully carried their message with all its real and imagined implications first to the Russian intelligentsia, then to the working classes through popular lecture series. Since scientists enjoyed overwhelming prestige and popularity among the Russian populace, they were quite successful in convincing many that man was not only a descendant of a lowly infusorium, but, in addition, was little more than a machine responding to external stimuli. There were apparently few, if any, biologists of national or international reputation that dared to question this fashionable concept of what constitutes a human being, and intellectual opposition thereto was provided only by the unpopular tsarist regime, the Orthodox Church, and some Russian humanists who considered man to be primarily a seat of the spirit or soul. A vitriolic public debate thus raged between the scientists and intelligentsia on the one hand, and the "establishment" on the other. Unfortunately, there was at that time nothing substantial in Russia that could take a middle position, to serve as a buffer, so to speak, between the two sides, and thus provide a meaningful alternative to both extremes. The result of this struggle turned out to be something that neither side wanted nor expected.

The social upheavals of 1917 resulted in the emergence of a Marxist government as the sole power in Russia. The views of the Russian scientific community in regard to the origin of man and the reflex nature of human behavior were in an apparent accord with the Marxist theoretical foundations that predicted that man would, within a short period of time, be adaptable to a radically altered way of life. With the support of the scientific community and the establishment of a new and monolithic social system, the stage was set for the creation of the *homo novus.* The environmentalist doctrines were first tested in the area of agriculture through the promotion of one Ivan M. Michurin (1855–1935). He was a high school dropout and a railroad clerk, and he cultivated a small garden as a hobby. On the basis of his experiences with this garden, he developed a theory of inheritance of acquired characteristics. His offers to solve all problems of agriculture in prerevolutionary Russia were apparently rebuked by the tsar's bureaucrats and the "official" (*kazennaya*) scientific establishment of that time, according to the Soviets. After the revolution, Michurin's ideas were championed by one T. D. Lyssenko, an agronomist, Communist Party functionary, and confidant of Joseph Stalin. His promotion of Michurinist ideas was accomplished not by scientific experimentation, but by persecuting those biologists and agronomists who subscribed to the principles of Mendelian genetics.

The most notable victim of the genetic "reform" movement in the Soviet Union was Nicholas I. Vavilov (1887–1942), the brother of a respected Soviet physicist. He was born in Moscow and graduated from the Moscow Agricultural Academy, then studied in England during the years of 1913 and 1914. He was then appointed to the faculty of Moscow University, and in 1917 became professor of botany in Saratov University. During the first twenty-five years of Soviet administration, Vavilov established and directed several agricultural research institutes and experimental stations, took part in many expeditions to collect specimens that could conceivably be grown in Russia, and did much for the Russian agriculture in general. He was a member of the Soviet Academy of Sciences and the Royal

Academy. Vavilov's scientific accomplishments were many, and were of course made during the Soviet era in Russia (68). He studied the different varieties of corn and potato in South America, especially the low-temperature varieties in Peru, which he called *S. Ajanhuiri* (2n=24 chromosomes) and *S. Jusepczukii* (2n=36 chromosomes). Vavilov formulated a theory whereby the geographic origin of a plant could be located by determining the area that contained the largest number of varieties of such a plant. He spent much time and effort describing similarities among the various species of plants, and proposed the so-called *Law of homologous series*, where he attempted to classify each species with respect to its chemical and physical properties. Each group of species with similar characteristics would then be classified into genera, similar genera into families, etc. The analogy that Vavilov drew for his proposal was the system of classifying organic compounds. He felt that only through such a systematic classification of plants could man be able to breed crops most suitable for any specific environment or need. Vavilov also systematized a large number of wheat hybrids with respect to their hardiness toward low temperatures, disease, draught, susceptibility to pests, and crop yields, and established what varieties of wheat were most suitable for cultivation in different parts of Russia.

Unfortunately, Vavilov and many other Russian geneticists were quite firmly committed to the gene theory of inheritance because of their previous publications, and were thus unable to shift with the winds of the party ideology. In spite of Vavilov's protests that he was a good party member and a loyal subject of Stalin, he was carted off to jail in the late thirties or early forties, and died in a concentration camp in 1942. Vavilov was of course not the only victim of the new genetics movement in the Soviet Union. Zirkle (69) states that I. J. Agol, L. P. Ferry, S. C. Levit, and N. A. Iljin, all repectable geneticists, were executed, and G. A. Levitzky, N. P. Ardoulov, G. D. Karpetchenko, and J. J. Kerkis died in concentration camps. Those who had a chance attempted to confess their errors, e. g., A. R. Zherbak's apology: "Now, however, since it has become clear to me that the fundamental aspects of Michurin's direction in Soviet genetics are approved by the Central Committee of the All-Union Communist Party (Bolsheviks), then I, as a member of the Party, do not consider it possible to retain those views that are recognized as erroneous by the Central Committee of our Party (69, p. 277). In spite of such recantations, most of the "erring" geneticists who were not arrested were relieved of their jobs. The Academy of Sciences abolished the cytogenetics laboratory, since the chromosomes no longer had anything to do with heredity. And the same academy, in accepting M. J. Muller's resignation from its ranks in 1948, pronounced that "There does not exist and cannot exist in the world a science divorced from politics." As Zircle states (69, p. 248), nothing like that has happened in the world in the preceding 300 years. As a result, Soviet biology fell back by some thirty years (70).

By 1963, however, things had somewhat normalized in the scientific establishment of the Soviet Union. The *Bol'shaya Meditsinskaya Entsiklopediya* (Large Medical Encyclopedia) now had an objective review on the function of chromosomes in heredity, and Gregor Mendel was accorded an honored place on its pages, though Michurinism was being retained as an important concept in the understanding of heredity. Lyssenko had become a nonperson. Vavilov, on the other hand, had become rehabilitated, and an extensive biographical sketch on him appeared in *Lyudi Russkoi Nauki* (Men of Russian Science). He was said to have been arrested on the basis of a "false accusation" in 1940, though no other details were given. The era of Lyssenkoism, or "Machiavellianism in scientific life" (68) had come to an end in Russia.

The question must now be answered as to whether or not the so-called Lyssenkoism of the Soviet era could trace its origins to the practices of the tsarist regime. In sifting through the biographies of Russian scientists, one does find instances of dismissals, forced retirements, or resignations under pressure. Yet most if not all such occurrences were the result of individual scientist's political activities or disputes with university administration regarding the rights of students and/or faculty members. The only direct vehicle for the control of scientifically-based thought was the tsar's censor, though even this method was not very effective as is evident from Sechenov's experience with his *Reflexes of the brain*, a book whose concept of what constitutes a human being was in opposition to that held by the official Orthodox Church and tsarist political theoreticians (e. g., Pobedonostsev). Of course, there were more subtle methods to "punish" erring professors, such as the case of a professor who was promoted over others because he was able to teach science without denigrating the religious establishment of Russia (71). At any rate, there does not appear to be any record of scientists being jailed, executed, sent to concentration camps, or even dismissed from their jobs because of their scientific views. The 1905–1906 constitution did away with both the subtle and more direct controls of scientific freedom. It would appear that the Stalinist persecution of classical geneticists was an unprecedented phenomenon whose primary origins are not to be found in the social order of Tsarist Russia.

REFERENCES

1. S. Borman. *Five Chemists who Should Have Won the Nobel.* Chem. & Eng. News 94 (15):19, 2016.
2. Sechenov. *Selected Works.* State Publ. Co., Moscow-Leningrad, 1935.
3. M. Peter Amacher. *Thomas Laycock, I. M. Sechenov, and the Reflex Arc Concept.* Bull. Hist. Med. 38:168, 1964.
4. I. Sechenov. *Autobiographical Notes.* American Inst. Biol. Sci., Washington, D. C., 1965.
5. N. E. Wedensky. *Are Excitation, Inhibition, and Narcosis Different Phases of the Same Activity?* Med. News 84:361, 1904.

6. N. E. Wedensky. *Die Erregung, Hemmung, und Narkose.* Archiv. f. d. gesammte Physiol. 100:1, 1903.
7. N. Wedensky. *Excitation prolongee du nerf sensitive et son influence sur le functionement au systeme nervoux central.* Compt. Rend. de Seances l'Acad. des Sciences 155:231, 1912
8. B. P. Babkin. *Pavlov. A. Biography.* Univ. of Chicago Press, Chicago, 1949.
9. I. P. Pavlov. *The Work of the Digestive Glands* (transl. by W. H. Thompson). Charles Griffin Co., London, 1902.
10. Ivan Petrovich Pavlov. *Lectures on Conditioned Reflexes.* Vol. 1 (transl. by Horsley Gantt). Internatl. Publ. Co., New York, 1928.
11. Ivan Petrovich Pavlov. *Lectures on Conditioned Reflexes.* Vol. 2 (transl. by Horsley Gantt). Internatl. Publ. Co., New York, 1941.
12. N. Ekk. *K voprosu perevyazke vorotnoy veny.* Voyenno-Med. Zh. 130 (II):1, 1877 (Natl. Med. Library, Bethesda, MD).
13. Solomon R. Kagan. *Elie de Cyon (1843–1912).* Med. Record 148:331, 1938.
14. E. von Cyon. *Die Nerven des Herzens. Ihre Anatomie und Physiologie.* Julius Springer, Berlin, 1907.
15. E. von Cyon. *Das Ohrenlabyrinth.* Julius Springer, Berlin, 1908.
16. J. Dogiel. *Die Ausmessung der stroemenden Blutvolumina.* Arbeiten aus der Physiol. Anstalt zu Leipzig 2:196, 1868.
17. J. R. Tarchanoff. *Die Bestimmung der Blutmenge an lebenden Menschen.* Pflueger's Archiv 23:548, 1880; 24:203, 1881; and 24:525, 1881.
18. Tupoumoff. *Vorlaeufige Mitteilung ueber eine neue Methode zur Bestimmung der Blutmenge am lebenden Menschen.* Pflueger's Archiv 26:409, 1881.
19. Editor. *Zur Aufklaerung ueber den (p. 409) in diesem Band von "Tupoumoff" publicirten Aufsatz.* Pflueger's Archiv 26:573, 1881.
20. J. Tarchanoff. *Ueber die galvanischen Erscheinungen in der Haut des Menschen bei Reizungen der Sinnesorgans und bei verschiedenen Formen der psychischen Thaetigheit.* Arch. f. d. gesammte Physiol. 46:46, 1890.
21. P. Owsjannikow. *Ueber den Stillstand des Athmungsprocesses waehrend der Exspirationsphase bei Reizung des centralen Endes von N. vagus.* Virchov's Archiv 19:221, 1860.
22. Nikolaus Kowalewsky. *Ueber die Epithelialzellen der Milzvenen.* Virchov's Archiv 19:221, 1860.
23. J. Dogiel and N. Kowalewsky. *Ueber den Blutstrombei unterbrochener Respiration. Pflueger's Archiv 3:489, 1870.*
24. Victor Robinson. *Pathfinders in Medicine.* Medical Life Press, New York, 1929, p. 749.
25. Paul de Kruif. *Microbe Hunters.* Blue Ribbon Press, New York, 1926.
26. Olga Metchnikoff. *Life of Elie Metchnikoff 1845–1916.* Haughton-Mifflin Co., Boston-New York, 1921.

27. Elie Metchnikoff. *The Prolongation of Life. Optimistic Studies.* G. P. Putnam's Sons, New York-London, 1908.
28. Gamaleia. *Ueber die Experimente zur Erforschung der Rinderpest.* Centralbl. f. Bakt. u. Parasitenk. 1:633, 1887.
29. N. Gamaleia. *Vibrio Metchnikovi et ses rappots avec le microbe du cholera asiatique.* Ann. Inst. Pasteur 2:482, 1888.
30. Raymond N. Doetsch. *Microbiology. Historical Contributions from 1776 to 1908.* Rutgers Univ. Press, New Brunswick, 1960.
31. S. Winogradsky. *Ueber Eisenbakterien.* Bot. Zeitung 46 (No. 17):261, 1888.
32. S. Winogradsky. *Clostridium pasteurianum, seine Morphologie und seine Eigenschaften as Buttersaeureferment.* Centralbl. f. Bakt. Parasitenk. u. Infektionskrankh.. Abt. II, 9:43, 101, 1902.
33. W. Omelianski. *Ueber die Trennung der Wasserstoff- und Methaengaerung der Cellulose.* Centralbl. f. Bakt., Parasitenk., u. Infektionskrankh. Abt. II, 11:369, 1904.
34. V. L. Omeliansky. *Fermentation methanique de l'alcohol ethylique.* Ann. Inst. Pasteur 30:56, 1916.
35. W. Omelianski. *Ueber die Zersetzung der Ameisensaeure durch Mikroben.* Centralbl. f. Bakt., Parasitenk., u. Infektionskrankh. Abt. II, 11:177, 256, and 317, 1904.
36. V. L. Omelyanski. *Noviy batsill: Bacillus flagellatus Omel.* Mikrobiologiya 1:24, 1914 (in Russian).
37. Dm. Iwanowsky. *Ueber die Mosaikkrankheit der Tabakpflanze.* Bull. St. Petersb. Acad. Sci. 35:67, 1892.
38. H. Lechevlier. *Dmitri Iosifovich Ivanovsky (1864–1920).* Bact. Revs. 36: 135, 1972.
39. H. S. Reed. *A Short History of the Plant Sciences.* Chronica Botanica Co., Waltham, 1942.
40. A. Famintzin. *Eine neue Bacterienform: Nevskia ramosa.* Bull. St. Petersb. Acad. Sci. 34:481, 1892.
41. Bergey's Manual of Determinative Bacteriology. R. S. Reed, E. G. D. Murray, and N. R. Smith, eds., 7-th ed., Williams and Wilkins Publ., Baltimore, 1957.
42. W. M. Wheeler. *Caspar Friedrich Wolff and the Theoria Generationis.* Biol. Lect. Woods Hole, Ginn and Co., Boston, 1899, p. 265.
43. W. Mayer. *Human Generation. Conclusions of Burdach, Dollinger, and von Baer.* Stanford University Press, Stanford, 1956.
44. Karl Ernst von Baer. *Das allgemeine Gesetz der Nature in aller Entwicklung.* In: Reder gehalten in wissenschaftlichen Versammlungen und kleiner Aufsaetze vermischten Inhalt. Schmitzdorff Publ. Co., St. Petersburg, 1864 (courtesy of Dr. G. F. Springer).

45. J. M. Oppenheimer. *Essays in the History of Embryology and Biology.* MIT Press, Cambridge, 1967.
46. 46, M. Kume and K. Dan. *Invertebrate Embryology.* NOLIT, Belgrade, 1968.
47. E. Rabinovitch. *Photosynthesis and Related Processes.* Vols. I and II. Interscience, New York, 1945.
48. T. Robinson. *Michael Tswett.* Chymia 6:146, 1960.
49. Supplement to ACS publications. *Luminaries of the Chemical Sciences.* Amer. Chem. Soc., Washington, D. C., 2002, p. 37.
50. R. L. M. Synge. *Tsvet, Willstaetter, and the Use of Adsorption for Purification of Proteins.* Arch. Biochem. Biophys. Suppl. I, 1962, p. 1.
51. M. Tswett. *Adsorptionsanalyse und chromatographische Methode. Anwendung auf die Chemie des Chlorophylls.* Ber. Deutsch. Bot. Gesellschaft 24:384, 1906.
52. M. Tswett. *Das neue System der sogenannten Chlorophyllderivate.* Biochem. Z. 10:426, 1908.
53. M. Tswett. *Beitraege zur Kenntnis der Anthocyane. Ueber kuenstliches Anthocyan.* Biochem Z. 58:225, 1914.
54. N. Gaidukov. *Weitere Untersuchungen ueber den Einfluss farbigen Lichtes auf die Farbung der Oscillarien.* Ber. Deutsch. Bot. Gesellschaft 21:484, 1903.
55. N. Gaidukov. *Die komplimentare chromatische Adaptation bei Porphyra und Phormidium.* Ber. Deutsch. Bot. Gesellschaft 24:1, 1906.
56. N. Gaidukov. *Zur Farbenanalyse der Algen.* Ber. Deutsch. Bot. Gesellschaft 22:23, 1904.
57. M. Woronin. *Plasmodiaphora brassicae. The Cause of Cabbage Hernia.* Phytopathol. Classics, No. 4, 1034.
58. Gustav Kuschinsky. *The Influence of Dorpat on the Emergence of Pharmacology as a Distinct Discipline.* J. Hist. Med. 23:258, 1968.
59. P. Peremeschko. *Die Entwicklung der quergestreiften Muskelfasern aus Muskelkernen.* Virchow's Archiv 27:116, 1863.
60. Arch. des sciences biol. publ. par l'Inst. imp. de med. exper. a St. Petersburg 6:1897 and 1898. Quoted by B. Morpurgo. *Ueber die kario-metrischen Untersuchungen bei Inanitionszustaenden.* Virchow's Archive 152:550, 1898.
61. S. M. Lukjanow. *Zur Frage nach der biologischen Autonomie des Zellkernes.* Virchow's Archiv 153:158, 1898.
62. S. M. Lukjanow. *Ueber den Einfluss partieller Leberexcision auf die Gallenabsonderung.* Virchow's Archiv 120:485, 1890.
63. M. G. Kurlov. *Alterations of the Blood in Animals in the Course of First Year After Removal of Spleen.* Vratch 10:515 and 538, 1889. Quoted by Index Medicus 11:407, 1889; and by J. C. G. Ledingham. *Sex Hormones and the Foa-Kurloff Cell.* J. Pathol. Bact. 50:201, 1940.

64. M. F. Dean and Helen Muir. *The Characterization of a Protein-Polysaccharide Isolated from Kurloff Cells in the Guinea Pig.* Biochem J. 118:763, 1970.

65. M. Lavdovsky. *Mikroskopische Untersuchungen einiger Lebensvorgaenge des Blutes .* Virchow's Archiv 96: 60, 1884.

66. M. Lavdovsky. *Von der Entstehung derchromatischen und achromatischen Substanzen in den tierischen und pflanzlichen Zellen.* Anat. Hefte 4:355, 1894.

67. R. M. Stecher und J. V. Klavins. *Charles Darwin and the Moscow Society of Naturalists.* J. Hist. Med. 20:157, 1965.

68. N. I. Vavilov. *The Origin, Variation, Immunity, and Breeding of Cultivated Plants. Selected Writings of N. I. Vavilov.* Chronica Botanica 13: Nos. 1–6 1949–1950.

69. C. Zircle. *Death of a Science in Russia.* Univ. Penna. Press, Philadelphia, 1949.

70. D. A. Pragay. *Lyssenko's Biology: Machiavellianism in Scientific Life.* Abst. No. 10, History of Chemistry Section, Abstracts of the 162nd meeting of the American Chemical Society, Washington, D. C. 1971.

71. H. M. Leicester. *The History of Chemistry in Russia prior to 1900.* J. Chem. Ed. 24, 438, 1947.

Illustrations

Ivan P. Pavlov (1849–1936), physiologist and Nobel laureate.

Elie I. Mechnikov (1845–1916), immunologist and zoologist, and Nobel laureate.

Nikolai F. Gamaleya (1859–1949), microbiologist.

Kliment A. Timiryazov (1843–1920), botanist.

Dimitry I. Ivanovsky (1864–1920), "father" of virology."

Mikhail S. Tswett (1872–1919), botanist and developer of chromatography.

Dimitry N. Pryanishnikov (1865–1948), botanist and agronomist.

Nicholas I. Vavilov (1887–1942), agronomist, victim of Lyssenkoism. Vavilov was quickly rehabilitated, though he was already dead for at least ten years after Stalin's death.

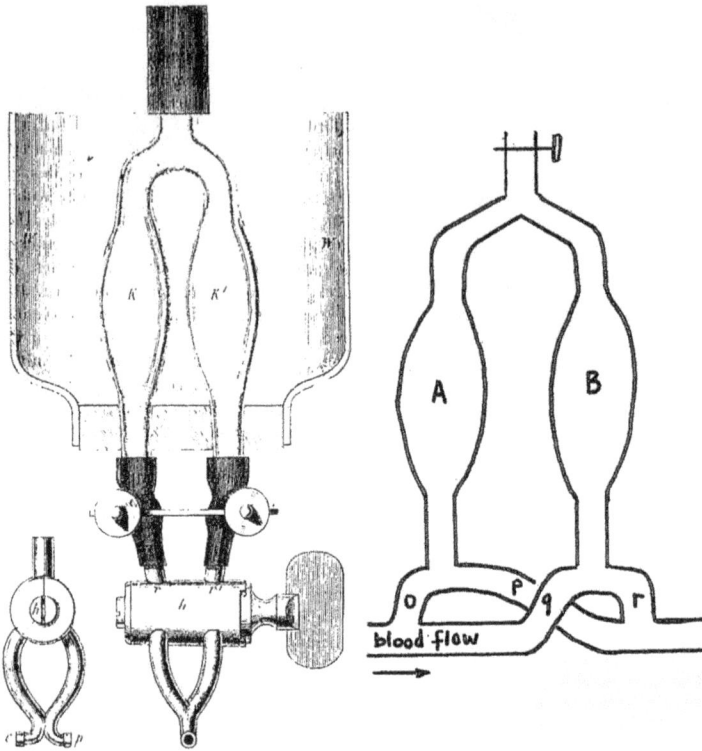

Dogel's Stromuhr. With respect to the schematic drawing, chamber A contained olive oil, and chamber B contained defibrinated blood. The stopcock was so turned that tubes o and r were open, and tubes p and q were closed. The blood flow displaced the olive oil from chamber A to B, and this maneuver was timed. Then the stopcock was turned so that tubesp and q were opened and o and r closed. The olive oil was replaced in chamber A.

Chapter IV

BIOLOGICAL CHEMISTRY

Introduction

B iological chemistry is a relatively new biological/chemical discipline. Important advances in our knowledge of what causes disease processes, for instance, could not have been made without understanding of the chemical processes occurring in animal organs; considerable knowledge of anatomy and even physiology was not enough to develop an understanding of how the different animal organs function and what goes wrong in their disease processes without understanding the complex chemical reactions occurring therein. As an example, one can mention the Russian scientists' obsession with the nervous system (Sechenov, Pavlov, de Cyon) as if it maintains an almost total control of human lives and behavior. In the early twentieth century, it would have been useless to look for some disorder of the nervous system that might be responsible for, say, diabetes. Nothing was known at the time, for instance, that endocrine glands existed for the purpose of emitting chemical substances that we now call hormones, which are crucial for normal human existence and whose disorders were the cause of serious diseases (e. g., insulin and diabetes). And in order to identify such hormones, advances in analytical and organic chemistry methodologies had to be available or had to be invented. At the end of the nineteenth and beginning of the

Anatoly Bezkorovainy, J. D., Ph. D.

twentieth centuries, many important facts about chemical compounds found in human organism were known, and some may have been utilized by physicians to diagnose diseases (again, e. g., blood sugar and diabetes). For instance, Emil Fischer studied amino acids and proteins and Michaelis/Menten developed the foundation for quantitative enzymology. And Russia too had its early physiological/biological chemists who participated in laying the foundations for today's biological chemistry. For instance, N. Lunin of Dorpat University in 1881 discovered the nutritional requirement by animals of organic compounds that were neither proteins, fats, nor carbohydrates (1). Casimir Funk later called them the *vitamins*. But it took another generation or two to put all that basic work together to understand how such a multitude of complex compounds works together, and, including the nervous system, organizes a living and thinking human being. There are several authors who have attempted to organize the beginnings of biological chemistry into a systematic narrative (2, 3, 4, 5); it was a difficult task and this author (A. B.) is happy to solute them for their effort. This author has used a dual approach to tell the story of Russian beginnings of biochemistry: firstly, the more far-reaching discoveries of Russian biochemists are described (e. g., Danilevsky's separation of "pancreatin" into three types of enzymes); and secondly, an attempt was made to review selected biochemical papers written by Russian scientists, papers representative of the type of research work that took place in Russian laboratories (or abroad) regardless of whether or not they represent significant breakthroughs in biochemistry.

The Early Period

Much of Russian biochemistry was imported from abroad, especially from Germany, where future Russian professors of biochemistry or medical chemistry learned their trade from men like Hoppe-Seyler, Abderhalden, Kuehne, Hofmeister, and Kossel. On the other hand, there were two centers in Russia where physiological chemistry developed to a large extent independently of Western Europe. The first such center was the Dorpat University, where physiological chemistry originated partly from its Pharmacological Institute, and in the later years was largely concerned with nutritional studies. The other early center of biochemical activity was St. Petersburg, where it developed as an outgrowth of the physiological investigations of Pavlov and his colleagues. Biochemical work originating from Pavlov's laboratories was largely concerned with proteolysis, enzymology, and protein metabolism.

The dean of Russian physiological chemistry was Carl Schmidt (1822–1894), a Dorpat professor of chemistry who, with Bidder the pharmacologist, is considered to have established the Dorpat school of biochemistry. Carl Schmidt was born in Mitau (now Jelgava), the capital of the Kurland province of Russia, was educated at Dorpat, and was on its faculty for forty years beginning with 1846. He was interested in a number of biochemical problems, including the chemistry of blood, the mechanism of alcohol fermentation, digestion, and the chemistry of carbohydrates and proteins. Schmidt discovered in 1847 that succinic acid was a product of yeast fermentation of fructose, sucrose, and grape juice (6). He also determined that yeasts could break up urea into NH_3 and CO_2, though only in the absence of sugar. When sugar was present, it was metabolized first, then urea was degraded. The yeasts gained weight when sugar was metabolized, whereas no weight gain took place when urea was metabolized in the absence of sugar. The weight gain was accounted for by an increase in the polysaccharide content of the cells rather than their protein (7).

Schmidt determined the densities of a number of biological materials such as egg albumen, red blood cells, muscle fibrils, and collagen. The values determined were 1.2617, 1.2507, 1.2678, and 1.2960, respectively. He proposed to use these numbers for the determination of serum protein and blood cell content using the following equations:

$$a = \frac{a' (p - p')}{1 - a'} \text{ for serum protein and } b = \frac{b' (p' - p'')}{1 - b'} \text{ for red blood cell content}$$

(equivalent to the modern hematocrit or grams of hemoglobin per 100 ml blood), where a was the absolute weight of the serum protein; a' and b' were the densities of albumin and red cells, respectively; p, p', and p" were the weights of the pycnometer filled with water, serum, and whole defibrinated blood, respectively; and b was the absolute weight of the red cells (8). Schmidt's other contribution to clinical biochemistry was the development of a method for the determination of uric acid in urine (9). The method involved the treatment of uric acid with iodine, whereby the uric acid was converted to alloxan and urea, then determining the unreacted iodine.

The phenomenon of peptic digestion of proteins was well-known among the early nineteenth century biologists, though the role of HCl was not completely appreciated. Schmidt's concept of peptic digestion of proteins involved the chemical combination of HCl and pepsin, where HCl was constantly to be replenished as the digestion proceeded, and could not be replaced by another acid (10). Though this theory was accepted for nearly fifty years following its formulation, we today know that Schmidt was only partially correct: instead of combining with pepsin chemically to activate it, HCl serves to activate the inactive zymogen, i. e., pepsinogen, into the active enzyme pepsin. HCl is also necessary to maintain an acidic pH in the stomach.

The nature of amyloid seen upon the autopsy of many human organs was elucidated by Carl Schmidt. It was accepted up to his time that amyloid was a polysaccharide, hence the term "amyloid." Schmidt was able to isolate very little carbohydrate from amyloid, but did find some 15.6% nitrogen. From this, he correctly concluded that amyloid was a protein in nature, and that the term "amyloid" was a misnomer (11). It is today known that amyloid fibrils consist of a number of proteins including a glycoprotein and the light chains of immunoglobulins.

Schmidt is credited with the introduction of the terms "carbohydrate," or "Kohlenhydrat" in German. He used this term in conjunction with his extensive study of a number of neutral polysaccharides and simple sugars from plant sources. He was able to arrive at a general formula for all neutral sugars, namely $C_{12}H_{10}O_{10}$ (12). Schmidt used atomic weights of 6 and 8 for carbon and oxygen, respectively, so that in modern terms, the formula would read $C_6H_{10}O_5$, not too different from the empirical formula of hexoses, $C_6H_{12}O_6$, or $C_6H_{10}O_5$ for glycosidically-bonded hexoses.

An important product of the Dorpat school of biochemistry was Alexander Schmidt (1831–1894), the founder of the modern theory of blood coagulation (13). He was born on the island of Oesel in the Estonian province of Russia, where his ancestors had migrated in search of religious freedom. He began his studies at Dorpat with the intention of following his family's wishes that he enter the ministry. However, he soon changed his interests to medicine, and after additional studies abroad was appointed to Dorpat's faculty. From 1885 to 1889 Schmidt served as the president of the Dorpat University. He conceived his theory of blood coagulation in 1861 while working with Hoppe-Seyler in Virchow's laboratory in Berlin. His entire research career was devoted to amplifying and improving his theory, which is viewed as an enzymatic denaturation of fibrinogen with the active enzyme arising from an inactive precursor. Schmidt proposed that essentially three substances were required for blood coagulation:

1. fibrinogen, which Schmidt called "Fibrinoplastische Substanz," 2. the "Fibrinogene Substanz," which was probably thromboplastin, and 3. a clotting enzyme, "das Fibrinferment," which was thrombin. Schmidt attempted to purify all three substances with varying degrees of success. The fibrinogen was obtained from plasma by acidification with either dilute acetic acid, or by the passage of CO_2 through plasma. It was insoluble in water, but was soluble in salt solutions and strong acids and alkalis. There were 0.7–0.8 grams of fibrinogen per 100 ml of bovine plasma, and 0.3 to 0.6 grams in horse plasma. He argued strongly against Bruecke's suggestion that fibrinogen was really serum albumin, and proved the difference via solubility and precipitability studies (14).

The tromboplastin-like substance was prepared from the pericardial fluid of horses by acidification or by saturating it with NaCl. Schmidt could not characterize it extensively because of its small yield. The clotting enzyme was, according to Schmidt, present in blood in an inactive form, and was to a large extent coprecipitated with the fibrinogen. The inactive state of the enzyme was demonstrated as follows: in one case, the blood was permitted to flow directly into alcohol from a cannulated blood vessel to be coagulated. The coagulum was then extracted with water and was tested for its ability to clot fibrinogen. No clotting took place. On the other hand, if blood was first vigorously defibrinated then coagulated with alcohol, the water extract of the coagulum was able to clot fibrinogen. Schmidt felt that the oxygen in air served to activate the enzyme. However, what probably had happened was that all the prothrombin in the first case had been coprecipitated with the fibrinogen and was not released from the fibrinogen denatured by the alcohol upon the extraction with water. In the second case, however, the two proteins may have become dissociated by the defibrinating procedure and was thus extractable from the alcohol coagulum. Thus, the inactive state of the coagulating enzyme was proposed for the wrong reason, though the basic idea of the absence of an active coagulating enzyme from the bloodstream was perfectly sound.

That blood coagulation was indeed an enzymatic process was shown by Schmidt in the following way: 1. Very small amounts of the purified enzyme were sufficient to initiate the clotting of fibrinogen. 2. Boiling the enzyme preparation destroyed its activity. 3. There was very little activity at $0^{o}C$. Its maximum activity was observed at $45^{o}C$. Furthermore, the material was apparently not destroyed during clotting, since the supernatant of the fibrin clot could initiate clotting in a fresh system. The purified enzyme was prepared by diluting plasma with 15–20 parts of "strong" alcohol, and permitting the precipitate to stand no less than 14 days. The precipitate was then extracted with water to obtain the active enzyme.

Nutrition and Its Disorders

It has already been stated that much of the work produced by Dorpat's physiological chemistry laboratories was concerned with nutrition. The force behind this type of approach was Georg Bunge, the son of a well-known Dorpat professor of botany, Alexander Bunge (1803–1897). Unfortunately, Georg Bunge left Dorpat in 1885 to accept the chair of physiological chemistry in Bern, Switzerland, so that his further influence upon the Russian scene was lost. Bunge's own interests included the absorption and utilization of iron by mammalian organisms (15). His theory of iron absorption provided for the lack of utilization of inorganic iron and the absorption of only organically-chelated iron by animals. This was based on the fact that ferric salts were found not to be absorbed by experimental animals, whereas the iron present in egg yolk was absorbed. Moreover, Bunge found that the iron of egg yolk was bound to an organic compound. He attempted to characterize the active form of iron by subjecting 200 egg yolks to peptic digestion, and obtaining a precipitate that contained all the iron originally present in the yolk. The precipitate was redissolved in dilute NH_4OH and reprecipitated with alcohol. Some 34 grams of a yellowish product were obtained, which contained 42.1% C, 6.1% H, 14.7% N, 0.6% S, 5,2% P, 31.1% O, and 0.3% Fe. It was probably phosvitin, or an impure fraction thereof. Bunge felt that the reason why anemic patients showed improvement after feeding on inorganic iron was the fact that the latter protected the organically-bound iron from digestion and release of free iron. It was eventually discovered that ferric iron is indeed not absorbed by the human organism, and that the form best suitable for absorption is ferrous iron. Apparently Bunge never bothered to make the distinction.

Another Dorpat researcher who studied iron metabolism in the mammalian system was G. Swirski. He followed the fate of ingested iron microscopically (16). He noted that iron entered the mucosal cells in the form of microscopic granules, though no iron was found in the wall of the stomach, the jejunum, and the ileum. Whereas the iron soon disappeared from the intestinal cells of iron-deficient animals, nothing happened to the iron granules in iron-loaded animals. In the latter case, the cells were eventually shed and the iron excreted in the stool. Swirski felt that the iron was taken away from the intestinal cells of iron-deficient animals by a circulatory protein (today known as transferrin), and when in circulation, was taken up by the macrophages to be transported to the liver and other tissues. Except for a few details, such as the involvement of macrophages, Swirski's view of iron metabolism was remarkably similar to that currently accepted by scientists (17).

An interesting iron intoxication disease, in some ways similar to hemochromatosis, was described in 1861 by N. Kashin (b. 1825), a medical officer in Siberia. According to Schipatschoff (18), Kashin observed a disease endemic to a 20,000 km^2 area in an isolated Siberian locale called the Urow Valley. The disease began already *in utero*, and progressed through rickets, scurvy, polyneuritis, and osteoarthritis. The bones of the affected individuals were so deformed and their muscles were so atrophied that they were complete invalids by the age of forty. Kashin felt that their drinking water was responsible for the disease, since moving inhabitants to a different area served to arrest and even cure the disease. He proposed an *en masse* resettlement of the population and this was partially carried out. In 1906, Beck, another Russian investigator, confirmed Kashin's original observations, but considered the disease to be primarily of the osteoarthritis type. In 1910, Velyaminow proposed that the disease was basically of thryrotoxic origin, since the thyroid gland was enlarged in most patients. He also termed the disease as *Kashin-Beck syndrome*. In 1931, Schipatschoff proposed that the Kashin-Beck syndrome was of the vitamin deficiency variety, since bread baked from crops of the affected area, when fed to laboratory animals, brought about symptoms of the disease. It was finally a Japanese investigator, K. Hiyeda, who in 1939 elucidated the etiology of the Kashin-Beck syndrome (19). The disease was found to be widespread in certain areas of Manchuria (then under Japanese occupation). Hiyeda found that the well water of the affected areas contained more than 0.3 mg iron per liter, with frequent readings of 10 mg/liter. All foods grown and raised in the area also contained high amounts of iron. Histologic examination of tissues obtained from those suffering from the disease showed large accumulations of iron granules. Hiyeda thus confirmed Kashin's implication of drinking water in the etiology of the disease, and suggested, as did Kashin, that the population be removed from the affected regions.

Perhaps the most important nutritional investigation carried out in Imperial Russia was that of N. Lunin (1854–1937), who, at the suggestion of Georg Bunge, made an attempt to evaluate the significance of common salts in the nutrition of animals. Lunin was a graduate of Dorpat University, and the work described below served as his doctoral thesis. Lunin subsequently practiced pediatrics in St. Petersburg until his death. Georg Bunge held the opinion that the intake of metal ions by the mammalian organisms was necessary in order to excrete sulfur present in proteins. To test this theory, Lunin divided his mice into four groups: the first was fed dried whole milk; the second received only distilled water; the third received ash-free casein, milk fat, and sugar; and the fourth was fed number 3 diet plus sodium carbonate in the amount equivalent to the amount of sulfur present in casein. The first group of mice lived for 2.5 months, then released from captivity. The second group survived 2–4 days, the third group lived for an averageof 15 days, and the fourth group survived for an average of 26 days. Replacing Na_2CO_3 by NaCl, or giving salt quantities greater than those of sulfur in casein resulted in shortened lifespans of animals. Lunin then fed the fourth group of mice a mixture of salts equivalent to that present in whole milk instead of just the sodium carbonate, namely, Ca, K, Na, Cl and F. The lifespan of these animals averaged 26 days. Lunin concluded that inorganic salts were indeed necessary for the proper nutrition of animals; however, the animals also required thereto unknown organic factors other than fat, carbohydrate, and protein for survival. He felt that perhaps such factors were like creatine, insofar as they were not metabolized by the organism (20). The organic chemical factors postulated by Lunin were, of course, vitamins, but it wasn't until 1906 that Hopkins made a similar suggestion, followed by Casimir Funk's definition of a vitamin in 1912.

Enzymology

The St. Petersburg school of physiological chemistry had its origins in the physiological investigations of Pavlov and his students. Understandably, such investigations were mostly concerned with the action of digestive enzymes on proteins. Much of the work originating from Pavlov's laboratory was published in the form of doctoral theses or in Russian periodicals, so that secondary sources must be often consulted to evaluate such contributions. A monograph by Babkin has been most helpful in obtaining such information (21).

Among Pavlov's collaborators in the field of digestive enzymes were N. P. Schepovalnikov, the discoverer of enterokinase; Noll, who studied secretory granules of the salivary glands and the stomach wall; and Savich, Babkin, and Rubashkin, who investigated the secretory granules of the pancreas. These investigators showed that the secretory granules were decreased in size following secretory activity of the pancreas, and eventually fused to form vacuoles. Babkin and Savich also showed that the secretion of protease, lipase, and amylase by the pancreas was parallel, and this finding, along with Schepovalnikov's discovery disposed of Pavlov's earlier theory to the effect that the pancreatic juice composition adapted itself to the nature of food ingested. Babkin (b. 1877) was probably Pavlov's most distinguished student. He worked with Pavlov from 1902 to 1912, then moved to the Novo-Alexandria Agricultural Academy and later to Odessa. After the Bolshevik coup, he emigrated to Canada, where he spent many fruitful years with the Dalhousie and McGill Universities.

Another collaborator of Pavlov, a Dr. Mett, devised a method for the measurement of proteolytic activity of enzymes. This involved the filling of glass columns with solutions of egg white, heat-coagulating the egg white, then inserting the column into the enzyme solution to be tested. The activity of the enzyme solution was then expressed in terms of millimeters of the column liquefied per unit time (22). Mett's method was a recommended procedure as late as 1948 (23). The development of Mett's method prompted Borissov, another co-worker of Pavlov, to check exhaustively the previous findings of Schuetz, who in 1885 proposed that pepsin acted on proteins in accordance with

$$v = k\sqrt{P},$$

where v was the velocity of the reaction, P was pepsin concentration, and k was a constant. Borissov established in 1901 an identical relationship for trypsin and amylase (using starch instead of egg white columns), and found, furthermore, that at very high enzyme concentrations, this relationship did not hold. For many years, the above equation was known as the Schuetz-Borissov rule (24).

Early attempts to quantitate enzymatic reactions were reviewed by Hans Euler in 1905 (25), and this review included the work of several other Russian biochemists, most notably those of Medvedev and Isayev. In looking over the contributions of the early enzymologists, Russian or otherwise, one is struck by the fact that experiments were done in practically unbuffered solutions, with no attempt to run the reaction at the pH that was optimal for the enzyme in question, and without any consideration for any cofactor requirements. In these pre-Michaelis-Menten times, there was no concept as to the role of substrate concentrations in the rates of enzymatic reactions. Nevertheless, new enzymes were being discovered, and modest successes in elucidating enzyme kinetics were being achieved.

As an example of the trials and tribulations of an early Russian enzymologist, one may mention the work of Medvedev from Odessa University, who between the years of 1896 and 1904 published a series of papers of increased degrees of sophistication (26, 27, 28). He discovered an enzyme in calf liver that oxidized salicylaldehyde into salicylic acid. As the reaction proceeded, the pH of the system decreased, and the reaction was soon enough arrested. In such a case, Medvedev found, the amount of product formed was inversely proportional to the square root of the aldehyde concentration and the volume of the vessel. He later discovered that the velocity was also proportional to enzyme concentration. He proposed that the initial step in the reaction was a "dissociation," or activation of the substrate by the enzyme, for which purpose an enzyme-substrate complex was formed:

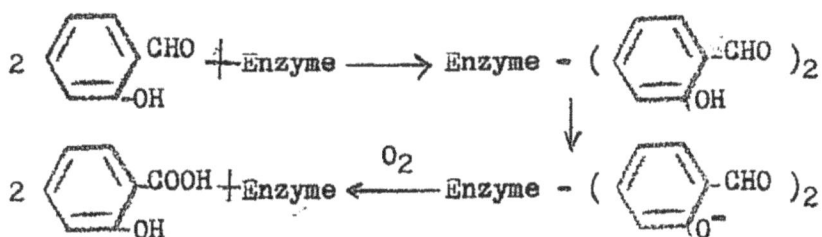

The function of the enzyme, according to Medvedev, was to bring about the dissociation of the phenolic group, which then favored a spontaneous oxidation of the carbonyl group by atmospheric oxygen.

Medvedev's most interesting paper was written partly in response to M. Jacoby's claim that Medvedev's oxidase was not a protein, since the enzyme failed to give positive Millon and biuret tests (29). Medvedev did a very simple thing (28): he preincubated his enzyme preparation with trypsin and found that the activity was lost. He concluded that his oxidase was indeed a protein. He did another unusual thing: he buffered his reaction mixture with sodium carbonate. As a result, the reaction velocity was no longer dependent on the volume of the reaction vessel, nor on substrate concentration when the latter was very high. There was only a direct dependence on the enzyme concentration. With small amounts of the substrate, the reaction obeyed the second order kinetics,

$$dx/dt = k (a - x)^2 ,$$

where a is the initial substrate concentration. In spite of the buffering by sodium carbonate, the enzymatic reaction eventually stopped, even though sufficient amounts of substrate were still present. Medvedev thought that the enzyme was inactivated by absorbing the protons from the phenolic group of salicylaldehyde during its activation. A more likely possibility was, however, a depletion of a cofactor such as NAD, which would be expected to participate in this type of a reaction.

Another enzymologist attempting to study the kinetics of enzyme action was Isayev of the Warsaw Polytechnic Institute. He was able to prepare a catalase from yeasts by first extracting the cells with water, then precipitating the enzyme with 50% ethanol (30). The enzyme had an optimal temperature of $40^{\circ}C$, and its action could be expressed by a first order kinetic equation,

$$k = (1/t)\ln(c_o/c_1),$$

where k was a kinetic constant, t was time of the reaction, and c_o and c_1 were concentrations of hydrogen peroxide at zero-time and t-time, respectively. Increased enzyme concentrations increased the rate of the reaction, but a simple relationship could not be established. The kinetic constant increased with increased enzyme concentration, whereas dialysis of the enzyme solution against water decreased the k (and hence the velocity of the reaction), indicating that there was a requirement for a dialyzable cofactor.

Isayev was also one of the first investigators to show the presence of respiratory enzyme systems in the yeast cells (31). The enzyme system was extracted from yeasts with glycerol, then precipitated from the extracts with alcohol. When incubated with pyrogallol in a manometric apparatus, oxygen was found to be absorbed. Another oxidizing enzyme was extracted by Isayev from malt (32). It oxidized a variety of compounds such as p-aminophenol, resorcinol, and pyrogallol with the uptake of oxygen and some production of carbon dioxide. The enzyme did not oxidize phenol, cresol, orcinol, formic acid, carbohydrates, or amino acids.

One of the most prolific Russian contributors to the German biochemical literature was E. S. London, a colleague of Pavlov at the Institute of Experimental Medicine in St. Petersburg and one of the first to use radiation for the treatment of malignancies (see chapter II). London's main interest lay in elucidating the digestive processes, and one of the problems he attempted to tackle was the relationships between the amounts of foodstuffs administered to the amounts digested. For instance, for the digestion of starch by the canine intestine, London and Kotchov (33) derived the following relationship:

$$x = k\sqrt[3]{M},$$

where x is the percentage of starch digested, M is the amount of starch administered, and k is a constant calculated to be 25 g. Though the number of London's publications in this field seems to be phenomenal, very little of fundamental importance seems to have been delivered from these studies.

Long before Pavlov's famous investigations on the function of mammalian digestive organs, which won him the Nobel Prize, there was another Russian investigator, A. Y. Danilevsky (1838–1923), who discovered the multienzyme nature of pancreatic secretions and did fundamental work on the action of trypsin. Danilevsky is even credited by some with the *discovery* of trypsin. Before Danilevsky's investigations, it was known that pancreatic juice caused the dissolution of fats, that it caused the inversion and degradation of sugar, and that it was probably responsible for the solubilization of coagulated protein, i. e., protein digestion. Yet there was a controversy in regard to the latter, since some investigators held that such solubilization was not due to the properties of pancreatic juice, but was rather due to "putrefaction." Moreover, among those who agreed that pancreatic juice had the power to digest proteins, there was a lack of consensus as to the optimal pH of the fermentation process. In addition, everyone agreed that all the enzymatic functions of the pancreatic juice were brought about by a single component called *pancreatin*, which was supposed to be a protein coagulable by heat and alcohol, and in most respects similar to serum albumin and milk casein.

For his investigations, Danilevsky used both the natural pancreatic juice obtained from a cannulated pancreatic duct of a large dog and the aqueous extract of excised dog pancreas (34). The pancreatic juice was mixed with a solution of collodion (soluble nitrocellulose) and separated into a precipitate and supernatant. The collodion was extracted from the precipitate with ether and alcohol, and the residue dissolved in water. It was capable of solubilizing fibrin clots, but not of inverting sugar or breaking down fats into glycerol and fatty acids. The reaction solubilizing fibrin worked in neutral and alkaline media, but not in acidic solutions. The activity was inhibited at 25 and 50°C, and was abolished at 60°. Optimal activity was seen at 37–40°C. The active material was always associated with protein material, but Danilevsky did not definitely state that the fibrin-solubilizing factor was indeed a protein. The active material did not give a xanthoproteic reaction, and it was precipitable with acid, though not with alkali.

The collodion supernatant was freed of ether and alcohol, and the precipitated collodion was removed by filtration. A large amount of inactive protein was precipitated from the supernatant by heating at 43–44°C, then adding a large quantity of absolute alcohol. The sugar-inverting factor was in the alcohol precipitate, whence it was extracted with a 2:1 water-alcohol mixture. It inverted sugar in an alkaline medium, and worked more slowly in an acidic environment. There was also a very small amount of fibrin-solubilizing activity in the invertase preparation, but no fat hydrolyzing activity. Danilevsky believed that the inverting factor was not a protein, since it did not give the xanthoproteic test. He did feel, however, that both the invertase and fibrin-solubilizing factor were colloid in nature. Danilevsky was never able to isolate the fat-splitting factor from pancreatic secretions. However, he did note that in preparing his pancreatic extracts, all endogenous fats were split into glycerol and fatty acids giving an acidic character to the extract. When he treated the extract with MgO to remove the fatty acids, the lipid-splitting factor was lost, probably being absorbed by the MgO. Danilevsky did believe that the lipid-splitting factor represented the third discrete component of pancreatic juice, the other two being the fibrin and sugar-inverting factors.

In 1886, Danilevsky described the phenomenon of plastein formation by gastric juice (34). He believed that this represented a *de novo* synthesis of protein from peptone, and proposed the presence of a specific enzyme, *chymosin* or *Labferment*, to carry out this reaction. It was later found that pepsin was actually responsible for plastein formation. This reaction was investigated more fully by Zavyalov from Dorpat (35), who found that the digests of many diverse proteins such as myosin, fibrin, and "Witte's peptone" could be converted into a gelatinous mass that was believed to represent a new protein. Zavyalov believed that the so-called common core theory of protein structure and the fact that a seemingly identical material (plastein) was formed from such a diverse group of proteins as myosin and fibrin represent proof-positive for this view. The plastein itself was studied by Lavrov of Dorpat (36), who found it to give a positive biuret, Millon, and xanthoproteic tests. In addition, the Molish test for carbohydrate was also positive. The material was insoluble in water and physiological saline, but was soluble in 0.25–5% HCl or H_2SO_4. It contained 57.6% C, 7.2% H, and 12.9% N when casein was used as its source. Kurayev (see below) was able to obtain plastein formation by the action of papain on the digests of several proteins, including casein and ovalbumin.

Danilevsky's work, in addition to its importance in understanding the function of the pancreatic juice, also represented one of the first successful attempts to separate mixtures of enzymes. Thus, Danilevsky was certainly one of the most gifted Russian biochemists, and probably ranks on the same level with the more famous Pavlov in terms of scientific ability and critical thinking. However, Danilevsky did not establish any "schools" of thought, nor was he able to gather a large group of students or other followers. One reason for this was perhaps that the peak of his activity was reached during the years when science was not yet being supported as generously as it later was, and another reason may have been the fact that Danilevsky made many moves during his most productive years. He graduated from Kharkov University in 1860, then went abroad, and in 1863 was appointed to a professorship at Kazan University. He was fired from his position in 1871 for having some strong differences of opinion with the administration, then went abroad once more to work with Virchow and other Western luminaries. In 1885, we find him on the faculty of Kharkov University, and in 1892 he moved to the Military-Medical Academy in St. Petersburg.

Danilevsky's work apparently inspired another Russian investigator, Victor Pashutin of the Military-Medical Academy, to improve upon the methods of separating the three types of enzymes in pancreatic juice (37). Pancreatic homogenates were extracted with sodium sulfite or ammonium nitrate solutions to produce the protease, with potassium arsenate solution to produce the carbohydrate hydrolase, and with potassium antimonate or sodium bicarbonate solutions to produce the lipase. Another example of enzyme separation was the work of Kazansky from Moscow University, where a separation of catalase from peroxidase was affected (38). A number of plant extracts were treated with pyrogallol up to a final concentration of 5%, whereby the catalase precipitated and the peroxidase remained in solution. The catalase of yeast extract was destroyed by the pyrogallol, though peroxidase could be recovered from the supernatant.

A noted Russian investigator in the area of catalase and peroxidase chemistries was A. N. Bach (1857–1946), whom the Soviets considered to be the father of Soviet biochemistry. He worked out an assay procedure for the determination of peroxidase using pyrogallol. With both peroxidase and catalase, he found that the speed of the reaction increased at a faster rate than could be accounted for by an increase in enzyme concentration: a twofold increase in enzyme concentration resulted in a threefold increase in the reaction speed. He likewise observed that in the presence of excess substrate, the eventual amount of the product formed was proportional to the amount of the enzyme used (39). Though Bach was certainly a competent biochemist, one can hardly ascribe any fundamental discoveries to his efforts. The title of "father of Soviet biochemistry" had apparently been bestowed upon him because of his revolutionary zeal and party loyalty: as a student, he belonged to the terrorist Narodnaya Volya organization, for which he was exiled to Siberia for three years. Upon his return, he went "underground" and supposedly agitated for a revolution among the workers of numerous Russian cities. After the tsar's police disbanded his organization in 1885, he fled to Switzerland, where he established a private laboratory to do his biochemical studies. There, he was apparently more concerned with science than revolution, until the Bolshevik coup of 1917 brought him back to Russia, where he assumed a high post in science administration. As such, he was apparently responsible for the development of biochemistry in the Soviet Union. One wonders how much he had to do with the persecution and demise of the Russian genetics scientists such as Vavilov (see chapter III). It is also interesting to note that Bach was not the son of an impoverished Russian peasant or proletarian, because even at the turn of the century, the financing of private laboratories was not a light matter. It is a certainty that the Russian revolutionary movement was not his benefactor in establishing a laboratory in Switzerland. They too experienced a chronic shortage of funds. It is to be remembered that the migration of Lenin and his group from Germany and Finland to Russia in 1917 was financed by Germany and other Western powers. A detailed study of Bach's activities in Switzerland would probably contribute an interesting chapter to the history of Russian revolution.

The inactivation and reactivation of enzymes by heat was studied by M. J. Gramenitsky, who worked with N. Kravkov, the pharmacologist, in St. Petersburg. He discovered that the heating of taka-diastase (an enzyme hydrolyzing starch) to 85 to 115° C resulted in loss of activity, but that could be reestablished by permitting its solution to cool slowly to room temperature (40). The hydrolytic activity of the enzyme disappeared already at 70°C, whereas its invertase activity was retained. In a similar experiment, if malt was heated to 100°C, the maltase activity was irreversibly destroyed, whereas the amylase and oxidase activities could be recovered by slow cooling. In another investigation (41), Gramenitsky permitted the taka-diastase, which he had heated to 85°C to recover its activity in neutral, acidic, and alkaline solutions. In the alkaline medium, activity was regenerated three times as fast as in water, and in an acidic medium, the recovery of activity was inhibited. Gramenitzky felt that the temperature sensitivity properties of enzymes could be used for enzyme purification purposes. In view of our present-day knowledge of enzyme chemistry, it is clear that Gramenitsky's experiments had important implications in the understanding of the denaturation phenomenon, as well as the three-dimensional aspects of protein structures. Gramenitsky, of course, did not recognize the significance of his findings, since, at that time, there was no agreement regarding the nature of enzyme structures, nor had it been settled as to what really constituted a protein.

The Structure and Metabolism of Proteins and Amino Acids

At the turn of the century, protein was considered to represent a substance that contained nitrogen, gave a biuret, the Millon, and the xanthoproteic reactions, and was coagulable by heat. Most workers also recognized that amino acids were sometimes components of proteins. Much information on protein structure was gathered on the basis of peptic or tryptic digestion, and since very little was at that time known about the specificity of these enzymes, erroneous conclusions were often reached on the structure of proteins. However, when properly carried out and interpreted, such experiments often did yield valuable information. One of the favorite proteins that was investigated in the laboratory was casein, because it was readily available in quantity and was considered by most investigators to be homogeneous. This view persisted well into the twentieth century until modern electrophoretic and ultracentrifugal techniques became available, which have shown that casein is a mixture of at least three components.

One of the first chemists to investigate casein was N. N. Lyubavin (1845–1917), who was at first a professor at St. Petersburg and then in Moscow (1886–1906). In 1871, he published a paper reporting his experiments with peptic digestion of casein. He was able to isolate a phosphorus-containing and water insoluble fraction and a water-soluble fraction from the digest. From the latter, Lyubavin was able to crystallize leucine and tyrosine (42). This work is remembered by many historians as providing conclusive proof for the existence of amino acids in proteins. In reporting the isolation of the two amino acids, he also communicated his thoughts on the process of protein digestion by enzymes. It had been noted for a long time that digested/hydrolyzed proteins contained more oxygen than did the undigested ones. This prompted Mulder, the "father" of the term *protein,* to propose that protein digestion is an oxidation process. Lyubavin argued that this was not the case, and that, instead, protein digestion represented hydrolysis, i. e., interaction of proteins with water. He proposed that the breakdown of proteins by acids, alkali, water over 100°C, enzymes, and bacteria (i. e., putrefaction or "Faeulniss") were identical processes in this respect. A significant increase in hydrogen content of digested proteins, Lyubavin argued, had not been noticed because of technical imperfections in the methods then in use. In a later communication (43), Lyubavin attempted to prove that casein was simply a phosphoprotein instead of a nucleoprotein, as had been proposed by several workers because of casein's phosphorus content. He did not exactly succeed in this, though he did show that the phosphorus-containing residue following pepsin digestion of casein was a heterogeneous substance.

One of the first investigators to suggest that casein was not a homogeneous protein was A. Danilevsky, who was able to separate casein into two fractions on the basis of solubility differences (44). The casein was extracted with hot 40–50% alcohol, yielding a precipitate he called "Caseoalbumin," and a soluble fraction called "Caseoprotalbstoff." The former, Danilevsky believed, was similar to coagulated serum albumin. He later changed the terminology to "Nucleoalbumin" and "Nucleoprotalbstoff," respectively, since both fractions contained phosphorus, and phosphorus, as stated above, was supposed to come from nucleic acids only. Danilevsky further believed that the two components were interconvertible, a conclusion he reached by comparing casein prepared by acid precipitation and that obtained by the action of rennin. Thus, whereas acid casein required a large quantity of alkali for neutralization, the rennin-coagulated casein required very little. Moreover, when the acid casein was extracted with hot alcohol, a large portion of it remained in solution as the "Nucleoprotalbstoff." On the other hand, very little of "Nucleoprotalbstoff" could be obtained from the rennin casein. He thus felt that the rennin was active against the "Nucleoprotalbstoff" only, converting it into "Nucleoalbumin." Such action of rennin, in Danilevsky's view, involved the attachment of calcium ions onto the phosphate groups of the "Nucleoprotalbstoff" (45). Danilevsky's "Nucleoprotalbstoff" was probably what we today recognize as kappa-casein, a glycoprotein stabilizing alpha-casein, which is the major component of total bovine milk casein. The action of rennin is now known to affect the splitting of kappa-casein into a glycomacropeptide and the paraprotein, thus destabilizing and causing the precipitation of alpha-casein. Acid casein is known to contain kappa-casein in the unaltered form. Needless to say, Danilevsky's views were severely criticized in the biochemical literature of his time, and were soon all but forgotten.

Another Russian investigator who attempted to elucidate the specificity of proteolytic digestion, and, at the same time, the structure of proteins, was D. Lavrov, who first worked with Danilevsky at the Military-Medical Academy, then moved to Dorpat. He digested fibrin with pepsin, and the digest was then fractionated with ammonium sulfate and alcohol. There were six fractions with varying ammonium sulfate precipitability, and six fractions not precipitated by ammonium sulfate but precipitated with alcohol. The various fractions were quite different with respect to optical rotation, the biuret and other typical protein color reactions, and precipitability with acids. The more soluble the fraction was in water, the fewer protein-positive reactions it tended to give. Extended digestion with large amount of pepsin yielded products that gave no reactions typical of proteins. Tryptic digestion yielded essentially similar results (46). In a later investigation, Lavrov was able to isolate crystalline leucine, valine, and aspartic acid following a prolonged (eight-week) digestion of proteins. In addition, he found cadaverine and putrescine in such digests, which, he felt, were produced from lysine and arginine via peptic digestion (47). More likely, the latter products were the result of bacterial contamination, though Lavrov claimed that the reaction mixtures were sterile. Lavrov and his students later developed a theory that the true function of pepsin was to hasten the hydrolysis of proteins by the HCl present in the stomach. One set of experiments in support of this hypothesis was provided by E. Svirlovsky, a collaborator of Lavrov at Dorpat who subjected a number of proteins to prolonged hydrolysis with 0.5% HCl at 36–37°C (48). He was able to isolate free amino acids such as histidine and arginine from all proteins studied except gelatin, and other products that upon further hydrolysis with 20% H_2SO_4 at 100°C for 6 hours yielded free amino acids. Similar results were, of course, obtained by peptic digestion of proteins. Lavrov and his students were thus first to suggest that peptides were products of proteolytic digestion of proteins.

The occurrence of arginine in animals was extensively investigated by a very accomplished Russian biochemist, Vladimir Gulevich (1867–1933). He isolated arginine from herring testicles and characterized both the free amino acid and a number of its derivatives. Its molecular weight, determined cryoscopically, was 166–185, hence it fit the formula $C_6H_{14}N_4O_2$. It melted with decomposition at 207°C and had a specific rotation of 9 to 10 degrees at the D-line of sodium.

At first he felt that his preparation was a stereoisomer of arginine isolated from plants by Schultze and Steiger, but upon recalculating his optical rotation data, he found that the two compounds were in fact identical. He later discovered arginine in the spleen of the ox, thus being the first to show the presence of arginine in higher animals (49). He also implied that arginine was important in the formation of urea by the mammalian organism (50).

Gulevich was a pioneer in the use of small-molecular weight compounds to determine the specificity of proteolytic enzymes (51). Since he was not very successful in defining the specificity of trypsin, mostly because the appropriate probes were not available, his work in this area is all but forgotten. He tested the following compounds for their ability to be split by trypsin: phenylethyl ether, biuret, acetanilide, sulfanilic acid, hippuric acid, salol, diphenylurea, diphenylthiourea, acetylsalicylic acid (aspirin), salicylic acid anilide, acetyl-p-O-ethylaminophenol, N-ethylaniline, acetylphenylhydrazine, and N, O-diacetylaminophenol. Only the latter was split by trypsin, though it was nor clear whether the ester-, the peptide-, or both bonds were broken. The above-mentioned investigations were carried out by Gulevich during his pilgrimage to Western European laboratories; however, his most productive period developed after he returned to Russia, first to Kharkov University (1899–1901), then Moscow University. In 1900, he and his student Amiradzibi described a new compound that they had isolated from meat extracts and which they called carnosine (52). As a nitrate, carnosine had a specific rotation of 22.3 degrees, a melting point of 211–212°C, and it gave an empirical formula of $C_9H_{14}N_4O_3$. HNO_3. The free base was prepared from the nitrate and was found to be strongly basic with a decomposition point of 239°C. Other derivatives prepared by Gulevich were its silver and copper complexes. The silver complex had an empirical formula of $C_9H_{14}N_4O_3$. Ag_2O, and a decomposition point of 220°C. The full structure of carnosine was established by Gulevich later (53). By the alkaline hydrolysis of the latter, he was able to isolate histidine as its silver salt and as a free amino acid. Its identity was deduced from its melting point of 253–254°C and its empirical formula of $C_6H_9N_3O_2$. By difference, the other component of carnosine was deduced to be alanine, and Gulevich proposed that carnosine was either histidyl alanine, or alanyl histidine. It is today known that carnosine is beta-alanyl histidine.

Another biological base isolated by Gulevich and his students was carnitine, which was prepared as a chloroplatinate salt from the by-products of carnosine isolation (54). The salt had a melting point of 214–218°C, and an empirical formula of $C_{14}H_{32}N_2O_6Cl_6Pt$. The free base was assumed to have a formula of $C_7H_{15}NO_3$. The function of both carnosine and carnitine were at that time unknown, though today it is recognized that carnitine transports fatty acyl CoA's across the inner mitochondrial membrane, but the function of carnosine remains unknown.

Further work on the composition of muscle extracts was carried out by Gulevich's associate, R. Krimberg (55). In order to dispose of the possibility that both carnitine and carnosine were artefacts produced by putrefaction or proteolysis, Krimberg prepared his extracts within half an hour of the animal's death. Both carnitine and carnosine were found in such extracts, and, in addition, there was methylguanine, which had been previously discovered in 1906 by both Gulevich and Kutscher. Krimberg was able to obtain 5.8 g. of pure carnosine from 4.5 kg of meat (i. e., 1.3% of total tissue weight), where creatine made up 1.5–2% of the tissue weight. Some 20 years later, when Krimberg was at the Riga University Physiological Institute in the then-independent Latvia, one of his coworkers, S. A. Komarov, discovered that muscle extracts, when injected into animals, would bring about both gastric and intestinal secretions (56). Creatine and creatinine had no effect, whereas carnitine, carnosine, and methylguanine did. Krimberg thereupon formulated a comprehensive theory to the effect that these bases were digestive hormones similar to secretin and gastrin, which, instead of, or in addition to the vagal system were responsible for the elaboration of appetite juice and the continuing activity of the secretory glands of the digestive tract (57). Komarov eventually moved to Canada to work with Babkin, and Krimberg did not further elaborate on his theory. This author (A. B.) published an article summarizing the work of Gulevich and Krimberg (58).

A base that was supposed to cure various kinds of diseases including insanity was isolated from mammalian testes by von Poehl and named spermine (59). Von Poehl (1850–1908) was a professor first at Dorpat, then in St. Petersburg. He prepared spermine as a platinum hydrochloride complex, and determined its empirical formula as $C_5H_{15}N_2PtCl_6$. The free base was given the formula $C_5H_{13}N_2$ or $C_{10}H_{26}N_4$. Its therapeutic effects were eventually proven to be nil, though it has been implicated in the biosynthesis of glycoproteins.

The isolation of new proteins was not as commonplace at the turn of the nineteenth century as it was at the mid- to end of the twentieth century, when almost every issue of a biochemical publication contained a description of a purified enzyme or another biologically active protein. Among the few proteins that had been purified at the end of the nineteenth century was a group of basic proteins called protamines, which were generally found in fish testes. One such protein was isolated by D. Kurayev (1869–1908) from the testes of the Baltic Sea mackerel (60). It contained 24% N and some 21% sulfate, and gave an empirical formula of $C_{30}H_{60}N_{16}O_6 \cdot 2\ H_2SO_4$. Kurayev called it *scombrine sulfate*, a name that is still in use. The protein was very basic, gave a strong biuret reaction, and had a specific rotation of -72 degrees. Histidine and arginine were found in its sulfuric acid hydrolysates.

Kurayev also proposed a novel method for the determination of protein molecular weights. He iodinated crystalline serum and egg albumins and found preparations that contained as much as 10–12% iodine (61). He assumed that the iodine reacted with the nitrogen atoms of proteins, and estimated that in his serum albumin preparations, there were 11 iodine atoms for every 116 nitrogen atoms. From this he deduced an empirical formula for serum albumin, $C_{450}H_{693}I_{11}N_{116}O_{132}S_4$, which yielded a molecular weight of 10,000–11,000. This value was quite higher than most values earlier suggested for serum albumin, but was, of course, far short of the 69,000-value that is the true molecular weight of serum albumin. Today, Kurayev's approach to protein molecular weight determination may be equated to that using end-group analysis and other number-average molecular weight determinations using chemical means.

One of the most influential biochemists in Russia was Marcel von Nencki (1847–1901), a Pole by nationality, who studied in Krakow, Berlin, and Jena, worked for many years in Switzerland, and in 1891 was appointed as head of physiological chemistry laboratories of the newly-created Institute of Experimental Medicine in St. Petersburg. Nencki was responsible for much of the classical work on hemoglobin, and probably introduced this area of research into Russia. With his professor, N. Sieber, Nencki crystaliized several hematins from animal red cells using amyl alcohol as the solvent and proposed the formula of $(C_{32}H_{30}N_4FeO_3.HCl)_4C_5H_{12}O$ for his product. The free heme was presumed to have the formula of $C_{32}H_{32}N_4FeO_4$. Upon fusing hematin with KOH, pyrrole was obtained, which represented one of the first indications as to the structure of the heme molecule (62). Nevertheless, reproducibility in regard to the empirical formula obtained from one preparation to another was wanting, and Nencki proposed that heme was susceptible to chemical modification during isolation procedures (63). Thus, if acetic acid was used in the procedure, there was a possibility of acetylating the heme, according to him. If ethanol was used, esterification was possible. He proposed that the hemin prepared by the classical Teichmann method (acetic acid-NaCl) should have the empirical formula of $C_{32}H_{31}N_4O_3ClFe$, and he prepared a number of derivatives of heme such as amyl and ethyl esters to prove his point. It is not exactly clear what compounds Nencki was dealing with here, since he claimed to have had free hydroxyl groups in his starting material (hemin). It is now known that there are no hydroxyl groups on protoporphyrin, but the vinyl groups can apparently be easily converted to hydroxyl residues to yield hematoporphyrin. It is thus possible that Nencki may have been working with mixtures of hematin derivatives. It may also be noted that Nencki coined the terms "hemin" and "hematin," but these had different connotations from those we use today. It is interesting to note that a formula for hemin derived by a certain Russian investigator named Shalfeyev (quoted in ref. 63) was correct to within one hydrogen atom: $C_{34}H_{33}O_4N_4ClFe$. Instead of using Teichmann's method, Shalfeyev crystallized his hemin from chloroform-quinine. Nencki criticized his work, and suggested that Shalveyev's compound was nothing but the acetylated form of the product that he, Nencki, had purified. Another interesting aspect of Shalfeyev's work was the isolation of what Shalfeyev thought was iron-free hemin, which he called the *carcasse*. This fraction accounted for some 6% of the total hemin isolated from a given batch of hemoglobin. Though Nencki felt that the carcasse was nothing more than proteinlike debris, the possibility remains that Shalfeyev had isolated protoporphyrin.

Quantitative aspects of hemoglobin structure were studied by Lavrov (64), who found that iron accounted for 0.47% of its total weight. He then separated the heme from the globin by treating hemoglobin with an ether-alcohol-H_2SO_4 mixture, whereby the globin precipitated and the heme remained in the organic phase. The former accounted for 94.09%, and the latter for 4.47% of the total hemoglobin weight. The rest was attributed to fat.

Russian contributions to forensic biochemistry may be exemplified by the work of Heinrich Struve of Tiflis (65). He devised methods for the detection of bloodstains by looking for red cells or red cell ghosts, instead of trying to crystallize hemin from suspected bloodstains. He eluted the red cells with either carbonic acid or dilute acetic acid, and was thus able to distinguish between human, sheep, goat, and avian cells through the use of microscope. He also described a complication encountered in such analyses: if the stained object was kept in a moist place for some time, fungi began growing on the stain, destroying both the cells and hemin. He then had to elute his stains with alkali and simply test for protein.

Marcel Nencki was also a pioneer in attempts to clarify the formation of urea in mammals. As early as in 1869, Nencki and Schultzen reported that urea could be formed from glycine (66). They kept dogs on low protein diets, so that their urea excretion dropped to 4–5 grams/24 hours. When acetamide was fed, the output immediately increased to 15 g. The amount of additional nitrogen excreted corresponded to the amount fed. The same finding was obtained with glycine: an equivalent of 12 g urea in the form of glycine was fed and 10–11 g were excreted. With leucine, a urea equivalent of 9.3 g was fed and 8 g was excreted. A more sophisticated experiment was later performed by Sergey Salaskin, who perfused dog livers with blood containing glycine, aspartic acid, and leucine. In all cases, an increae in blood urea was obtained, with 100% of glycine and leucine being converted to urea (67). These experiments not only furnished conclusive evidence to the effect that urea was formed from amino acids, but also implicated the liver in this process. Salaskin later attempted to amplify the latter finding by measuring blood and urine urea content in hepatectomized dogs (68). He found that blood and urine ammonia were increased and the amounts of urea were decreased, but the production of urea in such animals did not cease completely. Both the urine and blood of hepatectomized animals were more acidic than those of normal animals, and large amounts of lactic acid in both biological fluids were also noted. The animals soon died of acidosis, thus obscuring in a major way the effects of hepatectomy on nitrogen metabolism.

The implication of ammonia in the formation of urea was suggested by Nencki in collaboration with Pavlov (69). It was found that dogs with the Eck fistula, i. e., the veno-portal shunt, could not tolerate meat and would exhibit acute toxic effects, primarily in the nervous system, when fed nitrogenous compounds. It was also found that the urine of such animals contained large amounts of what was thought to be carbamic acid. The toxic symptoms seen in such animals could be reproduced in healthy animals by injection of carbamic acid (it is not clear exactly what was meant by "carbamic acid," since free carbamic acid is unknown; it is possible that acidified ammonium carbamate served as a source of "carbamic acid"). A number of organs were then analyzed for ammonia, and its presence was found to predominate in the lining of the stomach and intestine. The ammonia content of the portal vein was high compared to the blood of other vessels: 3.5–8.4 mg/100ml in the portal vein vs. 0.5–1.8 mg/100 ml in the hepatic vein. The urea content of both veins was almost identical. The ammonia content of arterial blood was 1.5 mg/100 ml, but in animals with the Eck fistula the ammonia level reached 5.5 mg/100 ml, i. e., a value identical to that of the portal vein. It was concluded that the ammonia, probably in the form of ammonium carbamate, was transported via the portal vein from the gastrointestinal tract to the liver, where it was converted to urea. The liver was thus assigned the role of a detoxification device with respect to ammonia (70). It is today known that large amounts of ammonia are produced in the gastrointestinal tract by the microorganisms, most of which is absorbed into the portal circulation and converted to urea in the liver. Nencki's group also suggested that uremia was nothing more than an accumulation of ammonia in the blood due to the failure to convert it to urea. However, Salaskin was not able to demonstrate the presence of excess ammonia in uremic dogs whose ureters had been tied, nor the presence of abnormal amounts of ammonia in the brains of patients who had died of uremia (71). Salaskin had previously found that dogs with the Eck fistula had high accumulations of ammonia in all tissues, especially in the brain. It is thus clear that the work of Russian laboratories provided much of the background for the formulation by Otto Folin of his classical theory on the endogenous-exogenous metabolism of nitrogen.

One of the great successes of modern biochemistry is the elucidation of a number of inherited diseases in terms of aberrations in enzymatic reactions. Among such diseases, one can mention the various porphyrias, the glycogen storage diseases, phenylketonuria, and a number of other rare disorders of amino acid metabolism. Alkaptonuria, a defect in the metabolism of tyrosine, has been known for a long time, and is characterized by the excretion of an abnormal substance in the urine, which turns dark upon exposure to air. The nature of this substance, which happens to be homogentisic acid, was discovered in 1891 by M. Volkov from St. Petersburg when he was working with Baumann in Germany (72). This compound was precipitated from urine by lead ions, and analyzed for 57.4% C and 5.3% H, giving an empirical formula of $C_8H_8O_4$. It formed a hydrate with some 9.5% water of crystallization. The melting point was 147°C. Functional group analysis gave one carboxyl group and two phenolic groups per molecule of acid. A number of derivatives were also characterized. Volkov felt that homogentisic acid was not a metabolic product of tyrosine, but was instead produced by microorganisms somewhere in the body. The intestinal microorganisms were apparently found not to be responsible. Today, it is known that homogentisic acid is an intermediate in the degradation process of tyrosine and is excreted in the urine because of absence of an oxidase that normally degrades homogentisic acid.

An early advocate of a rapid state of protein metabolism and general turnover was N. Tichmenev from Moscow, who in 1914 showed the presence of a high rate of protein biosynthesis in mice (73). He starved mice for two days, then sacrificed one group and weighed their livers. The other group was fed a high protein diet for 3 days and then sacrificed. The livers of the second group were some 20% heavier than those of the first group, and the increase in liver weights was accounted for by their protein content. However, the Folin theory of endogenous-exogenous protein metabolism, which held that proteins turned over very slowly, remained the dominant dogma until the early 1940s, when isotope tracer experiments of Schoenheimer and Rittenberg conclusively showed that proteins in the living organism are present in a dynamic equilibrium, i. e., a state of rapid turnover.

Carbohydrate Chemistry and Metabolism

It has already been mentioned that Carl Schmidt of Dorpat University was one of the founders of modern carbohydrate chemistry. Another Russian biochemist who is familiar to every student of biochemistry was F. F. Selivanov (b. 1859), who devised a test for keto-sugars (e. g., fructose) using HCl and resorcinol. The first laboratory synthesis of a sugar was performed by M. A. Butlerov in 1861 using formaldehyde and alkali (74). Formaldehyde was at that time a relatively new compound, having only recently been prepared from methylene iodide and silver oxalate by Butlerov, who believed that he had synthesized methylene glycol. He visualized the reaction between methylene glycol and alkali as follows:

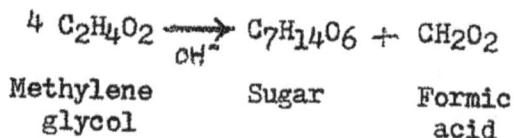

$$4 \ C_2H_4O_2 \xrightarrow{OH^-} C_7H_{14}O_6 + CH_2O_2$$

Methylene glycol   Sugar   **Formic acid**

The product was sweet to the taste, was strongly reducing, but did not rotate plane polarized light. It analyzed for 41–44% C, 7% H, and 49–52% O. The general formula of $C_nH_{2n}O_{n-1}$ was proposed by Butlerov for this type of sugars. The second compound obtained in the above reaction was named "methylenitan" by him, and was believed to be similar to mannose.

The study of carbohydrate metabolism received its greatest impetus from the discovery of Buchner in 1897 that cell-free yeast extracts were able to metabolize sugar into ethanol. Another milestone in this field was the discovery of Harden and Young in 1905 that during glycolysis, inorganic phosphate was converted to organic phosphate via esterification of sugar hydroxyl groups. However, few biochemists realize that the esterification of inorganic phosphate during glycolysis was simultaneously reported by a Russian botanist, Leonid Ivanov (b. 1871), a professor at the St. Petersburg Forestry Institute. Ivanov published a paper to this effect in 1905 in the 34th volume of *Proceedings of the St. Petersburg Society of Naturalists* (75), a publication hardly known in the West. He showed that inorganic phosphate was converted to organic phosphate, and that inorganic phosphate stimulated the process. Ivanov was able to precipitate the organic phosphate as a copper salt, to hydrolyze off the phosphate group, and to obtain a phenylhydrazine reaction with the product. He concluded that the phosphate group was attached to an aldo- or keto-group, and from the melting point of its phenylhydrozone, as well as its phosphorus content (19–20%), he concluded that the phosphate he had observed was a triose phosphate.

The scramble to discover glycolysis intermediates involved many Russian scientists, such as Kostichev (1877–1931) and A. N. Lebedev (1881–1938). Kostichev, who was with the St. Petersburg Technological Institute, discovered the participation of acetaldehyde in glycolysis by inhibiting the pathway with Zn-ions, then recovering the acetaldehyde by distilling the reaction mixture (76). The compound was identified through color reactions with sodium nitroprusside and diethyamine, and by preparing the acetaldehyde p-nitrophenyhydrazone (m. p. 127.5°C). The derivative also gave the expected elemental analysis of 23.6% N, 54.1% C, and 5.6% H. No other volatile aldehydes were found. Kostichev then proved that acetaldehyde was indeed a glycolytic intermediate by starving yeast cells, then giving them acetaldehyde. The starved yeast cells then produced a much greater quantity of alcohol than did the control yeasts. The addition of formic acid did not improve the yield (77). Kostichev believed that the production of acetaldehyde took place through the formation of pyruvic acid from glucose, though pyruvate had at that time not been unequivocally implicated as an intermediate in the glycolytic pathway. He proposed the following set of reactions leading to acetaldehyde:

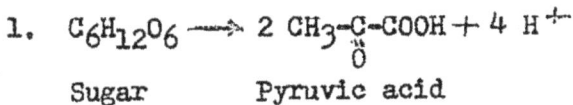

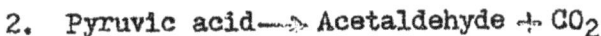

$$1. \quad C_6H_{12}O_6 \longrightarrow 2\ CH_3\text{-}\underset{O}{\overset{}{C}}\text{-COOH} + 4\ H^{+}$$

Sugar   Pyruvic acid

$$2. \quad \text{Pyruvic acid} \longrightarrow \text{Acetaldehyde} + CO_2$$

$$3. \quad \text{Acetaldehyde} \longrightarrow 2\ H^{+} + \text{Alcohol}$$

The 4 hydrogen ions produced in the first reaction were to be transferred to some receptor *via* "reductases." The utilization of pyruvate by a variety of tissues, presumably *via* the glycolytic pathway, was demonstrated by Chernorutsky (78), and the presence of the "reductase" system in the yeast had been previously shown in 1908 by Kostichev's professor named Palladin (79). He had shown that yeast cells could bring about a reduction of methylene blue and sodium selenite, and that glucose inhibited this reaction. Kostichev's alcohol formation proposal indicated above is recognized in today's biochemistry texts as shown below (80):

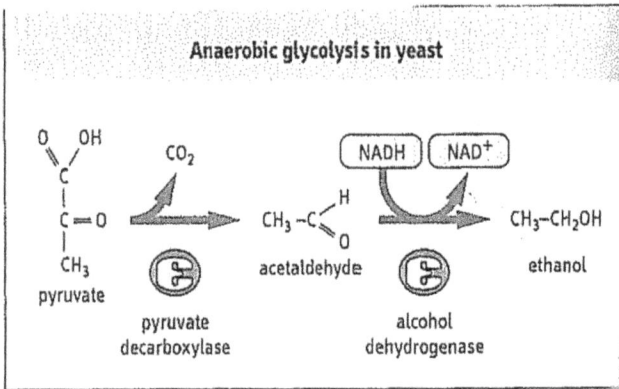

Fig. 11.6 **Anaerobic glycolysis in yeast.** Formation of ethanol by anaerobic glycolysis during fermentation. Pyruvate is decarboxylated by pyruvate decarboxylase, yielding acetaldehyde and CO_2. Alcohol dehydrogenase uses NADH to reduce acetaldehyde to ethanol, regenerating NAD^+ for glycolysis.

The great Russian synthesizer of the glycolytic pathway of the meager number of data available in his time was A. N. Lebedev, who spent some time working at the Pasteur Institute in Paris, and was later on the faculty of the Don Technological Institute of Novocherkassk. He was one of the discoverers of dihydroxyacetone phosphate as a glycolytic intermediate, and showed, using kinetic arguments, that the disappearance of glucose during fermentation was independent of carbon dioxide evolution (81). When dihydroxyacetone was added to the fermentation mixture, inorganic phosphate was first taken up to form organic phosphate. After a time interval, the organic phosphate was reconverted to inorganic phosphate. The dihydroxyacetone under Lebedev's conditions was converted to hexose diphosphate. This compound was isolated by Lebedev and identified as glucose diphosphate. More likely, Lebedev had isolated fructose-1,6-diphosphate, which is readily formed from triose phosphates *via* the aldolase reaction, because the equilibrium constant is now known to favor the condensation. This was, of course, not known to Lebedev, and he proposed the following series of reactions for the glycolysis process:

1. Glucose $\longrightarrow$ 2 triose
2. 2 Triose $+$ 2 H_3PO_4 $\longrightarrow$ 2 triose phosphate
3. 2 Triose phosphate $\longrightarrow$ hexose diphosphate
4. Hexose diphosphate $+$ H_2O $\longrightarrow$ alcohol $+$ CO_2 $+$ triose phosphate $+$ P_i

   or

   Hexose diphosphate $+$ 2 H_2O $\longrightarrow$ 2 alcohol $+$ 2 CO_2 $+$ 2 P_i

Proceeding with his investigations. Lebedev found that glyceraldehyde was fermented to carbon dioxide by the yeast, but did not give hexose diphosphate . By that time, Lebedev had also seen Kostichev's work on acetaldehyde, and proceeded to propose the following scheme for glycolysis (82):

Glucose $\longrightarrow$ Glyceraldehyde $+$ Dihydroxyacetone
   $\downarrow \longrightarrow$ 2 H^+ $\downarrow \longleftarrow$ H_3PO_4
   Enol pyruvate   Dihydroxyacetone phosphate
   $(C_3H_4O_3)$ $\downarrow$
   $\downarrow$ Hexose diphosphate
   Acetaldehyde $+$ CO_2 $\downarrow \longrightarrow$ H_3PO_4
   $\downarrow \longleftarrow$ 2 H^+ Hexose
   Ethanol

It may be noted that the above scheme has included a compound $C_3H_4O_3$, which is reminiscent of phosphoenolpyruvate, an important glycolytic intermediate. As may have been expected, Lebedev came under a severe criticism from his colleagues in regard to his glycolysis scheme. Especially adamant were Harden and Young, Buchner, and Meisenheimer. It apparently got so bad that Lebedev had to beg his colleagues to desist and give the experimenter a chance to either disprove or confirm his theory: "*Wir moechten die Fachgenossen bitten, uns die noetige Zeit zu lassen, um diese Untersuchung, die von dem einen von uns* (Lebedew) *fortgesetzt wird, ungestoert zu Ende zu fuehren.*" And furthermore, he proposed that criticisms not based on experimental data be avoided, and that, instead, more data be gathered: "*Uebrigens, werde ich mich sehr freuen, wenn die HHrn. Buchner und Meisenheimer meine Arbeit fortsetzend, etwas neues finden und die Wissenschaft damit bereichern.*" In view of the modern knowledge of the glycolytic scheme, Lebedev's main error lay in his proposal that the condensation of triose phosphate was a *bona fide* step in alcoholic fermentation. Since the equilibrium constants of the various glycolysis steps were unknown at that time, it is difficult to imagine how Lebedev could have made a different conclusion.

In the early 1900s, there was a great interest in the role of protein breakdown during glycolysis. This problem was investigated by L. Ivanov (83), who found that in the presence of ample amounts of sugar, there was no change in the protein content of yeast during glycolysis. This was, of course, a demonstration of the now well-known protein-sparing action of carbohydrate. In fact, protein synthesis ensued when asparagine was added to such systems. The protein was degraded when the yeasts were starved. Ivanov also discovered that yeasts that had undergone a vigorous glycolytic episode did not destroy protein as well as yeasts that had not, when both were starved. The factor that inhibited proteolysis was ethanol, where such inhibitory effect was noted when alcohol concentration had reached 4%. A similar study was performed by Gromov and Grigoryev of Palladin's laboratory at the St. Petersburg Women's College (84), who also found a large loss of total protein in the yeasts in the absence of sugar. In the presence of glucose or sucrose, there was only a small loss of protein. A similar though smaller effect was observed with glycerol and glycine. A potent stimulant of proteolysis was potassium nitrate, which at a concentration of 5% was able to bring about the destruction of 85% of the endogenous protein present in the yeast. Glucose was able to negate these effects to a large extent. It may be remembered that Carl Schmidt obtained similar results with the destruction of urea by yeasts some 60 years earlier.

That anaerobic fermentation does not necessarily lead to alcohol production was demonstrated by Palladin and Kostyichev in plant seedlings (85). It was shown that under normal conditions, both CO_2 and ethanol were produced by the seedlings, but when the seedlings were frozen and then thawed, only the CO_2 production was retained. Acetone was found in the wheat seedlings under these conditions, though it could not be demonstrated in any other plant. It is not clear exactly where the glycolytic pathway was inhibited by freezing. Perhaps it was stopped at the acetaldehyde step, perhaps acetoin was formed instead of alcohol. The normal wheat seedling was able to produce products of alcoholic fermentation with the production of CO_2 and uptake of O_2, without producing alcohol (86). Pyruvate was just as effective.

Lipids and Lipid Metabolism

Research on lipids in Russia, as it was in other lands, was restricted largely to composition studies, cleavage by enzymes, and intestinal absorption, because appropriate methodology was behind that of other areas of biochemistry. One of the first Russian researchers in the field of lipid chemistry was C. Diakonov from Kazan University, who, while working with Hoppe-Seyler in Berlin, was first to provide a quantitative analysis of lecithin shortly after its discovery by his boss (87). Egg lecithin was analyzed to contain 3.8% P, 1.8% N, 11.4% H, and 64.3% C. The empirical formula was calculated to be $C_{44}H_{90}NPO_9$. Upon alkaline hydrolysis, lecithin was split into two moles of stearic acid, one mole of glyceryl phosphate, and 1 mole of choline. The same type of lecithin was isolated from the brain. Much of the work on lipid metabolism was uninspiring, such as the report by Yushchenko, who found that the thyroid gland contained a lipase that split monobutyrin, tributyrin, ethyl butyrate, and neutral fats. Extirpation of the thyroid gland resulted in a decline of the serum lipase, thus suggesting to the author that the thyroid gland was its source (88).

The most interesting work on lipid metabolism performed by Russian investigators seems to have been done by Subbotin and by Lebedev. Subbotin's work was published in the form of a monograph, and probably represented a thesis. It was later published abroad (89). Subbotin set out to answer two questions: did tissue fat originate directly from dietary fat, and was tissue fat synthesized from protein and carbohydrate? The first question was answered by feeding starved dogs whale oil (Spermacet). The oil was efficiently absorbed, but no deposition of its fatty acids was observed in dog tissues, even though a net fat deposition was taking place. Dietary fat was thus found not to be directly deposited in tissues. The possible transformation of protein into fat was investigated by feeding starved dogs a mixture of protein and palm oil (the latter consisted only of palmitates and oleates). After such feedings, the animals were shown to have synthesized stearates. In addition, protein was fed with a soap consisting of palmitates and stearates of sodium. The tissues showed a net deposition of oleic acid in addition to palmitic and stearic acids. Subbotin concluded that fats could be synthesized from proteins. He apparently assumed that fatty acids were not interconvertible to each other! He was not able to show directly that fat was either synthesized or not synthesized from carbohydrate. However, indirect data were, in his opinion, suggestive of lack of carbohydrate participation in lipid biosynthesis.

Subbotin's work was extended by A. Lebedev, whose investigations were aided by the availability of a method for the separation of saturated from unsaturated fatty acids, which was developed by Lebedev himself (90). This method involved the saponification of the fat with KOH or NaOH, extraction of the unsaponifiable fraction with ether, and the precipitation of fatty acids with lead acetate. The precipitated salts were then extracted with aqueous but alcohol-free ether, whereby the lead salt of oleic acid was found in the ether phase, whereas stearic and palmitic acids remained insoluble. Using this procedure, Lebedev investigated the composition of human and goose tissues. The geese were fed corn oil containing 77 to 80% oleic acid and 12 to 14% of saturated fatty acids. The tissue neutral fats contained 61–69% oleic acid and 21 to 32% of stearic-palmitic acids depending on the organ source of the neutral fat. The fatty acid fraction from the fatty liver of a deceased patient had 60–68% oleic acid and 27–32% saturated acid; a similar composition was observed in a lipoma fatty acid fraction. The oleic acid content was higher in the fatty acid fraction of fat infiltrating lung tissue following a lung embolism episode, in subcutaneous fat, and in intestinal fat. The fatty acid fraction of these structures contained 73–80% oleic acid and 14 to 22% saturated fatty acids. In an experiment similar to that of Subbotin's, Lebedev fed starved dogs some tributyrin that he had synthesized. The compound was almost completely absorbed by the animals, but in spite of a net increase in the animal's fat content, no deposition of tributyrin in tissues was observed. In an additional experiment, Lebedev fed geese diets with and without lipid. Those receiving fat in their diets had larger fat deposits in their tissues, but the lecithin fractions of their fat deposits were much lower than those in animals not receiving fat in their diets. Lebedev concluded, as did Subbotin, that animal fat does not directly originate from dietary fat, but is instead sunthesized from dietary fat by some as yet unknown mechanism.

The origin of fat present in fatty livers of fasted animals and in patients with phosphorus poisoning was also elucidated by Lebedev (91). He found that the composition of fat from fatty livers was identical to that of subcutaneous fat: the fatty acid fraction from both sources contained 55–56% oleic acid, and 37–39% palmitic and stearic acids; for rabbit fat, this was 50–52% oleic acid and 42–43% stearic-palmitic acids. He concluded that during starvation and phosphorus poisoning, the subcutaneous fat deposits were transported to the liver rather than originating from protein, as had been previously proposed by Voit. Lebedev also analyzed the fat fraction of milk from a number of sources, including human milk. He found that the fatty acid fraction from human milk lipids was richest in saturated fatty acids, where stearic and palmitic acids accounted for 29% of all fatty acids present.

Endocrinology

It is generally believed that modern endocrinology was born with the discovery of *secretin* in 1902 by Bayliss and Starling, who also defined a hormone as a substance that was produced in one organ to exert its effect on another organ. Secretin could have been easily discovered in Pavlov's laboratory in St. Petersburg, and Pavlov and his students were indeed ridiculously close to doing so. However, Pavlov's school, as well as most other Russian physiologists, were so much committed to the nervous system as the sole regulator of the mammalian organism that they could hardly imagine that there was another equally important control mechanism in the living system. This rigidity of thought is well demonstrable by the work of Popielski, a student of Pavlov, and later, director of the Pharmacologic Institute in Lemberg (Austrian Poland). Popielski was studying the mechanism of pancreatic secretion that was elicited by the introduction of 0.5% HCl into duodenal fistulas of dogs (92). Pancreatic secretion was then studied after cutting the spinal chord, the vagi, the splanchnics, and after the destruction of the spinal chord and medulla (dogs were kept under anesthesia with an artificial respirator). Pancreatic secretion still persisted. He then cut the stomach and small intestine at several locations thinking that a local nervous network may be involved, but this either had no effect on pancreatic secretion. The celiac plexus was then destroyed also without any effect. Papielski all but eliminated the nerves as controllers of pancreatic secretions, yet he did not recognize that and, instead, proposed that it is controlled by as yet unrecognized nerve center. Even after the publication of Bayliss and Starling's work, Popielski refused to believe that secretin played a major role in controlling pancreatic secretions (93).

The presence of endocrine glands in the mammalian organism was well known in the nineteenth century, but their function was not well understood until after secretin was discovered. Research on endocrine glands was largely limited to studying their composition and to physiological effects brought about by the injection of their extracts. Such physiological effects, e. g., changes in blood pressure, were interpreted in various ways, and the Russian investigators made such interpretations invariably in terms of changes in the nervous system activity (e. g., de Cyon's view on the function of thyroxine). The situation did not improve even after Baumann's discovery of the role of iodine in the function of the thyroid gland.

The first Russian investigator to study the composition of the thyroid gland was N. Bubnov, a student of Botkin at the Military-Medical Academy and a classmate of Pavlov. He participated in the Russo-Turkish war of 1876–1878 under Pirogov, then continued working with Botkin until he contracted diphtheria from a patient and died in 1884 at the age of thirty-three. Bubnov was particularly interested in the nature of the "colloid" substances of the thyroid glands (94). He extracted both the bovine and human glands first with water, then with 10% NaCl, and finally with two portions of 0.1 % KOH. The water extract contained largely small molecules such as lactic acid, hypoxanthine, and guanine, although it also gave a positive test for proteins. The next three fractions consisted of proteins that Bubnov called "thyreoprotin." These proteins were precipitated from the 10% NaCl and the two KOH extracts with dilute acetic acid, and represented what Bubnov felt were homogeneous thyroproteins 1, 2, and 3. The bovine thyroproteins gave ash values of 1.4, 1.7, and 1.5% respectively; 16.0, 16.1, and 16.7% nitrogen, respectively; and 49.4, 50.2, and 49.4% carbon, respectively. Similar results were obtained from human materials. Bubnov did not do iodine analyses on his protein fractions. Nevertheless, Bubnov is considered to be the discoverer of thyroglobulin, a large protein of the thyroid glands that is intimately involved in the biosynthesis of thyroxine from tyrosine. The thyroglobulin was most likely present in Bubnov's 10% NaCl extract (thyroprotein 1).

A very interesting, though in retrospect incorrect, proposal on the function of the thyroid gland was provided by Notkin of the Kiev University (95). He noted that in some animals, total thyroidectomy brought about a tetanic state, in others only myxedema. In human subjects, thyroidectomy produced a tetanic state (kachexia) with myxedema. The myxedema could be cured by the administration of thyroid gland extracts, whereas the tetanic state could not. He concluded that these were two completely unrelated symptoms, and that both were caused by two different toxins. It is possible that the tetanic state observed by Notkin in theyroidectomized patients was brought about by the inadvertent extirpation of the parathyroid glands, whose existence at that time was unknown. Notkin prepared thyroprotein by extracting water-extracted thyroid glands with 5% NaCl, then precipitating the protein with ammonium sulfate. At first, his preparations contained as much as 0.6% iodine, but a continued purification of the product (redissolution and reprecipitation) resulted in the loss of iodine. The material thus prepared contained carbohydrate, but was largely protein in nature and precipitable by dialysis against water and by heat. When injected into dogs, the thyroprotein caused a myxedemalike disorder similar to that observed in thyroidectomized dogs. He proposed that the thyroprotein was a toxin that brought about the symptoms of myxedema, that it was a product of general metabolism of the animal, and that the thyroid gland then acted as a trap or detoxifier of the thyroprotein. This detoxification reaction was supposedly accomplished by an enzyme that split the thyroprotein into carbohydrate and peptide portions, the peptide portion then combining with thyroxine, and the complex then being used by the organism for some as yet unknown physiological function. To prove this point, Notkin was actually able to show the cleavage of a small quantity of the thyroprotein by thyroid extracts into a precipitable protein fraction and a water-soluble carbohydrate portion. In Notkin's view, the tetanus-producing toxin was also a product of the general metabolic activity of the organism, and it too was supposed to be inactivated by the thyroid gland. However, this toxin was not supposed to be a protein in nature, and its inactivation was supposedly accomplished by a specific antitoxin previously isolated from the thyroid glands by a Vienna investigator named Fraenkel. In retrospect, Fraenkel may have prepared a parathyroid gland extract, which then relieved the tetanus symptoms of the thyroidectomized animals.

Notkin was undoubtedly dealing with thyroglobulin, since the latter is known to contain carbohydrate. However, it is not clear why he found no iodine in his preparation. It is to be noted that Notkin did not do the iodine analysis himself, but had these done by a colleague who may not have taken the best care in performing this delicate analysis. In 1899, a Swiss investigator, A. Ostwald, showed that both Bubnov's and Notkin's thyroproteins contained iodine, which could be released by mild acid hydrolysis. Notkin did, nevertheless, succeed in showing that thyroxine in the thyroid gland is associated with a larger protein molecule, and did establish the glycoprotein nature of thyroglobulin.

Elie de Cyon (see chapter III) was another Russian investigator who attempted to elucidate the function of the thyroid gland. He did most of this work after he had left Russia, and was temporarily associated with a Swiss institution in Bern. De Cyon felt that the thyroid gland had essentially two functions, one mechanical and the other—biochemical (96). In his extensive studies of the gland, he discovered its innervation and connection with the vagi. The purpose of such innervation, in his view, was to dilate the blood vessels of the thyroid gland, which indeed showed a remarkable capacity to become dilated. The purpose of such dilation was to shunt large quantities of blood away from the brain, should the blood pressure get too high, according to de Cyon. This was supposed to represent an antistroke device in the human organism.

De Cyon's view of the function of the thyroid gland was in the role of an iodine-trapping device and, at the same time, to make thyroxine. Thyroxine, de Cyon found, was a valuable cofactor in the regulation of heart function and vasomotor activity by the vagi, where the thyroxine served to stimulate both the vagi and the depressor nerves so as to lower the frequency of heartbeat, to increase the force of contractions, and to cause the lowering of blood pressure. Iodine, on the other hand, was supposed to be a thyroxine antagonist, and have an inhibitory effect upon the vagus nerves. He noted that the thyroidectomized animals showed the same symptoms as animals poisoned by NaI. De Cyon felt that the thyroproteins were not toxic (as had been proposed by Notkin), and represented merely a stage intermediate between iodide and thyroxine (this view was entirely correct). In collaboration with A. Ostwald, de Cyon showed the nontoxic nature of thyroglobulin by injecting both the iodinated and iodine-free thyroglobulins into experimental animals (97). Only the iodinated thyroglobulins showed a depression of the blood pressure and lowering of the heart rate. It was concluded that thyroxine was a physiologically active substance.

The function of pituitary glands was also studied by de Cyon during his stay in Switzerland (98). He made pituitary gland extracts and injected these into dogs. He observed a small rise in blood pressure, a decline in the frequency of heartbeat, and an increase in the force of cardiac contractions. He assumed that the extract had stimulated the vagi, since the pituitary extracts, like thyroxine, were able to negate the effects of atropine on the nerves. Electrical or mechanical stimulation of hypophysis gave the same results. Hypophysectomy resulted in an accelerated heart rate, an effect of cutting the vagi. Having, in his mind, established a relationship between the vagi and the pituitary gland, de Cyon formulated a theory to the effect that hypophysis was capable of sensing increases in the intracranial pressure (99). When such pressure became higher than normal, the hypophysis was stimulated, mechanically resulting in the stimulation of the vagi. This in turn resulted in the dilation of the thyroid gland veins and a constriction of arterioles. The blood pressure was thus increased, yet the blood was shunted through the thyroid to relieve the pressure on the brain. Thus, de Cyon's view was that the pituitary organ was primarily a pressure sensing device for the protection of the brain. The chemical (hormonal) function of this structure was, in his view, merely as a backup system for its primarily mechanical role. De Cyon's proposals on the function of the endocrine glands were thus primitive and, in most cases, erroneous, though not unimaginative. After all, he was one of the leaders of the Russian school of the organism's total control by its nerve networks. Nevertheless, his work, carried out with precision and technical perfection, undoubtedly served as an important stimulus for further research in the area of endocrinology, especially since de Cyon was one of the first authors to propose the interrelationship between hypophesis and other internal secretion organs.

The adrenal gland did not escape de Cyon's attention (100). He found that blood pressure was increased after the injection of adrenal gland extract, which was brought about because of a massive vasoconstriction. As in case of the thyroid and hypophysis, de Cyon felt that these effects were due to the action of vagi and depressor nerves: adrenal extracts merely served to depress the activity of these nerves, since a stimulation of them could not bring about a reversal of the effects of adrenal extracts. On the other hand, adrenal extracts served to stimulate the accelerator nerves and, of course, the sympathetics. He also noted a momentary decline in the rate of heartbeat after injecting the adrenal gland extract, and thought that this was due to the response of the hypophysis to increased blood pressure in the cranial cavity. Another aspect of the adrenal gland action was studied by Gramenitsky, who gave adrenalin to rabbits and found a marked hyperglycemia in such animals (101). The intensity of hyperglycemia was directly proportional to the amount of adrenaline administered. When blood glucose concentration exceeded 0.2%, a glycosuria was observed. Gramenitsky was thus one of the first to observe the glycogenolytic effect of epinephrine.

It would be interesting to speculate as to who were the most distinguished biochemists in Russia of the tsarist era. In view of modern biochemical knowledge, the most progressive investigators were probably Lunin, Gulevich, Danilevsky, and Alexander Schmidt. These would be closely followed by such synthesizers as Lebedev and de Cyon. Among the developers of methodology, Tswett, mentioned in Chapter III, is of course, the most outstanding. The men who probably did most to introduce biochemistry into Russian universities and scientific institutions were Nencki, Carl Schmidt, and after the Bolshevik coup, A. N. Bach. Nencki was best known abroad (102). In summary, it is apparent that biochemistry was rather well developed in Russia before the First World War, and had the revolutionary turmoil with its war on classical genetics and DNA not interfered, Russian biochemistry would have achieved the same degree of excellence as that of the Western world.

REFERENCES

1. H. Wain. *A history of preventive medicine.* Charles C. Thomas, Springfield, IL, 1970, p. 369.
2. J. T. Edsall. *Some personal history and reflections from the life of a biochemist.* Ann. Rev. Biochem. 40, 1971.
3. R. Hill. *The chemistry of life. Eight lectures on the history of biochemistry.* Cambridge Univ. Press, Cambridge, 1970.
4. M. Florkin. *The history of biochemistry.* Elsevier Press, New York, 1972.
5. H. M. Leicester. *Development of biochemical concepts from ancient to modern times.* Harvard Univ. Press, Cambridge, 1974.
6. C, Schmidt. *Zur Geschichte der Gaerung.* Ann. Chem. Pharm. 126:126, 1863.
7. C. Schmidt. *Gaerungsversuche.* Ann. Chem. Pharm. 61:168, 1847.
8. C. Schmidt. *Ueber das specifische Gewicht des Albumins, Muskelfibrins, der Blutkoerperchen, und Sehnen.* Ann. Chem. Pharm. 61:156 and 165, 1847.
9. A. Vogel and C. Schmidt. *Volumetrische Bestimmung der Harnsaeure im Harne.* Centralbl. F. med. Wissensch. 6:385 and 420, 1868.

10. C. Schmidt. *Ueber das Wesen des Verdauungsprocesses.* Ann. Chem. Pharm. 61:311, 1847.

11. C. Schmidt. *Ueber sogenannte "thierische Amyloid" (Substanz der corpuscular amylacea).* Ann. Chem Pharm. 150:250, 1859.

12. C. Schmidt. *Ueber Pflanzenschleim und Bassorin.* Ann. Chem. Pharm. 51:29, 1844.

13. E. Jorpes. *Alexander Schmidt.* J. Chem. Ed., 28:578, 1951.

14. A. Schmidt. *Neue Untersuchungen ueber die Faserstoffgerinnung.* Pflueger's Archiv 6:413, 1872.

15. G. Bunge. *Ueber die Assimilation des Eisens.* Hoppe-Seyler's Zeitschr. 9:49, 1885.

16. G. Swirski. *Ueber die Resorption des Eisens im Darmkanale der Meerschweinchen.* Pflueger's Archiv 74:466, 1899.

17. A. Bezkorovainy. *Biochemistry of nonheme iron.* Plenum Press, New York, 1980.

18. W. G. Schipatschoff. *Die Kaschin-Becksche Krankheit.* Deutsch. Arch. F, kiln. Med. 170:133, 1931.

19. K. Hiyeda. *The cause of Kashin-Beck disease.* Jap. J. Med. Sci. Sect. 5, Pathology, 4:91, 1939.

20. N. Lunin. *Ueber die Bedeutung der anorganischer Saltze fuer die Ernaehrung des Thieres.* Hoppe-Seyler's Zeitschr. 5:31, 1881.

21. P. Babkin. *Secretory mechanism of the digestive glands.* Paul B. Haeber, Publ., New York, 1950.

22. A. Samojloff. *Einige Bemerkungen zu der Methode von Mett.* Pflueger's Archiv 85:86, 1901.

23. F. C. Koch and M. E. Hanke. *Practical methods in biochemistry.* Williams and Wilkins, Publ., Baltimore, 1948, pp. 130–132.

24. J. Schuetz. *Zur Kenntniss der quantitativen Pepsinwirkung.* Pflueger's Archiv 80:1, 1900.

25. H. Euler. *Katalyse durch Fermente.* Hoppe-Seyler's Zeitschr. 45:420, 1905.

26. A. Medwedew. *Ueber die oxydativen Leistungen der thierischen Gewebe.* Pflueger's Archiv 74:193, 1899.

27. A. Medwedew. *Ueber die oxydativen Leistungen der thierischen Gewebe. Zweite Mitteilung.* Pflueger's Archiv 81:540, 1900.

28. A. Medwedew. *Ueber die oxydativen Leistungen der thierischen Gewebe.* Pflueger's Archiv 103:403, 1904.

29. M. Jakoby. *Ueber das Aldehyde oxydirende Ferment der Leber und Nebenniere.* Hoppe-Seyler's Zeitschr. 30:135, 1900.

30. W. Issayew. *Ueber Hefekatalase.* Hoppe-Seyler's Zeitschr. 42:102, 1904.

31. W. Issayew. *Ueber die Hefeoxydase.* Hoppe-Seyler's Zeitschr. 42:132, 1904.

32. W. Issayew. *Ueber die Malzoxydase.* Hoppe-Seyler's Zeitschr. 45:331, 1905.

33. E. S. London and A. P. Korchow. *Zur Kenntnis der Verdauungs- und Resorptionsgesetze. IX. Zur Verdauung der Kohlenhydrate.* Hoppe-Seyler's Zeitschr. 68:363, 1910.

34. Y. Danilewsky. *Ueber specifish wirkenden Koerper des natuerlichen und kuenstlichen pancreatischen Saftes.* Virchow's Archiv 25:279, 1862.

35. W. W. Sawjalow. *Zur Theorie der Eiweissverdauung.* Pflueger's Archiv 85:171, 1901.

36. D. Lawrow. *Ueber die Wirkung des Pepsins resp. Labferments auf konzntrierte Loesungen der Produkte der peptischen Verdauung der Eiweisskoerper (Reaktion von A. Danilewski).* Hoppe-Seyler's Zeitschr. 51:1, 1907.

37. V. Paschutin. *Ueber Trennung der Verdauungsfermente.* Centralbl. F. d. med. Wissensch. 10:97, 1872.

38. A. Kasanski. *Ueber die Abtrennung der Peroxydase und Katalase.* Biochem. Z. 39:64, 1912.

39. A. Bach. *Zur Kenntnis der Katalase.* Ber. 38:1878, 1905.

40. M. J. Gramenitzki. *Der Einfluss verschiedener Temperaturen auf die Fermente und die Regeneration fermentativer Eigenschaften.* Hoppe-Seyler's Zeitschr. 69:286, 1910.

41. M. J. Gramenitzki. *Ueber den Einfluss von Saeuren und Alkalien auf das im Stadium der Regeneration befindliche diastatische Ferment.* Biochem. Z. 56:78, 1913.

42. N. Lubavin. *Ueber die kuenstliche Pepsin-Verdauung des Caseins und die Einwirkung von Wasser Eiweisssubstanzen.* Med.-Chem. Untersuch. V. Hoppe-Seyler 4:463, 1871.

43. G. Wagner. *Sitzung der russischen chemischen Gesellschaft am 1./13. December 1877.* Ber. 10:2237, 1877.

44. A. Danilewsky and P. Radenhausen. *Forschungen auf dem Gebiete der Vienhaltung und ihrer Erzeugnisse.* No. 9, 1880. Quoted by O. Hammersten. *Zur Frage, ob das Casein ein einheitlicher Stoff sei.* Hoppe-Seyler's Zeitschr. 7:227, 1882-1883.

45. A. Danilewsky. *Zur vorlaeufigen Abwehr.* Hoppe-Seyler's Zeitschr. 7:427, 1882-1883.

46. D. Lawrow. *Zur Kentniss der Chemismus der peptischen und tryptischen Verdauung der Eiweissstoffe.* Hoppe-Seyler's Zeitschr. 26:513, 1898–1899.

47. D. Lawrow. *Zur Kentniss der Chemismus der peptischen und tryptischen Verdauung der Eiweisskoerper.* Hoppe-Seyler's Zeitschr. 33:312, 1901.

48. E. Swirlowsky. *Zur Frage nach der Einwirkung von verduennten Salzsaeure auf die Eiweissstoffe.* Hoppe-Seyler's Zeitschr. 48:252, 1906.

49.   W. Gulewitsch. *Ueber das Arginin.* Hoppe-Seyler's Zeitschr. 27;178 and 368, 1899.

50.   W. Gulewitsch and A. Jockelsohn. *Zur Frage nach dem Chemismus der vitalen Harnstoffbildung. II. Ueber das Vorkommen von Arginin der Milz.* Hoppe-Seyler's Zeitschr. 30:533, 1900.

51.   W. Gulewitsch. *Ueber das Verhalten des Trypsins gegen einfache chemische Verbindungen.* Hoppe-Seyler's Zeitschr. 27:540, 1899.

52.   W. Gulewitsch and S. Amiradzibi. *Zur Kenntniss der Extraktivstoffe der Muskeln.* Hoppe-Seyler's Zeischr. 30:565, 1900.

53.   W. Gulewitsch. *Zur Kenntniss der Extraktivstoffe der Muskeln. VIII. Mitteilung. Ueber die Bildung des Histidins bei der Spaltung von Carnosin.* Hoppe-Seyler's Zeitschr. 50:535, 1907–1908.

54.   W. Gulewitsch and R. Krimberg. *Zur Kenntniss der Extraktivstoffe der Muskeln: II. Ueber das Carnitin.* Hoppe-Seyler's Zeitschr. 45:326, 1905.

55.   R. Krimberg. *Zur Kenntniss der Extraktivstoffe der Muskeln: IV. Ueber das Vorkommen des Carnosins, Carnitins, und Methylguanidins im Fleisch.* Hoppe-Seyler's Zeitschr. 48:412, 1906.

56.   S. A. Komarow. *Zur Frage nach dem Mechanismus der Darmsekretion.* Biochem Z. 147:221, 1924.

57.   R. Krimberg. *Zur Frage nach der Bedeutung der Muskelhormone im Sekretionsprozesse der Verdauungsdruesen.* Biochem. Z. 157:187, 1925.

58.   A. Bezkorovainy. *Carnosine, carnitine, and Vladimir Gulevich.* J. Chem. Ed. 51:652, 1974.

59.   A. Poehl. *Ueber Spermin.* Ber. 24:359, 1891.

60.   D. Kurajeff. *Ueber das Protamin aus den Spermatozoen der Makarele.* Hoppe-Seyler's Zeitschr. 26:524, 1898–1899.

61.   D. Kurajeff. *Ueber Einfuehrung von Jod in das krystallisierte Serumund Eieralbumin.* Hoppe-Seyler's Zeitschr. 26:462, 1898–1899.

62.   M. Nencki and N. Sieber. *Untersuchungen ueber den Blutfarbstoff.* Arch. F. Exper. Pathol. U. Pharmakol. 18:401, 1884.

63.   M. Nencki and J. Zaleski. *Untersuchungen ueber den Blutfarbstoff.* Hoppe-Seiler's Zeitschr. 30:384, 1900.

64.   D. Lawrow. *Quantitative Bestimmung der Bestandtheile des Oxyhaemoglobins des Pferdes.* Hoppe-Seiler's Zeitschr. 26:348, 1898–1899.

65.   H. Struve. *Beitrag zur gerichtlich-chemischen Untersuchung von blutverdaechtigen Flecken.* Virchow's Archiv 79:524, 1880.

66.   O. Schultzen and M. Nencki. *Ueber die Vorstufen des Harnstoffs im Organismus.* Ber. 2:566, 1869.

67.   S. Salaskin. *Ueber die Bildung von Harnstoff in der Leber der Saeugethiere aus Amidosaeuren der Fettreihe.* Hoppe-Seyler's Zeitschr. 25:128, 1898.

68. S. Salaskin and J. Zaleski. *Ueber den Einfluss der Leberextirpation auf den Stoffwechsel bei Hunden.* Hoppe-Seyler's Zeitschr. 29:517, 1900.

69. M. Hahn, O. Massen, M. Nencki, and I. Pawlow. *Die Eck'sche Fistel zwischen der unteren Hohlvene und der Pfortader und ihre Folgen fuer den Organismus.* Arch. F. exper. Pathol. U. Pharmakol. 32:161, 1893.

70. M. Nencki, J. P. Pawlow, and J. Zaleski. *Ueber Ammoniakgehalt des Blutes und der Organe und die Harnstoffbildung bei den Saeugethieren.* Arch. F. exper. Pathologie u. Pharmakol. 37:26, 1896.

71. S. Salaskin. *Ueber das Ammoniak in physiologischer und pathologischer Hinsicht und die Rolle der Leber im Stoffwechsel stickstoffhaltigen Substanzen.* Hoppe-Seyler's Zeitschr. 25:449, 1898.

72. M. Wolkow and E. Baumann. *Ueber das Wesen der Alkaptonurie.* Hoppe-Seyler's Zeitschr. 15:228, 1891.

73. N. Tichmeneff. *Ueber die Eiweissspeicherung in der Leber.* Biochem. Z. 59:326, 1914.

74. M. A. Boutlerow. *Formation synthetique d'une substance sucree.* Comp. Rend. 53:145, 1861.

75. L. Iwanoff. *Ueber die Synthese der phosphoorganischen Verbindungen in abgetoeteten Hefezellen.* Hoppe-Seyler's Zeitschr. 50:281, 1906–1907.

76. S. Kostytschew. *Ueber Alkoholgaerung. I. Mitteilung. Ueber die Bildung von Acetaldehyd bei der alkoholischen Zuckergaerung.* Hoppe-Seyler's Zeitschr. 79:130, 1912.

77. S. Kostytschew and E. Huebbenet. *Ueber Alkoholgaerung. II. Mitteilung. Ueber Bildung von Aethylalkohol aus Acetaldehyd durch lebende und getoetete Hefe.* Hoppe-Seyler's Zeitschr. 79:359, 1912.

78. M. Tschernorutzky. *Ueber die Zerlegung von Brenztraubensaeure durch tierische Organe.* Biochem. Zeitschr. 43:486, 1912.

79. W. Palladin. *Beteiligung der Reduktase im Prozesse der Alkohogaerung.* Hoppe-Seyler's Zeitschr. 56:81, 1908.

80. Baynes, J. W. and M. H. Dominiczak. *Medical Biochemistry.* Elsevier, Philaderlphia, 2005, p. 148.

81. A. v. Lebedew. *Ueber den Mechanismus der alkoholischen Gaerung.* Ber. 44:2932, 1911.

82. A. v. Lebedew and N. Griazunoff. *Ueber den Mechanismus der alkoholischen Gaerung.* Ber. 45:3256, 1912.

83. L. Iwanoff. *Ueber das Verhalten der Eiweissstoffe bei der alkoholischen Gaerung.* Hoppe-Seyler's Zeitschr. 42:464:1904.

84. T. Gromow and O. Grigoriew. *Die Arbeit der Zymase und der Endotryptase in den abgetoeteten Hefezellen unter verschiedenen Verhaeltnissen.* Hoppe-Seyler's Zeitschr. 42:299, 1904.

85. W. Palladin and N. Kostytschew. *Anaerobe Atmung, Alkohogaerung, und Acetonbildung bei den Samenpflanzen.* Hoppe-Seyler's Zeitschr. 48:214, 1906.

86. S. Kostytschew. *Ueber den Eifluss vergorener Zuckerloesungen auf die Atmung der Weizenkeime.* Biochem. Z. 23:137, 1910.

87. C. Diakonow. *Ueber die chemische Constitution des Lecithin.* Centralbl. F. d. med. Wissensch. 6:2, 97 and 434, 1868.

88. A. Juschtchenko. *Ueber die fettspaltenden und oxydierenden Fermente der Schilddruese und den Einfluss letzterer auf die lipolytischen und oxydierenden Prozesse im Blute.* Biochem. Z. 25:49, 1910.

89. V. Subbotin. *Beitraege zur Physiologie des Fettgewebes.* Zeitschr. F. Biol. 6:73.

90. A. Lebedeff. *Ueber die Ernaehrung mit Fett.* Hoppe-Seyler's Zeitschr. 6:139, 1882.

91. A. Lebedeff. *Woraus bildet sich das Fett in Faellen der akuten Fettbildung?* Pflueger's Archiv 31:11, 1883.

92. L. Popielski. *Ueber das peripherische reflectorische Nervencentrum des Pancreas.* Pflueger's Archiv 86:215, 1901.

93. L. Popielski. *Die Sekretionstaetigkeit der Bauchspeicheldruese unter dem Einfluss von Salzsaeure und Darmextrakt (des sogenannten Sekretins).* Pflueger's Archiv 120:451, 1907.

94. N. A. Bubnow. *Beitrag zur Untersuchung der chemischen Bestandtheile der Schilddruese des Menschen und des Rindes.* Hoppe-Seyler's Zeitschr. 8:1, 1883-1884.

95. J. A. Notkin. *Zur Schilddruesen-Physiologie.* Virchow's Archiv 144 (Suppl):224, 1896.

96. E. v. Cyon. *Beitraege zur Physiologie der Schilddruese und des Herzens.* Pflueger's Archiv 70:126, 1898.

97. E. v. Cyon and A. Ostwald. *Ueber die physiologischen Wirkungen einiger aus der Schilddruese gewonnener Produkte.* Pflueger's Archiv 83:199, 1901.

98. E. v. Cyon. *Die Verrichtungen der Hypophyse.* Pflueger's Archiv 71:431, 1898.

99. E. v. Cyon. *Die physiologische Herzgifte.* II. Theil. Pflueger's Archiv 73:339, 1898.

100. E. v. Cyon. *Ueber die physiologische Bestimmung der wirksamen Substanz der Nebennieren.* Pflueger's Archiv 72:370, 1898.

101. M. Gramenitzki. *Blut-und Harnzucker bei kontinuierenden Adrenalinfusion.* Biochem. Z. 46:186, 1912.

102. 102. M. Hahn. *Marcel v. Nencki.* Ber. (Vol. 4 Suppl.):289, 1902.

Illustrations

A. N. Bach (1857–1946), biochemist and Communist Party functionary. He is said to be the "father of Soviet biochemistry."

The St. Petersburg University in the 19th century.

352

Chapter V

CHEMISTRY

The Early Period

M odern chemistry based on theoretical principles came to Russia by edict through the establishment of the Academy of Sciences in 1725. Chemical knowledge on the practical level certainly existed in Russia prior to this time. In fact, well before the Tatar invasions in the thirteenth century, ancient Russians were familiar with the manufacture of enamel, of earth and plant pigments, of glass, of medicines from plants, and were expert in metallurgy involving the handling of gold, silver, lead, arsenic, iron, tin, copper, and antimony. Common chemicals such as sulfur, lime, acetic acid (vinegar), salts, acids, and alkalis were also known and used. Much of such practical knowledge apparently came from Byzantium, which had an active trade and spiritual ties with ancient Russia. The Tatar invasions of Russia destroyed most of its culture and arrested the normal exchange of ideas between it and the rest of the world for the next 200 years thereafter (1).

Practical chemical knowledge made a slow recovery in Russia following the emergence of Muscovy as the dominant Russian principality. Theoretical chemistry in the form of alchemy and iatrochemical concepts was absent from the Russian soil, because such ideas, being West European in origin, were apparently regarded as corrupting to the soul and on par with religious heresy. However, there is an account of a native alchemist who, in 1596, claimed to have been able to transform an "ore" to gold and silver. Having heard of this, the tsar summoned the man to Moscow for a demonstration, but the latter was unable to produce gold. Thereupon, the hapless alchemist was tortured and poisoned to death with his own chemicals (1). The chemists of the seventeenth-century Russia were chiefly physicians and apothecaries who had been summoned from abroad to provide medical care to the tsar's court and to those nobles who could afford it.

The druggeries connected with the Aptekarsky Prikaz (see Chapter II) were apparently the first chemical laboratories in Russia. These performed drug assays, and often did analyses for the military or the industry. The first formal chemical laboratory was established near Moscow in 1707 by Peter the Great, and its primary purpose was to analyze ores and industrial products. Subsequently, a chemical laboratory was established for military needs in St. Petersburg in 1716, and a large civilian laboratory was constructed in St. Petersburg in 1719–1725. The first research laboratory was established in 1748 by Lomonosov under the auspices of the Academy of Sciences (2).

Organization of the Academy of Sciences provided for a chair of chemistry, and the man invited to occupy it was Johann Georg Gmelin, Jr (1709–1755) of Tuebingen, Germany. He moved to St. Petersburg in 1730, but since no laboratory was available for research, he instead joined the Academy's expedition to Siberia (1733). He remained with the expedition for the next ten years, as a result of which he published a multivolume treatise on Siberian flora. He is thus better known as a botanist than a chemist. Gmelin returned to Tuebingen in 1747. It is interesting to note that Gmelin's nephew, Samuel Gottlieb Gmelin (1744–1774), was also a professor of chemistry in St. Petersburg, and like his uncle, was sent on an expedition to Russian frontiers. He was there captured by a nomad Tatar tribe and held for ransom. He died in captivity while negotiations for his release were taking place (3). Other early chemists of prominence associated with the St. Petersburg Academy of Sciences were the foreign-born Lehmann (1700–1767), Laxmann (1738–1796), and Lovitz (1757–1804); and the native Russians Lomonosov (1711–1765), Severgin (1765–1826), Musin-Pushkin (1760–1805), and Petrov (1761–1834) (4). Of these, Lomonosov was probably the most accomplished theoretician, whereas Lovitz excelled in practical laboratory work.

Mikhail V. Lomonosov is considered by many to be the father of Russian science (5). The son of a simple fisherman, he was able to obtain an education by pretending to be a nobleman. He was eventually sent to Marburg to study physics, chemistry, and metallurgy, and upon his return to Russia, he was admitted to the Academy of Sciences. Lomonosov's accomplishments were many and varied. He was a poet as well as a linguist who had written a treatise on Russian grammar; and a chemist who wrote on the structure of matter and questioned the existence of phlogiston well before the work of Lavoisier was published (6). He divided all matter into the so-called basic substances and complex substances (also called corpuscles). The latter were to consist of various combinations of the basic substances. In terms of modern terminology, basic substances may be equated to elements and corpuscles to compounds or molecules. This concept of structure of matter was incorporated into Lomonosov's book entitled *Principles of Mathematical Chemistry,* which was never finished nor published in his time. He was apparently also one of the first to propose what we today recognize as the law of conservation of mass or energy. In 1748, in a letter to Euler, Lomonosov expressed the belief that if a substance loses weight through a chemical reaction, another substance must then gain an equivalent amount of weight. Similarly, if one object causes another to move, the first loses a force equal to the force imparted upon the second object. Again, these thoughts were not published in a formal scientific journal of that time.

Among the chemists, Lomonosov is best known for his opposition to the phlogiston theory (7). He apparently published such criticism in 1750 in the St. Petersburg Academy proceedings, and it is believed that Lavoisier had seen the paper before performing his own famous experiments on combustion. Lomonosov's work was based on the experiments of Boyle who had previously reported that if lead was heated in a closed vessel, it acquired weight. The increase in weight was ascribed to an acquisition by lead of a "substance of fire." Boyle had also noted that air rushed into the vessel in which the lead had been heated, but Boyle was at a loss to explain this. Lomonosov repeated Boyle's experiment and noted that the vessel containing the burned lead did not actually gain weight, though the weight gain took place when the air had rushed into the vessel upon its opening. He proposed that the lead had combined with a substance from the air rather than with phlogiston or any "substance of fire." It is generally agreed, however, that Lomonosov's influence in this respect was not great outside of Russia.

Lavoisier's theory of combustion, published in 1773, did not find an extensive opposition in Russia, probably because Lomonosov had already laid the ground for it. One of the more prominent followers of Lavoisier in Russia was Vassily V. Petrov (1761–1834), who not only repeated Lavoisier's experiments but also made important additions to them. He attempted to burn several substances in the vacuum, or an inert atmosphere, and found in 1801 that only those substances were oxidized which contained endogenous oxygen such as sugars, wood, and cotton. Others were not affected in the absence of air. Petrov is also said to have been the first Russian electrochemist, and is credited with the discovery of the electric arc and construction of a powerful storage battery. Other important backers of Lavoisier in Russia were

A. N. Scherer (1771–1824), V. M. Severgin, and Y. D. Zakharov (1775–1836). Both Petrov and Scherer were professors at the Military-Medical Academy in St. Petersburg. There were, of course, adherents to the phlogiston theory in Russia, and among these the most prominent ones were Tobias Lovitz and Ferdinand Reuss (1778–1852). The latter was the first professor of chemistry at Moscow University, chemistry having been previously taught by an assortment of physicians and pharmacists. However, these men were apparently quite content with using the phlogiston theory in their research and lectures without mounting a crusade for its preservation as was done by Priestley. In addition to his efforts in the laboratory, Lomonosov was active in improving the educational opportunities for Russian youth, and many give him credit for promoting the establishment of the Moscow University. In the academy, he was apparently in constant conflict with its German "party," which resisted his efforts to improve academic opportunities for Russian population.

The most successful experimental chemist in the early period of the St. Petersburg Academy was probably Tobias Lovitz (8). He was born in Germany in 1757, and as a child of ten moved to St. Petersburg when his father joined the academy in 1767. He was educated in various apothecaries of St. Petersburg, then received formal education in Goettingen. He accompanied his father on one of the academy's expeditions to the Russian interior, where his father was captured by Pugachov's rebels and hanged as a government representative. This made such a strong impression on the young man that he subsequently seldom ventured beyond the city limits of St. Petersburg. In 1785, Lovitz discovered the decolorizing action of charcoal while trying to decolorize a tartaric acid preparation. The brown substance in Lovitz's tartaric acid was believed by him to be rich in phlogiston, whereas he believed that charcoal was poor in phlogiston. Thus, in his mind, mixing the brown tartaric acid with charcoal (to decolorize it) served to transfer the phlogiston to the charcoal! He applied the new method of decolorization to the purification of sugar, honey, acids, alcoholic beverages, oils, and even water for the army in the field. Lovitz was also the first to prepare glacial acetic acid from vinegar by first decolorizing it with charcoal, then freezing it. The melting point of acetic acid was determined to be $16°C$. He also noted that glacial acetic acid could, under certain conditions, be cooled to $3°C$ without solidifying. He induced crystallization in such solutions by seeding with crystals. Lovitz was also first to crystallize NaOH and KOH, and to prepare the hydrated alkali, $NaOH.2H_2O$. Working on freezing mixtures, he was able to prepare anhydrous $CaCl_2$, which, in turn, enabled him to prepare anhydrous ether and alcohol. His work on acetic acid also resulted in the preparation of trichloroacetic acid. It is said that Lovitz independently discovered chromium and strontium, but was beaten to publication of his results by Klaproth and Vanquelin. Lovitz died in 1804 as a result of a laboratory accident. His assistant, C. G. S. Kirchhoff (1764–1833) discovered the hydrolysis of starch in 1811 and the action of diastase in 1814.

Practical chemistry in Russia showed a growth parallel to that of laboratory and theoretical chemistry in the eighteenth and beginning of the nineteenth centuries. This may well be characterized by the developments in metallurgy, brought about in part by the discovery of mineral deposits in the Ural Mountains (9). An early investigator of Russia's minerals and ores was A. A. Musin-Pushkin, who was a government administrator and a founder of the mining institute in St. Petersburg. He was first to prepare the transparent colorless phosphorus, and discovered sodium tungstate and chrome alum. He started his research activity by working on platinum ores, and was able in 1805 to report the production of malleable platinum metal. This was done by preparing an amalgam from mercury and platinum sponge, then heating the amalgam until the mercury had evaporated, burning off the wooden molds, and finally, heating the platinum to white heat for about 2 hours. After the discovery of large platinum deposits in the Ural Mountains in 1824, the process of rendering the metal malleable was simplified by the royal assayer V. V. Lyubarsky and the head of the mining institute, P. G. Sobolevsky (1781–1841). The platinum ore was first dissolved in *aqua regia*, then the platinum was precipitated as chloroplatinate with NH_4Cl. The chloroplatinate was calcinated (heated) to produce platinum sponge, which was compressed in a mould by a screw press, heated to white heat, and again compressed while still hot. The process was later modified by omitting the second pressing operation. Both Lyubarsky and Sobolevsky were decorated for their discovery by Tsar Nicholas I. In addition to platinum, osmiridium was also discovered in the Ural Mountains. Lyubarsky reported that the ore contained 60% iridium, 30% osmium, 5% iron, 2% platinum, and 0.7% gold. There were at first few uses for the large quantity of platinum mined in Russia, but the problem was solved when it was decided to use the metal for coinage. The production of platinum coins continued in Russia from 1828 to 1845, when a total of 1.4 million such coins were issued in three denominations, using a total of 14.8 tons of platinum. Eventually, as uses other than the production of jewelry and coins became known for platinum (e. g., in various alloys), the minting of platinum coins was discontinued.

The role of the Academy of Sciences in chemical research probably reached its zenith in the early nineteenth century through the activities of men like Fritzsche, the organic chemist, and G. H. Hess, the physical chemist (10). Hess (1802–1850) was born in Switzerland, but at the age of three was taken to Russia, where he remained for the rest of his life. In 1825, he graduated from Dorpat University's medical school, then studied for a short while with Berzelius before going to Irkutsk in Siberia to set up his medical practice. In his spare time, Hess investigated the geology of that region, and was honored for his discovery of various mineral deposits by being appointed member of the Academy of Sciences. He thereupon abandoned his medical practice, and in 1830 moved to the Academy's laboratories in St. Petersburg to devote the rest of his life to research and public service. In addition to his research activities at the academy, Hess participated in the teaching of chemistry at a number of institutions in St. Petersburg, most notably at the newly organized Technological, Mining, and Pedagogical Institutes. At the latter, he was the chemistry instructor for young V. V. Voskresensky, who later became an important force in the popularization of chemistry in Russia. Hess was also chosen to tutor Crown Prince Alexander (later Tsar Alexander II) in chemistry, who upon Hess's suggestion visited Berzelius' laboratory during his tour of Sweden in 1838. Hess completed the work on Russian chemical nomenclature in the 1830s, which had been initiated earlier by Scherer, Severgin, and Zakharov.

Hess' main contribution to chemistry is his thermochemical law, or the so-called law of heat summation: the change in heat content experienced by transforming one substance into another is constant regardless of the path taken. He established this fact by calorimetric measurements of heat evolved (as a function of rise in temperature) when different proportions of water and H_2SO_4 and alkalis were mixed. This principle was published in 1840 in the St. Petersburg Academy's journal some two years before the law of conservation of energy was enunciated. Abbreviated accounts of Hess' work also appeared in *Annalen der Chemie und Pharmacie* (vols. 31, 36, 40, and 52) and in *Poggendorf's Annalen* (11). As is well known, Hess' law is the foundation for the concept of the heat of formation, and probably served as an important experimental criterion for the formulation of the first law of thermodynamics. Other contributions of Hess included the determination of cobalt oxide structure, Co_3O_4, and the discovery of the law of thermoneutrality, which states that no heat change occurs when salts exchange their ionic components.

Of the early organic chemists associated with the academy's laboratories, Carl Julius Fritzsche (1808–1871) was probably its most distinguished member (12). He was born in Germany, had his training in pharmacy, and took his doctorate in botany in Berlin in 1833. He thereupon moved to Russia and spent his entire life at its Academy of Sciences. He never participated in teaching, since he did not wish to be distracted from his research activities. He published mostly in the academy's publication, though some accounts of his work did appear abroad in an abbreviated form. Fritzsche is credited with the rediscovery and naming of aniline, the discovery of nitrophenol isomers, and the discovery of chrysanilic and anthranilic acids.

A great portion of Fritzsche's work was concerned with the degradation of products of indigo. When this compound was heated with concentrated bases such as KOH and NaOH, an oily product was obtained which Fritzsche called aniline (13). It boiled at 228° C, and its formula was deduced from elemental analysis to be $C_{12}H_{14}N_2$ (C_6H_7N, using modern values for the atomic weights of C, H, and N). He also prepared oxalic acid anilide and the hydrochloride of aniline. Though aniline had been previously prepared by Erdmann in 1826 and was known under the name of *crystallin,* extensive investigations of its properties and potential uses did not begin until Fritzsche's work was published. When indigo was heated to $150^{\circ}C$ with KOH, its blue color disappeared and was replaced by an orange coloration. Upon cooling, a copious brown crystalline precipitate appeared, whose main component was the potassium salt of an organic acid (14). Fritzsche named this compound *crysanilic acid.* Its lead salt had the empirical formula of $C_{28}H_{20}N_5O_5.PbO$ ($C_{14}H_{10}N_2O_{2-3}$. PbO). He had some difficulty reproducing his elemental analyses. The true structure of crysanilic acid was not elucidated until 1910, when Friedlaender published his interesting findings on this compound. When crysanilic acid was hydrolyzed by mineral acids, a new compound and another new acid were obtained by Fritzsche. He called the latter *anthranilic acid*. Its melting point was 135°C, and its empirical formula was $C_{14}H_{12}N_2O_3$ ($C_7H_6N_{1-2}$), which was only slightly off the true empirical formula of anthranilic acid ($C_7H_7NO_2$). The acid was also isolated in its hydrated form, and when it was heated beyond its melting point, it decomposed into aniline and CO_2. Some seventy years later, Friedlaender and Schwenk finally elucidated the connection between indigo blue, crysanilic acid, and anthranilic acid (15). They found that alkali actually splits indigo into anthranilic acid and an aldehyde, whereupon these two compounds condensed to give crysanilic acid. The latter could then be hydrolyzed by acid to give back the anthranilic acid:

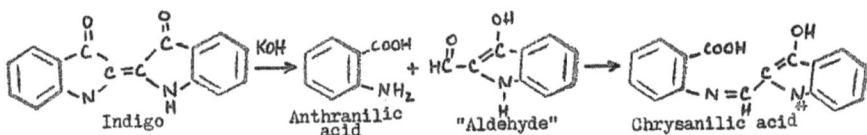

As early as in 1839, Fritzsche published a paper in the St. Petersburg Academy's journal dealing with the preparation of nitrophenol by treating indigo with nitric acid. He later synthesized the compounds by nitrating phenol with fuming nitric acid (16). The product had a melting point of 45°C, a boiling point of 214°C, and apparently represented o-nitrophenol. A by-product, which was a dinitrophenol, was obtained by steam-distilling the residue. The residue remaining after the distillation of dinitrophenol was another nitrophenol; Fritzsche called it isonitrophenol, and it apparently was p-nitrophenol. Fritzsche is also the discoverer of the ability of polynitro compounds to form complexes with aromatic substances (17). Using picric acid as the polynitro compound, he prepared the benzene-picrate complex containing the two reactants in a 1:1 ratio and having a melting point of 85 to 90°C. It was decomposed to its constituents by water. A similar complex was made with naphthalene, whose melting point was 149°C and which was also decomposed by water. Fritzsche prepared two other such complexes: one apparently from anthracene that he had isolated from coal tar distillate and whose picric acid complex melted at 170°C, and another from a compound he had isolated from wood tar and which formed a picric acid complex melting at 125°C (18).

Chemistry at Russia's Universities: The University of Kazan

With the emergence of the universities as a significant force in the cultural mainstream of Russia, the importance of the Academy of Sciences as the primary center of chemical research was diminished. For a number of reasons, the "provincial" Kazan University (est. 1804–1805) assumed the leadership position in both the training of chemists and the quality and quantity of chemical research produced in Russia. Because of Kazan's chemistry faculty and their initiatives, it was possible for Nathan M. Brooks of New Mexico State University to write the following statement almost two hundred years later in an American Chemical Society journal (19): "During the nineteenth century and up to 1917, Russian chemists produced a significant number of 'cutting-edge' advances in all branches of chemistry. Indeed, one could plausibly argue that—considering the size of the chemical community—Russian chemists were among the most productive chemists at that time." Dr. Brooks then proceeded to tell the story of Nikolai Zinin, one of Kazan's early chemistry professors, whom we will mention again further below.

A major reason why Kazan became a magnet for talented chemistry students and superior teaching and research professors was probably the presence of Karl Karlovich Klaus (1796–1864), discoverer of the ruthenium element (20). He was born of Baltic Germanic parents in Dorpat, in the Russian Empire at that time, and today in Estonia. His parents died in his early childhood, and in 1811 he left Dorpat for St. Petersburg to become an apprentice with a pharmacist. After he became a licensed pharmacist, he moved to Kazan (on the Volga River) in 1826 to open his own pharmacy business there. In 1827, Klaus took part in an expedition to explore medicinal plants in the Volga region, and a year later, he participated in another expedition to examine Ural Mountain mineral deposits, as a result of which he became interested in learning more about the platinum metals that were being mined in the Urals. He eventually sold his pharmacy business and joined the Dorpat University to study chemistry. There, he came under the influence of G. F. Osann (1797–1866), a Dorpat professor of chemistry. Osann was an expert on platinum family metals, and even claimed that he had discovered three new metallic platinumlike elements from Ural Mountain mines, one of which he called *ruthenium* (Russia). However, this turned out to be an incorrect conclusion. In 1837, Klaus was offered a professorial position in the pharmacy department of Kazan University and he accepted it. As an aside, the Russians did not know what to do with all the platinum they produced from the Ural Mountain ores. As indicated above, it was decided to mint coins from it, and in 1820s and 1830s, platinum coins were circulating in Russia.

At Kazan, Klaus' original research project was to devise a procedure (based on Osann's methodologies) for the extraction of residual platinum, osmium, iridium, palladium, and rhodium from platinum ore residue. As a result, he isolated a pure new metal that he called *ruthenium* (21, 22). He took that portion of platinum ore that was insoluble in aqua regia (the so-called osmiridium fraction) and heated it with potash and KNO_3 to bright redness. This was then poured into water and allowed to stand for 4 days. The resulting solution was treated with HNO_3, thus precipitating OsO_2 and RuO_2. This was distilled with aqua regia, whereby OsO_2 was obtained from the condensate. The residue contained the chlorides of ruthenium. To this, Klaus added NH_4Cl and precipitated ammonium chlororuthinate, $(NH_4)_2RuCl_6$. Upon calcination, spongy ruthenium was obtained. Klaus named his newly discovered metal *ruthenium* in honor of Russia (his native land) and Osann, his erstwhile mentor at Dorpat. He first reported on his discovery in the *Scientific Notes of Kazan University,* a local science journal, in 1844, and later in other journals. Klaus returned to live in Dorpat in 1852 for family reasons. He was offered, and he accepted, Dorpat's chair of the pharmacy department. In 1861, he was appointed member of the St. Petersburg Academy of Sciences. He passed on in 1864 after a short pneumonialike lung illness.

Klaus was followed at Kazan University by a large number of talented professors and students with names of A to Z (Arbuzov to Zaitsev), who made advances in all branches of chemistry, especially organic chemistry. David D. Lewis writes that "Kazan University had, by the end of the nineteenth century, established a school of chemistry that was preeminent in Russia, supplying many department chairs and professors of chemistry to Russian and foreign universities" (23). Their accomplishments are described below. But chemistry was, of course, not the only branch of the sciences that were advanced by Kazan's eminent professors. One such person was Nicholas I. Lobachevsky (1793–1856), a professor of mathematics and inventor of the non-Euclidian geometry (24). He was associated with Kazan University since its founding in the years of 1804–1805, and became a professor of mathematics there in 1816. His faculty honored him in 1827 by electing him their "rector," i. e., university president. He served in this capacity for nineteen years.

The second chemist to follow Klaus' footsteps in making Kazan a powerhouse of chemical research was Nikolai N. Zinin (1812–1880), who entered Kazan U. in 1830 (19, 25). He graduated in 1833 and received the magister degree in 1836. He then went abroad to work with European chemists, including Liebig, returned to Russia in 1840, and a year later, he passed his dissertation exam for the doctorate degree. He then joined the faculty of Kazan University as professor of technology, where he remained until 1847. In that year, he joined the University of St. Petersburg as professor of chemistry and physics. Zinin's research work involved primarily aromatic compounds, his most famous work being the synthesis of aniline and naphthylamine from the corresponding nitro compounds (26). He treated nitronaphthalene with H_2S or $(NH_4)_2S$ in alcoholic solution, and upon the removal of alcohol, obtained a crystalline product that he called "naphthalidam." The product was very basic, forming salts with organic as well as mineral acids, it melted at 50- and boiled at 300°C, and had the empirical formula of $C_{20}H_{18}N_2$ ($C_{10}H_9N$). He characterized both the sulfate and hydrochloride salts of the naphthylamine. Likewise, the reduction of nitrobenzene yielded a base that was crystallized as a hydrosulfide. The removal of sulfide yielded a water-insoluble basic oil with a boiling point of 200°C. It formed salts with various acids, of which the sulfate and hydrochloride were extensively characterized. Its empirical formula was $C_{12}H_{14}N_2$ (C_6H_7N), and Zinin called it "benzidam." Fritzsche, however, immediately recognized it as aniline which he had prepared at about the same time from indigo and appended Zinin's paper with a statement to that effect. Zinin did this work while he was in Kazan, and apparently had no idea that it could be extremely helpful for the synthesis of various dyes. A. W. von Hoffman, who was working in London on synthesizing dyes at that time, stated in 1880 that "if Zinin had done nothing more than to convert nitrobenzene to aniline, even then his name should be inscribed in gold letters in the history of chemistry" (19).

Another area of Zinin's interest was the chemistry of benzyl and its derivatives. He treated benzyl, $C_{14}H_{10}O_2$, with PCl_5 and obtained a product with a formula of $C_{14}H_{10}OCl_2$, which he called "chlorobenzyl" (27). It melted at $71°C$, and reacted with KOH to give potassium benzoate and what was most likely benzaldehyde. He also described conditions for the reduction of benzyl to benzoin with zinc and HCl (28), and found that the reaction proceeded beyond benzoin if the conditions were made more drastic. He isolated a new compound whose melting point was $45°C$, and which supposedly contained more hydrogen than did benzoin. In a later paper, however, Zinin could not confirm the increase in hydrogen, and determined the empirical formula of the new compound to be $C_{14}H_{12}O$ (29). He called this compound desoxybenzoin, and its structure is today known to be as follows:

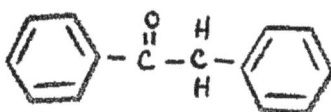

By the middle of the nineteenth century, chemical research at the Russian universities thus came of age and retained its dominant position until after the Bolshevik coup. The dissemination of chemical research in Russia and the interaction of Russian chemists was greatly aided by the establishment of the Russian Chemical Society in 1864, and the *Journal of the Russian Chemical Society* in 1869. The latter publication became second to none in quality. It should be noted that this journal was not the first chemical publication in Russia. This honor belongs to the *Khimicheskii Zhurnal Sokolova I Engel'kharta* (Chemical Journal of Sokolov and Engelhart), which was issued in

1859 and 1860, and appeared in the form of 24 issues grouped into 4 volumes.

The years of 1830–1870 were, according to Leicester (4), a period of the most rapid growth of chemistry in Russia. After 1870, "reaction" supposedly set in, and the quality of science teaching and research allegedly declined. In fact, Leicester notes, there was a decline in the number of first-rate chemists at the turn of the century. Vucinich (30) and his reviewer, David Joravsky (31), are apparently in agreement. Yet, their verdict is difficult to accept if one takes a look at the late nineteenth and early twentieth-century issues of *Berichte* or the *Journal of the Russian Chemical Society*. The contributions of such individuals as Wagner, Zelinsky, Walden, Chugaev, Chichibabin, Kurnakov, Ipatieff (see below) and others are anything but of second-rate quality. It is possible that instead of declining, chemistry of this period experienced a transition from the scientific empire system that was controlled by a few luminaries such as Markovnikov and Butlerov to a more democratic arrangement, where a larger number of workers found the opportunity to do independent research. Credit for discoveries in the field of chemistry during that period was thus diluted to a larger number of individuals than was the case in mid-nineteenth century. In addition, chemically-oriented research was being done in other scientific fields not traditionally associated with chemistry. For instance, the field of plant pigment chemistry was enjoying a golden era in Russia at the beginning of the twentieth century, and this is commonly associated with botany. And then the emerging field of biological chemistry was of great interest to aspiring chemists, who practiced their trade in the departments of microbiology, botany, forestry, and the medical schools. As an example, see the description of Michael Tsvett's story above, in chapter IV.

It was a pleasure to see Dr. Leicester's mention of superior chemistry teaching in Russia in the 1830–1870 era. Not much is generally written in science history publications about the quality of science teaching in colleges and universities. Success in research is the Great Master. This author will, however, end the story of this section by mentioning a Russian chemist of the 1830–1870 era who was a principally a teacher and had such research luminaries as Mendeleyev as students. His name was A. A. Voskresensky (1809–1880), a graduate of the St. Petersburg Pedagogical Institute (1836) and a student of Hess (32). He devoted his entire professional life to teaching and administration, except for a short period in Liebig's laboratory and a year or so of research in Russia. Yet during his short-lived stay in the research laboratory, he managed to isolate quinic acid from cinchona bark and synthesize quinine therefrom (1838). He was also able to isolate a new alkaloid from the cocoa bean, which he called *theobromine* (1841). It is always a good idea for aspiring teachers to first accomplish something in the research laboratory, and then undertake the much harder (for some) job of teaching chemistry. Success in research increases the teacher's authority among his/her students and provides them with a model that they may want to emulate. Voskresensky's teaching activities were carried out principally at the Pedagogical Institute and the St. Petersburg University.

Theoretical Organic Chemistry

Significant discoveries toward the elucidation of the nature of organic compounds are generally ascribed to C. F. Gerhardt (1816–1856), and later to F. A. Kekule (1829–1892). It is only recently that the contributions of the Russian chemist Alexander M. Butlerov (1828–1886) have been brought to light (32), and that only through the now well-known attack of the Soviet establishment on the resonance theory during the last years of Stalin's regime. Butlerov was a student of Klaus at the Kazan University, and upon graduation in 1849, remained there as an assistant while working on his doctoral dissertation. His thesis was not accepted at Kazan, but Butlerov was luckier at Moscow University, which conferred the degree upon him in 1854. Thereupon, he became a professor of chemistry at Kazan. In 1857 and 1858, he visited the laboratories of Mitscherlich, Kekule, and Wurtz in Western Europe, and upon his return to Kazan, he initiated research on one-carbon compounds. As a result, he made important discoveries in the field of formaldehyde chemistry and was able to prepare a synthetic sugar (see Chapter IV). In 1861, he again visited Western Europe, where he presented his now-classical paper on chemical structure at a naturalists' convention in Spyer, Germany. He traveled again to Europe in 1867 and 1868, and upon his return to Russia was appointed to the chair of chemistry at St. Petersburg University. Butlerov was a member of the St. Petersburg Academy of Sciences, and was an honorary member of a number of foreign chemical societies, including the American Chemical Society, to which he was elected in 1876.

Butlerov's most important work was presented at the Spyer meeting and was subsequently published in Germany (33). He noted that a given organic compound can undergo many types of chemical reactions, and since formulas previously proposed to represent such substances (e. g., those of Kekule) were based on the functional status of such compounds in a given reaction, then it would follow that such organic compounds could have several "rational" formulas depending on the number of chemical reactions that they could enter. He rejected the idea of such a multiplicity of formulae for a single organic compound, and proposed that each compound must have a single formula, which he termed "chemical structure." Such a structure would show the relationships among all atoms present in the compound, the valences of all constituent atoms, and the effects of the constituent atoms upon each other. Such a definite chemical structure should predict, in Butlerov's view, both the chemical and physical properties of the compound in question. The chemical structures of compounds were to be determined through both the degradative and synthetic reactions involving the compound of interest. However, the claim to the effect that Butlerov advanced the concept of the tetravalency of the carbon atom (34) cannot be substantiated, since Butlerov himself freely acknowledged the priority of Kekule and Couper in this respect (35).

As an illustration of the significance of his structural theory of organic compounds, Butlerov presented the problem of isomerism, more specifically, the properties of the various butanes, butenes, and butyl alcohols, including trimethyl carbinol, which Butlerov was the first to prepare. The latter served as a starting compound for the preparation of a number of other substances used by Butlerov to illustrate his concept of chemical structure. Trimethyl carbinol was prepared by the action of acetyl chloride on dimethyl zinc (30, 36, 37) as follows:

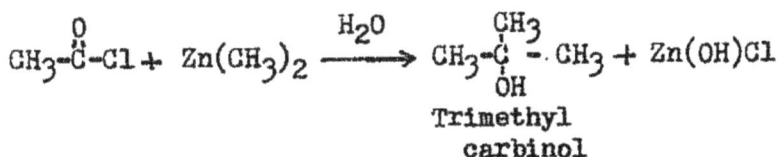

$$CH_3\text{-}\overset{O}{\overset{\|}{C}}\text{-}Cl + Zn(CH_3)_2 \xrightarrow{H_2O} CH_3\text{-}\overset{CH_3}{\underset{OH}{C}}\text{-}CH_3 + Zn(OH)Cl$$

Trimethyl
carbinol

The trimethyl carbinol served as a starting material for the preparation of tert-butyl iodide, which, in turn, served to prepare isobutene:

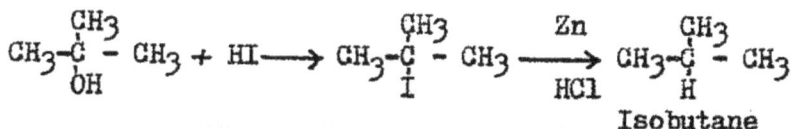

$$CH_3\text{-}\overset{CH_3}{\underset{OH}{C}}\text{-}CH_3 + HI \longrightarrow CH_3\text{-}\overset{CH_3}{\underset{I}{C}}\text{-}CH_3 \xrightarrow[HCl]{Zn} CH_3\text{-}\overset{CH_3}{\underset{H}{C}}\text{-}CH_3$$

Isobutane

The isobutene was a colorless gas that burned with a white flame and was more volatile than n-butane that had previously been prepared by Frankland. Butlerov also noted that the trimethyl carbinol was more volatile than n-butanol. He was thus pointing out physical differences among compounds of identical empirical formulae.

Another set of isomers studied by Butlerov were the butylene types, C_4H_8. The butylenes obtainable from n-butanol and sec-butyl alcohol were already known, and Butlerov prepared a third type from trimethyl carbinol by first converting it to isobutyl iodide, then treating the iodide with alcoholic KOH. The product, isobutylene, $CH_2=C(CH_3)_2$, was a gas liquefiable at a pressure of 2.5 atm. At 15–18°C, with a boiling point of -7 to -8°C at atmospheric pressure. It reacted with HI to give tert-butyl iodide, and reacted with hypochlorous acid to give a chlorohydrin. The latter gave trimethyl carbinol upon reduction with sodium amalgam. Butlerov thus described the isomers of butane, butyl alchohol, and butylene, all compounds in each group having identical empirical formulae. The differences in the physical and chemical properties of the compounds within each group could, according to Butlerov, be understood only in terms of his chemical structural theory.

Butlerov's work on formaldenyde has been summarized by Walker (38). Butlerov had attempted to prepare methylene glycol, but prepared formaldehyde and various polymers thereof instead as follows:

$$I-\overset{H}{\underset{H}{C}}-I \; + \; 2 \; Ag\text{-acetate} \longrightarrow CH_3-\overset{O}{\overset{\|}{C}}-O-\overset{H}{\underset{H}{C}}-O-\overset{O}{\overset{\|}{C}}-CH_3 \; \xrightarrow{H_2O} \; \text{Formaldehyde} + \text{acetic acid}$$

Methylene acetate

$$I-\overset{H}{\underset{H}{C}}-I \; + \; Ag\text{-oxalate} \longrightarrow (CH_2O)_n \quad (\text{paraformaldehyde})$$

$$I-\overset{H}{\underset{H}{C}}-I + Ag_2O + NH_3 \longrightarrow \text{Hexamethylenetetramine}$$

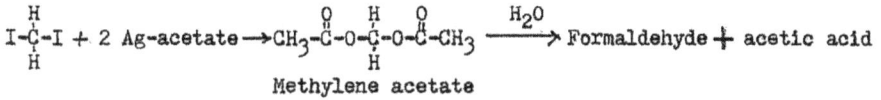

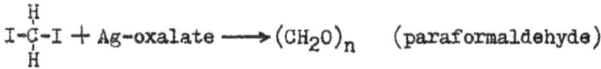

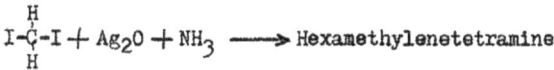

Butlerov was thus one of the first to prepare paraformaldehyde and hexamethylenetetramine. He did not actually isolate formaldehyde from the above reaction system, though he did note its peculiar odor and recognized that he had prepared a then-new substance.

The discovery of tautomerism is also credited to Butlerov's laboratory, though Butlerov apparently did not realize the importance of this phenomenon (39). His main interest in connection with this project was to study the properties of isobutylene, which was prepared from trimethyl carbinol and used to prepare isodibutylene as follows:

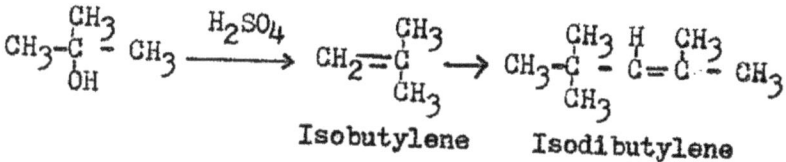

$$CH_3-\overset{CH_3}{\underset{OH}{C}}-CH_3 \; \xrightarrow{H_2SO_4} \; CH_2=\overset{CH_3}{\underset{CH_3}{C}} \longrightarrow CH_3-\overset{CH_3}{\underset{CH_3}{C}}-\overset{H}{C}=\overset{CH_3}{C}-CH_3$$

Isobutylene   Isodibutylene

The new olefin, isodibutylene, could be oxidized by potassium dichromate and sulfuric acid to trimethyl acetic acid and acetone on the one hand, and to a lesser extent, to an octanoic acid with a boiling point of 215°C, and a ketone with a boiling point of 125–130°C on the other. The structures of the latter two products were as follows:

$$CH_3-\underset{\underset{CH_3}{|}}{\overset{\overset{CH_3}{|}}{C}} - CH_2-\underset{\underset{H}{|}}{\overset{\overset{CH_3}{|}}{C}} - COOH$$

$$CH_3-\underset{\underset{CH_3}{|}}{\overset{\overset{CH_3}{|}}{C}} - CH_2-\overset{\overset{CH_3}{|}}{C} = O$$

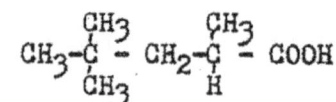

"Octanoic acid"   **"Ketone"**

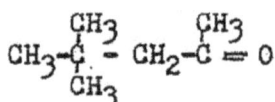

The formation of the octanoic acid was explained by Butlerov by the following set of reactions:

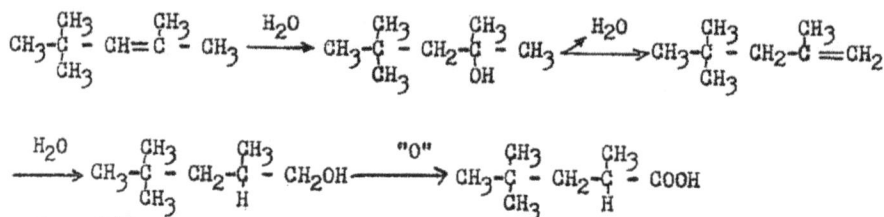

$$CH_3-\underset{\underset{CH_3}{|}}{\overset{\overset{CH_3}{|}}{C}} - CH = \overset{\overset{CH_3}{|}}{C} - CH_3 \xrightarrow{H_2O} CH_3-\underset{\underset{CH_3}{|}}{\overset{\overset{CH_3}{|}}{C}} - CH_2-\underset{\underset{OH}{|}}{\overset{\overset{CH_3}{|}}{C}} - CH_3 \xrightarrow{H_2O} CH_3-\underset{\underset{CH_3}{|}}{\overset{\overset{CH_3}{|}}{C}} - CH_2-\overset{\overset{CH_3}{|}}{C} = CH_2$$

$$\xrightarrow{H_2O} CH_3-\underset{\underset{CH_3}{|}}{\overset{\overset{CH_3}{|}}{C}} - CH_2-\underset{\underset{H}{|}}{\overset{\overset{CH_3}{|}}{C}} - CH_2OH \xrightarrow{"O"} CH_3-\underset{\underset{CH_3}{|}}{\overset{\overset{CH_3}{|}}{C}} - CH_2-\underset{\underset{H}{|}}{\overset{\overset{CH_3}{|}}{C}} - COOH$$

The first and third compounds in the above series are today known as tautomeric forms of the same compound; however, as pointed out above, Butlerov did not realize the significance of this discovery. The second compound, a tertiary alcohol, was prepared by Butlerov by treating isodibutylene with HI, then treating the latter with silver oxide.

Butlerov's laboratory was also responsible for the discovery of a novel method for the synthesis of alcohols from iodohydrin and zinc alkylates (40):

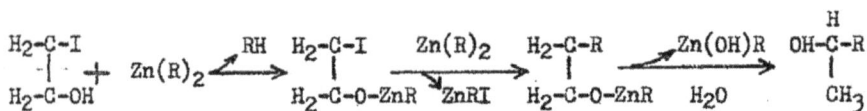

$$\underset{H_2-C-OH}{\overset{H_2-C-I}{|}} + Zn(R)_2 \xrightarrow{RH} \underset{H_2-C-O-ZnR}{\overset{H_2-C-I}{|}} \xrightarrow[ZnRI]{Zn(R)_2} \underset{H_2-C-O-ZnR}{\overset{H_2-C-R}{|}} \xrightarrow[H_2O]{Zn(OH)R} \underset{CH_3}{\overset{OH-C-R}{|}}$$

For structure proofing the alcohols, an oxidation with potassium chromate was performed, whereby isopropyl and isobutyl alcohols gave acetic acid rather than propionic and butyric acids as would be expected from n-alcohols.

The method for structure proofing of alcohols and ketones by their oxidation to acids with chromic acid was developed by Alexander N. Popov (1840–1881), a Butlerov associate at Kazan who later became a professor at the Warsaw University. Popov prepared methylamylketone by two methods: first by using caproyl chloride and dimethyl zinc, and by using acetyl chloride with diamyl zinc. Oxidation of both preparations with potassium chromate and sulfuric acid yielded a mixture of valeric and acetic acids:

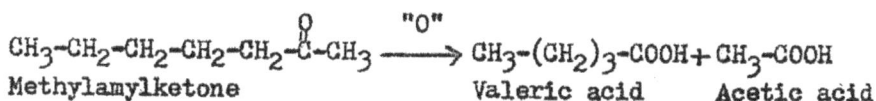

$$CH_3-CH_2-CH_2-CH_2-CH_2-\overset{O}{\overset{\|}{C}}-CH_3 \xrightarrow{\text{"O"}} CH_3-(CH_2)_3-COOH + CH_3-COOH$$

Methylamylketone   Valeric acid   Acetic acid

In a similar fashion, methylethylketone was prepared from propionyl chloride and dimethyl zinc, and from acetyl chloride and diethyl zinc. Oxidation of the ketone prepared by both pathways yielded acetic acid only and no formic acid. Popov was also first to synthesize ethyldimethylcarbinol, an isomer of amyl alcohol (41, 42). He used propionyl chloride and dimethyl zinc for this purpose. Oxidation of this compound by chromic acid yielded acetic acid only. Popov's method of oxidation with chromic acid thus served as a useful tool in the chemist's laboratory for many subsequent years. Butlerov has been called the best organic chemist of Russia. With his introduction of dialkyl zinc into synthesis of organic compounds (often called the Butlerov reaction), he basically initiated the formation of carbon-to-carbon bonds in the area of organic compound synthesis (e. g., tertiary alcohols) (23). Today, dialkyl zinc compounds have been replaced by alkyl magnesium compounds in the so-called Grignard reactions.

The man who is believed to have been the first to recognize the influence of neighboring atoms on the reactivities of functional groups in organic compounds was Vladimir Markovnikov (43), who is best known to students of organic chemistry for the famous Markovnikov rule. He was born in Nizhnii Novgorod (Gorkii was its Soviet name) (1838–1904), studied finance and economics at the Kazan University, but later switched to chemistry through the influence of Butlerov. His first degree was in economics (1860), but his magister degree was in chemistry (1865). Markovnikov then worked abroad with Erlenmyer, Bayer, and Kolbe in Germany, and returned to Kazan in 1867. His doctorate was awarded him in 1869. He succeeded Butlerov as the professor of chemistry at Kazan when the latter moved to St. Petersburg, but shortly thereafter he moved to Odessa, and then to Moscow in 1873. In 1893 he was supposedly dismissed from his position as professor for supporting student rioters, though he was permitted to stay at the university to continue his research.

Markovnikov's main contribution to chemistry was his work on the substitution of hydrogen atoms in organic compounds by other atoms; however, his work went unnoticed until 1899 when Mayer recognized its importance. Leicester states (43) that this happened because Markovnikov was a Russian supernationalist and refused to publish his work in anything but Russian journals. Hence, few non-Russian scientists supposedly had access to his papers. In addition to his work on the fundamentals of organic reactions, Markovnikov was also concerned with the properties of Russian crude oil, being aided financially by Russian petroleum industry. He was able to isolate a number of compounds from petroleum, and was one of the first to isolate and identify cycloheptane from this source. He advocated a strong relationship between industry and the academe, and urged the establishment of research laboratories in the relevant industrial organizations. He placed many of his students in industrial positions.

The Markovnikov rule, which states that a substitution on an organic compound will take place at that carbon with the fewest number of hydrogen atoms, was formulated after an extensive investigation of reactions of various types of organic compounds with halogens, halides, and nitrates (44, 45, 46). He found that halides could be introduced with the greatest ease into tertiary carbon atoms, such as trimethylcarbinol, which had previously been prepared by Butlerov, and with greatest difficulty into primary carbon atoms. Halogenation of carbons already containing halide groups (e. g., chloroethanol) took place on that carbon atom that was already bonded to a halide atom. Nitration of hydrocarbons with nitric acid showed similar behavior: tertiary hydrogens were most labile, followed by secondary ones, and ending with the primary hydrogens. The addition of halides to olefins followed the same rule, where the halide atom was added to that carbon containing the smallest number of hydrogen atoms. Thus, the addition of hydrogen iodide to propylene resulted in the formation of 2-iodopropane. The addition of hydrogen iodide to isobutylene gave tert-iodoisobutane. The addition of hydrogen bromide to vinyl bromide produced 1,1-dibromoethane, and the addition of hydrogen iodide to 2-chloropropylene resulted in the formation of unsaturated compounds, where the atoms abstracted originated from neighboring carbon atoms rather than from the same carbon.

Markovnikov's discovery of the effects of neighboring groups on the reactivity of hydrogen atoms was based on the halogenation of the butyric acids. He found that the reaction between bromine and butyric acid resulted in the formation of alpha-bromobutyric acid. He then proposed that the introduction of the halide atom into a saturated hydrocarbon occurs on that carbon atom which is most under the influence of a carbon atom containing oxygen. In the process of his studies on the rules of substitution of one radical by another, Markovnikov had the opportunity to prepare several new derivatives of butyric acid. The alpha-hydroxyisobutyric acid was prepared by reacting acetone with HCN, then hydrolyzing the resulting nitrile. The hydroxyl group could be substituted by a bromide atom to yield alpha-bromoisobutyric acid, the same compound as that obtained by bromination of isobutyric acid. The alpha-hydroxybutyric acid was prepared by hydrolyzing the alpha-bromobutyric acid. Beta-hydroxybutyric acid was synthesized from propylene chlorohydrin and KCN, followed by hydrolysis of the resulting nitrile. It was prepared at the same time by Wislicenus from acetoacetic acid.

Markovnikov's work on both the addition of acid halides to olefins and the rules governing the dehydrohalogenation was continued by his Kazan associate, A. M. Zaitsev (Saytseff) (1841–1910), who posed basically two questions (47): when the elements of an acid halide are removed from a halogenated alkane, which neighboring carbon atom loses the hydrogen; and, in a symmetrical alkene of the type

$$CH_3\text{-}CH\text{=}CH\text{-}CH_2\text{-}CH_3 \text{ or } CH_3\text{-}CH\text{=}CH\text{-}CH_2\text{-}CH_2\text{-}CH_3,$$

where does the addition of an acid halide take place? The first question was answered by what we today know as the Zaitsev rule: when a halo-alkane is dehydrohalogenated, the hydrogen is lost preferentially from that carbon which is least hydrogenated. Thus, 2-bromobutane lost HBr to form 2-butene, and 1-bromobutane lost HBr to form 1-butene. The second question was answered by Zaitsev in collaboration with G. Wagner (48, 49). When the above two structures were treated with HI, the iodine atoms were found on the second carbon atoms, giving 2-iodopentane and 2-iodohexane respectively. From this, another rule was proposed stating that the halide atom adds to that carbon of an ethylene group which is attached to a methyl residue.

Wagner and Zaitsev were also first to use organometallic compounds of the Grignard type for organic synthesis, except that instead of Mg they used Zn (50). They used this procedure to synthesize diethylcarbinol, an isomer of amyl alcohol unknown at that time, and other secondary alcohols such as isobutyl alcohol (51). For the preparation of diethylcarbinol, one equivalent of ethyl formate was mixed with four equivalents of ethyl iodide over zinc pellets. The reaction mixture was heated, the crystalline product was added to water, and the water solution was distilled. The final product, obtained in a 26% yield, had an empirical formula of $C_5H_{12}O$ and had a boiling point of 116–117°C at 736.3 mm Hg pressure. The reaction sequence was visualized as follows:

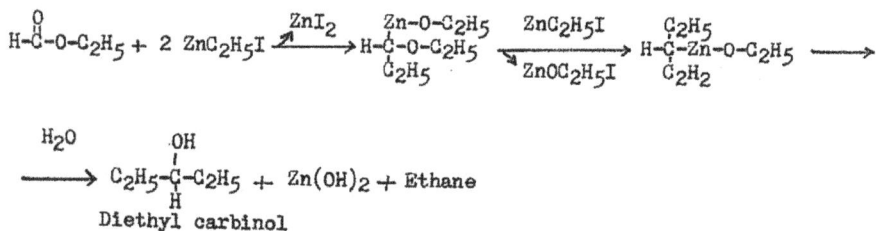

$$H-\overset{O}{\underset{}{C}}-O-C_2H_5 + 2\ ZnC_2H_5I \xrightarrow{ZnI_2} H-\overset{Zn-O-C_2H_5}{\underset{C_2H_5}{C}}-O-C_2H_5 \underset{ZnOC_2H_5I}{\overset{ZnC_2H_5I}{\rightleftharpoons}} H-\overset{C_2H_5}{\underset{C_2H_2}{C}}-Zn-O-C_2H_5 \longrightarrow$$

$$\xrightarrow{H_2O} C_2H_5-\overset{OH}{\underset{H}{C}}-C_2H_5 + Zn(OH)_2 + Ethane$$
$$\text{Diethyl carbinol}$$

It may be noted that Zaitsev's brother Michael (b. 1845), an industrial chemist, discovered the reduction of acyl chlorides to aldehydes using a palladium catalyst.

A corollary to Zaitsev's rule was provided by Alexander Eltekov (b. 1847) of Kharkov University, who studied the dehydrohalogenation of alkyl dihalides (52). He proposed that in such reactions, the loss of the hydrogen atom is experienced by that carbon which holds the other halogen. Thus, 1,2-dibromopropane gave two products when dehydrobrominated: 1-bromopropene and 2-bromopropene. This rule was valid even if there was an adjacent carbon atom with fewer hydrogens. Thus, vigorous dehydrobromination of 1,2-dibromo-3-methylbutane resulted in the formation of isopropylacetylene, CH_3-$CH(CH_3)$-$C=CH$. Eltekov is also known for the so-called Eltekov reactions, which involve the synthesis of olefins using lead oxide as the catalyst (53). For instance, when amylene was treated with methyl iodide and PbO, the following compounds were formed:

$$2 \; CH_3\text{-}\overset{\underset{\displaystyle CH_3}{|}}{C} = \overset{\underset{\displaystyle H}{|}}{C}\text{-}CH_3 \; + \; 3 \; CH_3I \; \xrightarrow[\text{heat}]{\text{PbO}} \; CH_3\text{-}\overset{\underset{\displaystyle CH_3}{|}}{\underset{|}{C}} \overset{\displaystyle CH_3}{\underset{|}{-}} \overset{\displaystyle CH_3}{C} = CH_2 \; \text{ and } \; CH_3\text{-}\overset{\displaystyle CH_3}{C} = \overset{\displaystyle CH_3}{C} \text{-} CH_3$$

Amylene

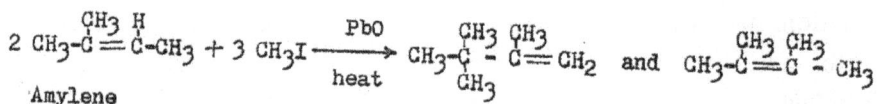

The same catalyst could convert alkyl halides into olefins, whereby ethyl iodide gave ethylene, and isobutyl iodide gave isobutylene. The Eltekov reaction provides one of the few available techniques for the synthesis of highly-branched olefins.

Isomerism of various types was of considerable interest to a number of Russian investigators. Butlerov was, of course, one of the first Russian chemists to synthesize and make a thorough study of the structural isomers of butyl alcohols and butylenes. In later years, the nature of the various isomers of amylene was also elucidated by Russian investigators (54). Thus, Flavitsky was able to determine the structure of amylene as $(CH_3)_2\text{-}CH{=}CH\text{-}CH_3$, which was confirmed by Vishnegradsky, who also showed that commercial amylene preparations were mixtures of several isomers. Eltekov was able to identify in such commercial amylene preparations the isomer studied by Flovitsky, and, in addition, to show the presence of ethylmethyl ethylene and isopropyl ethylene therein. Trimethyl ethylene was prepared from amyl alcohol by Ermolayev.

The most accomplished Russian researcher in the field of isomerism was undoubtedly Paul Walden (55). He was born in the Livonian province of Russia in 1863 and graduated from the Riga Polytechnicum in 1882, where he was student of Ostwald before the latter moved to Leipzig in Germany. Walden later followed to Leipzig, where he received his doctorate in 1891. After a brief visit to other German laboratories, Walden returned to Riga to join the faculty at the Polytechnicum, where his research interests changed from physical chemistry to organic chemistry. During the period of Walden's return to Riga and the First World War, Walden provided the scientific community with a thorough description of what we now know as the

Walden inversion phenomenon, which, in turn, formed the basis for the understanding of the Sn_2, or the bimolecular type mechanism of organic reactions. In addition, he made extensive contributions to the understanding of electric conductivity of nonaqueous solvents, the discovery of the Walden viscosity rule, where the product of the viscosity and equivalent conductivity at infinite dilution is a constant, and the discovery of the so-called Walden-Ostwald rule for the determination of valences of acids and bases *via* conductimetric measurements. Walden was the recipient of numerous honors from his *alma mater* as well as his country's government. He was president of the Riga Polytechnicum for some years, was chosen along with Mendeleyev and Menshutkin to plan the organization of the St. Petersburg Polytechnic Institute; he was knighted by the tsar and elected to the St. Petersburg Academy of Sciences in 1906 following the departure of Beilstein. During the First World War, Walden and the Polytechnicum were evacuated to Moscow. When Walden returned to Riga in 1918, he found more turmoil as the Bolsheviks, the Germans, and the nationalist Latvians were struggling for the control of the former Russian Livonian and Kurland provinces. So, as Walden himself had said, he had had enough of fighting and revolutions, and fled to Rostock, Germany, where he became the head of the chemistry department. He had already received offers to occupy the chemistry chairs vacated by Skraup in Switzerland and by Van't Hoff in Holland before the First World War, but he had chosen at that time to remain in Riga, Russia. War once again drove Walden from his home when Allied bombardments during the Second World War destroyed Rostock. He then settled in Tuebingen, where he taught chemistry practically until his death in 1957.

Walden's work on stereochemistry was initiated by the observation that the treatment of optically active organic acids with phosphorus halides produced halogen-substituted acids with an optical rotation opposite to that of the initial material (56). He produced from 1-malic acid the d-chlorosuccinic- and d-bromosuccinic acids by the action of PCl_5 and PBr_5; d-bromopropionic acid ethyl ester from 1-lactic acid ethyl ester and PBr_5; d-chlorophenylacetic acid from 1-mandelic acid and PCl_5; and 1-diethylchlorosuccinate and 1-diethylbromosuccinate from d-diethylmalate and PCl_5 and PBr_5, respectively. He later devised chemical cycles for the interconversion of optical isomers into each other (57, 58), as for instance the following:

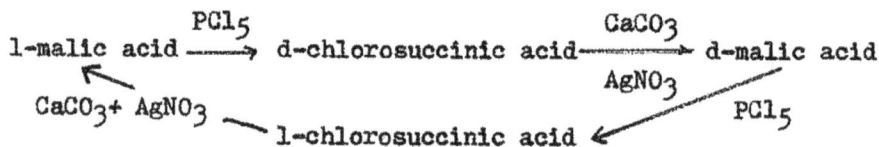

$$1\text{-malic acid} \xrightarrow{PCl_5} d\text{-chlorosuccinic acid} \xrightarrow{CaCO_3} d\text{-malic acid}$$

$$CaCO_3 + AgNO_3 \nwarrow \qquad \xleftarrow{AgNO_3} \qquad \nearrow PCl_5$$

$$1\text{-chlorosuccinic acid} \xleftarrow{\qquad}$$

Walden was also one of the pioneers in developing the kinetic method for the separation of racemates (59). For this purpose, he used 1-amyl alcohol to show preferential esterification with d-bromopropionic acid: d-bromopropionic acid was reacted with half the molar equivalent of 1-amyl alcohol. The product (an ester) had an optical rotation of 3.3 degrees; if a four-fold excess of amyl alcohol was used, the product had specific rotation of 2.99 degrees. Hence, the d-isomer of the acid reacted preferentially with the 1-amyl alcohol. Similarly, if d,1-methylmalonic acid was reacted with 1/3 the theoretical equivalent of 1-amyl alcohol, a product with specific rotation of 1.5 degrees was obtained. When an almost equimolar amount of 1-amyl alcohol was used, the ester had a specific rotation of 3.5 degrees. Hence, the 1-amyl alcohol reacted preferentially with 1-methylmalonic acid. At about the same time, Marckwald and McKenzie published a more successful method using 1-methol instead of 1-amyl alcohol. Finally, it may be mentioned that Walden was familiar with the technique of optical rotatory dispersion, a technique that was quite novel at the end of the nineteenth century. Walden reported the optical dispersion values for tannin using the red, green, yellow, and blue filters in addition to the standard sodium lamp (60). No structural conclusions were, however, made from these measurements.

The field of chemical documentation owes its very existence to the untiring efforts of some twenty years' duration of Friedrich Beilstein (61). He was born in St. Petersburg in 1838, and studied chemistry in Germany from 1853 to 1858, when he received his doctorate in Goettingen. His teachers included Bunsen, Liebig, and Woehler. Beilstein then remained in Germany, first as an assistant in Breslau and Goettingen, then as a professor in Goettingen. In 1866, he left Germany to accept the chair of chemistry at the St. Petersburg Technological Institute, which had been vacated by Mendeleyev. He remained there for the next thirty years. Beilstein apparently did not enjoy much working in Russia, firstly because of the relatively tight governmental controls, and, on the other hand, because of the obstreperous nature of Russian students, who would rather protest and demonstrate than learn sulfur chemistry. Yet he refused to move back to Germany, because he was being well paid and because in Germany he would have to spend much time in research. The latter he wanted to avoid in order to devote his time to the compilation of his "Handbuch." In 1881, Beilstein was elected to the Academy of Sciences over the objections of the "Russian faction," which wanted to elect Mendeleyev. Beilstein was also a consultant to numerous governmental and industrial agencies in Russia, and executed his duties with competence but little enthusiasm. He was an honorary member of the German Chemical Society, the British Chemical Society, the German (Prussian) Academy of Sciences, and the Uppsala Scientific Society.

Beilstein's research work was done largely in Goettingen, though some projects were carried over to St. Petersburg. His interests included the chemistry of cresol isomers, the composition of crude oil, and analytical chemistry problems, such as flame tests for halogens, the separation of Mn from Fe, and methods for the determination of Cd, Zn, and Sb. His greatest contribution to science was, of course, the compilation of the *Handbuch der organischen Chemie*, the first volume being published in 1880 by Leopold Voss in Hamburg. The work was a detailed list of all known organic compounds with their physical and chemical properties as well as literature references. The second edition appeared in 1890 with twice the amount of information, and the third edition appeared between the years of 1892 and 1899. At that time it became clear to Beilstein that he alone with an assistant or two could not handle the work required to keep up the "Handbuch." Consequently, he relinquished his authorship rights to the German Chemical Society, which formally assumed its publication in 1897.

Though Beilstein was born in Russia and lived there most of his life, he nevertheless remained a German in his heart, avoiding contact with native Russians as much as he could, and remaining in Russia only because of the privileges and honors accorded to him there. He chose his assistants from among the German graduates of the Dorpat University. He deeply resented the Slavophile mood in Russia at that time, and was in general exceedingly critical of Russian customs and beliefs. He was not liked by his colleagues other than those of his Germanic orientation. Beilstein died in St. Petersburg in 1906 of an apparent heart attack.

During the Soviet era of Russian history, the field of organic chemistry underwent a painful period reminiscent of the purges in the ranks of the biologists in the 1930s. In the late 1940s, the political structure of Soviet Union directed its community to reexamine the ideological implications of chemical theories, so as to "struggle against the alien reactionary ideas of bourgeois science." The result of such "critical examination" was a paper in *Uspekhi Khimii* (a chemistry journal) (62), which singled out the resonance theory as having "neither experimental nor theoretical justification," and furthermore, that it acted as a "brake upon the further development concerning chemical structure." Instead of the resonance concept, the authors of this position paper proposed to return to the writings of Butlerov, whose concept of the chemical structure was supposedly most compatible with the doctrine of dialectical materialism (63). As was indicated above, Butlerov had proposed that each organic compound ought to be represented by a finite structural formula showing the relative positions of all constituent atoms, from which most of the compound's properties could be predicted. What apparently was attractive about Butlerov's concepts was the supposed immutability of chemical structures, whereas modern resonance theory stressed the flexibility of structures of unsaturated compounds. Such flexibility is apparently inconsistent with the principles of dialectical materialism. It is doubtful, however, that Butlerov had political ideology in his mind when he read his paper at the Spyer convention in 1861. Considering the state of the art at that time, Butlerov's proposals were a significant advance toward the understanding of the basic principles of organic chemistry. They in no way excluded the possibility of refinements in the years to come.

The attacks on the resonance theory also subjected to abuse its Western proponents such as Pauling, Wheland, and Ingold, as well as those Soviet scientists who used the resonance theory in their work. Surprisingly, even the social scientists got into the act, and their views of the resonance were especially amusing in view of their meager knowledge of organic chemistry (64). As was the case with biology, a Lyssenko of theoretical organic chemistry appeared on the scene in the person of one C. V. Chelintsev (63), who proposed a structural theory that was both consistent with Butlerov's ideas and the then current political doctrines. Chelintsev's theories were, however, denounced as worthless both in the West, and, surprisingly, by some fearless Soviet organic chemists. In contrast to the situation with geneticists in the 1930s in the Soviet Union, no organic chemists were arrested and sent to the Gulags, and the attacks on the resonance theory subsided after Stalin's death. It can be said that the only benefit that was derived from that whole Soviet resonance theory affair was that the West finally became well aware of Butlerov's invaluable contributions to organic chemistry.

Organic and Name Chemical Reactions

Organic chemistry is unique among the sciences in that it immortalizes its practitioners by retaining their names with the reactions they discovered. These are known as the name reactions, and may be exemplified by such well-known reactions as the Cannizaro reaction, the Strecker synthesis, the Hoffman degradation, or the Wagner rearrangement. Among the names associated with specific organic chemistry reactions, the Germans tend to predominate, though Russian names seem to occupy a second place in this sort of priorities. The following section provides a description of the more important organic chemistry reactions discovered by Russian investigators, which are significant from either the theoretical point of view or their utility for analytical or synthetic purposes. Many of them carry their discoverers' names.

The Russian chemist best known to the American public is undoubtedly Alexander Borodin, though few realize that beside being a great composer, he was also a great chemist (65, 66). He was born in 1834, supposedly being an illegitimate son of a Caucasian prince, and obtained his education at the Military-Medical Academy in St. Petersburg. In 1859, he worked abroad with Kekule and Erlenmyer, and in 1862, he joined the faculty of Military-Medical Academy as a professor of organic chemistry. All his life, Borodin attempted to divide his time between science and music, succeeding admirably in both, but, undoubtedly, also hastening his demise, which occurred in 1887. Whereas his professor, N. Zinin, urged him to devote more time to chemistry, his music teacher recommended that he spend his time developing his music talents. That, of course, he did, eventually composing two symphonies, a number of chamber music pieces, the popular symphonic poem "In the steppes of Central Asia," and one major opera, the *Prince Igor.* In his musical activities, Borodin was closely associated with a number of other Russian part-time composers: Rimsky-Korsakov, a naval officer; Cui, a military engineer; Mussorgsky, an army officer; and Serov and Glinka, government bureaucrats. Together, they are known as the "mighty five," who pioneered the Russian National music school. It may be noted that another Russian composer, Peter Tchaikovsky, was as nationalistic as any of his contemporaries (the "1812 Overture," March Slav," etc.), yet he preferred to work alone and not to belong to any specific group. The Soviet regime preferred to remember Borodin as a chemist rather than composer.

Borodin's scientific interests included a number of areas of organic and inorganic chemistry. One of his first papers dealt with aromatic nitrogen compounds, whose structures he wanted to determine. These compounds were hydrobenzamide, amarine, and benzoylanilide (67, 68):

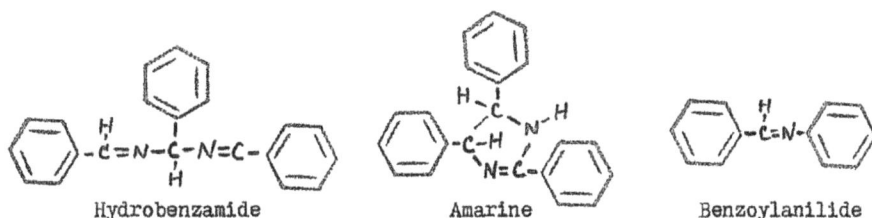

Hydrobenzamide   Amarine   Benzoylanilide

Borodin used ethyl iodide as a reagent, obtaining from hydrobenzamide an oil with an empirical formula of $C_{25}H_{28}NO$ (using modern atomic weights). He concluded that hydrobenzamide accepted two ethyl groups, and proposed that the product he viewed as being a derivative of ammonium oxide, whereas the original reactant (hydrobenzamide) be viewed as a derivative of ammonia with all its hydrogen atoms substituted. Amarine was also able to react with ethyl iodide taking up two ethyl groups with relative ease, and a third ethyl group with some difficulty. He concluded that there were three replaceable hydrogen atoms in amarine. This was, of course, incorrect, since it is proposed that amarine has a free amino group. Benzoylanilide took up one molecule of ethyl iodide giving a product with an empirical formula of $C_{15}H_{16}NI$ (modern atomic weights). He correctly concluded that benzoylanilide was an amine, where one of the hydrogen atoms of the nitrogen was substituted by the phenyl residue, whereas the other two hydrogens were substituted by the divalent radical C_7H_6. Though Borodin's structures were indeed primitive by modern standard (he used the accepted organic chemistry designations before Butlerov's introduction of atomic structures of compounds at the Spyer convention in 1861), he is, nevertheless, recognized as a pioneer in viewing nitrogen-containing compounds as substituted ammonias.

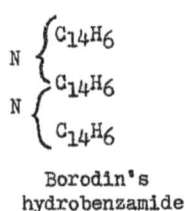

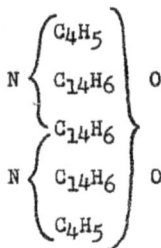

Borodin's hydrobenzamide coupled with ethyl iodide

Note that in both structures, atomic weights assumed for for C, O, and N are 6, 8, and 7 respectively.

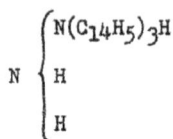

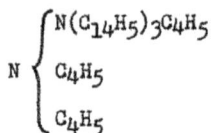

The reaction of amarine with ethyl iodide, as viewed by Borodin. Atomic weights as above.

388

Borodin later changed his interests to reactions of aliphatic compounds, one of his better known contributions being the so-called Borodin reaction (also called Hunsdiecker reaction) for the shortening of the carbon chain in aliphatic acids (69). This involves the treatment of a silver salt of the acid with bromine:

$$RCOOAg + Br_2 \rightarrow R\text{-}Br + CO_2 + AgBr$$

He also devised a method for the preparation of ethylene from diethyl zinc and chloroiodoform (70). The latter compound was prepared by Borodin from iodoform and mercurous chloride, boiling point being at 131°C. Simultaneously with Wurtz, Borodin discovered the aldol condensation, the results of both workers being reported in the same volume of the *Berichte* (71). Whereas Wurtz used acetaldehyde for this purpose, Borodin showed the same reaction with valeraldehyde, which, upon treatment with alkali and heating, polymerized into a higher compound with the loss of water. He was able to isolate from such reaction mixtures an aldehyde ($C_{10}H_{18}O$) and a "neutral" substance ($C_{20}H_{38}O_3$), which he considered to be a hydrate of the aldehyde, ($C_{10}H_{18}O)_2.H_2O$. Borodin's reaction probably ran as follows:

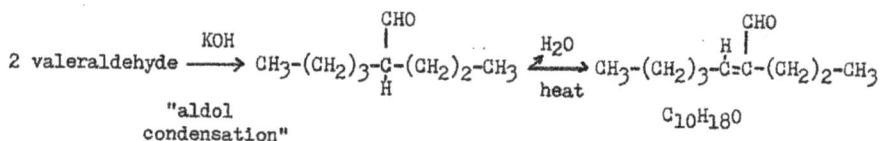

Borodin's contribution to inorganic chemistry involved the elucidation of some properties of hydrogen fluoride (72). He was interested in why hydrofluoric acid existed in the form of H_2F_2 and potassium fluoride in the form of KF . HF. He reacted acetic acid with the anhydrous potassium fluoride, KF, and found the products to consist of potassium acetate and KF. HF. He thereupon proposed that HF acts much like water, which forms HF complexes very much like the formation of hydrates with water. The KF. HF was used by Borodin in the first synthesis of benzoyl fluoride from benzoyl chloride. The chemical properties of the fluoride were very similar to those of the chloride.

And finally, the field of clinical chemistry remembers Borodin for his invention of a method for the determination of urea in urine, which was reported in 1876 at a Russian Chemical Society meeting. This method involves the treatment of urine with NaOBr, then measuring the nitrogen released by volumetric (manometric) means.

The various rearrangements frequently observed during chemical reactions of organic compounds are a fascinating subject to study, and a number of Russian investigators were the discoverers of new rearrangements. The best known among these is the Wagner-Meerwein rearrangement, originally described by Georg (Egor Egorovich) Wagner in 1899 in a relatively complicated group of compounds called the terpenes. A more general case of this type of rearrangement was studied by Zelinsky and Zelikov in 1901 (73), who treated pinacolyl alcohol with dry oxalic acid, obtaining unsaturated hydrocarbons. A similar rearrangement was seen with cyclohexanol:

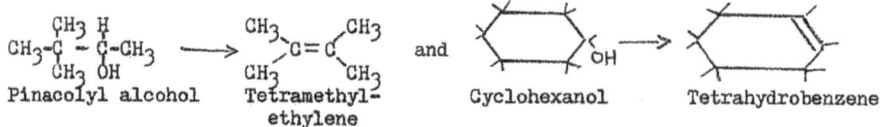

| Pinacolyl alcohol | Tetramethyl-ethylene | and | Cyclohexanol | Tetrahydrobenzene |

Wagner's original observation involved the rearrangement of borneol to camphene (74):

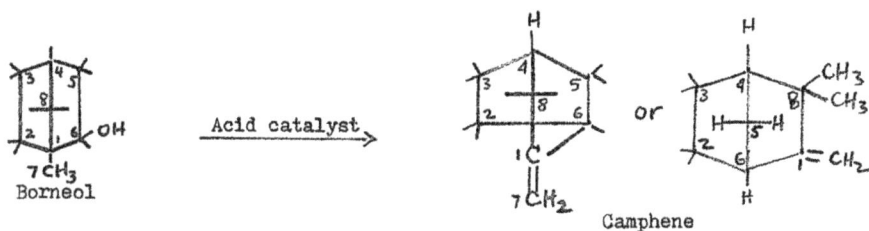

Borneol → Acid catalyst → Camphene

No less important were Wagner's contributions to the structural chemistry of terpenes, where he was the first to propose the correct structures for limonene, borneol, pinene, and a number of other compounds. He did this mainly on the basis of experimental work performed by Wallach and his students, performing only a few key experiments in his own laboratory. It is probably safe to say that had Wagner been alive in 1910, he would have shared the Nobel prize with Wallach, which the latter received for his work on the chemistry of terpenes. The formula for pinene was established from the reaction products of pinene and hypochlorous acid. These products were presumed to be chlorohydrins of sobrerol, whose structure was consequently also determined in these series of experiments (75):

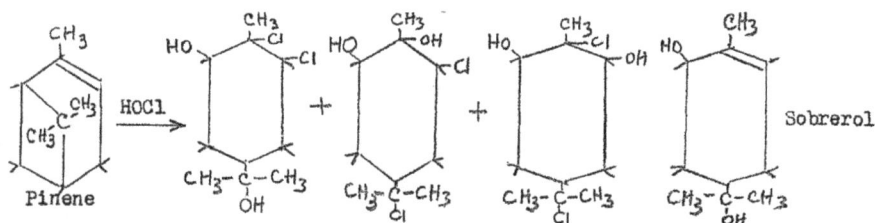

Having determined the structural formula for pinene, Wagner proceeded to investigate some of its reactions, which, in turn, led to the determination of correct structures for limonene and borneol (76). The reactions performed were as follows:

391

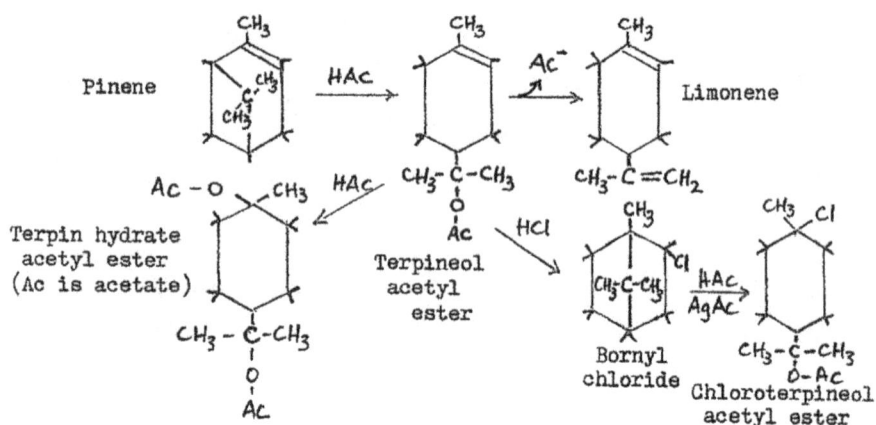

Georg (Egor Egorovich) Wagner was the product of the Kazan school of chemistry. He was born in 1849 in Kazan in a family of a civil servant who was apparently of German origin. After graduating from Kazan University in 1874, where he was principally under the guidance of Zaitsev, Wagner spent some time with Butlerov in St. Petersburg, then taught at the Novaya Alexandriya Agricultural Academy, and finally in 1886 settled in Warsaw, where he taught at both the University and the Technological Institute. He died in 1903 at the peak of his intellectual activities (77).

An interesting rearrangement of the Wagner type, i. e., one involving the formation of carbonium ions, was reported by N. Demyanov (1861–1938), a graduate of the Moscow University and a professor at the Moscow Agricultural Academy. The Demyanov rearrangement involves both the ring expansion (78, 79) and ring contraction (80) in cyclic alcohols and amines:

$$CH_2 \atop CH_2 \!\!\!\diagdown\!\! CH\text{-}CH_2\text{-}NH_2 \xrightarrow{\ HNO_2\ } \ {CH_2 \atop CH_2}\!\!\!\diagdown\!\! CH\text{-}CH_2\text{-}OH + {CH_2 - CH_2 - OH \atop CH_2 - CH_2} \ + \ N_2$$

Cyclobutyl
alcohol

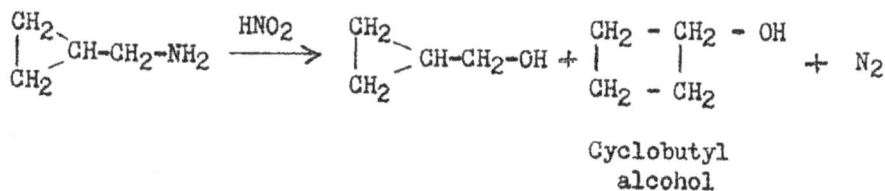

The cyclobutyl alcohol was also prepared from cyclobutylamine, and upon oxidation with nitric acid yielded succinic acid. Ring expansion was also accomplished without the presence of the amino group:

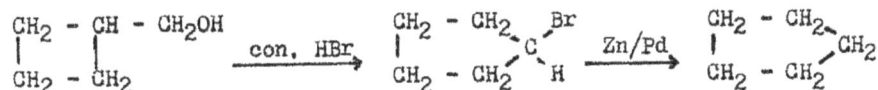

$$\underset{CH_2 - CH_2}{CH_2 - CH - CH_2OH} \xrightarrow{\ con,\ HBr\ } \underset{CH_2 - CH_2}{\overset{CH_2 - CH_2}{\diagdown}}\!\!C\!\!\underset{H}{\overset{Br}{\diagup}} \xrightarrow{\ Zn/Pd\ } \underset{CH_2 - CH_2}{\overset{CH_2 - CH_2}{\diagdown}}\!CH_2$$

Ring contraction proceeded with cyclic amines via the formation of a bicyclic intermediate:

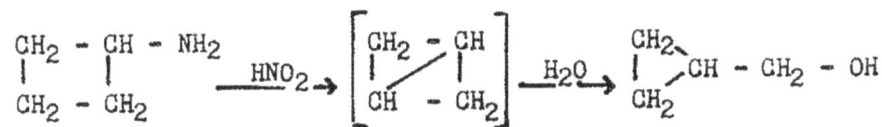

$$\underset{CH_2 - CH_2}{CH_2 - CH - NH_2} \xrightarrow{\ HNO_2\ } \left[\underset{CH - CH_2}{CH_2 - CH}\right] \xrightarrow{\ H_2O\ } \underset{CH_2}{\overset{CH_2}{\diagdown}}\!CH - CH_2 - OH$$

Phosphoorganic chemical investigations in Russia may be exemplified by the discovery of the so-called Arbuzov rearrangement (81). Its discoverer, A. E. Arbuzov (1877–1968), was first a professor at the Novo-Alexandriya Agricultural Academy, and later at Kazan University. Arbuzov prepared a number of trivalent phosphorus esters, $P(OR)_3$, then treated these with alkyl halides. The result was a rearrangement into pentavalent phosphorus esters:

393

$$P(OC_2H_5)_3 + {}^*C_2H_5I \longrightarrow \left[{}^*C_2H_5\overset{I}{\underset{}{P}}(OC_2H_5)_3 \right] \longrightarrow {}^*C_2H_5\overset{O}{\underset{\parallel}{P}}(OC_2H_5)_2 + C_2H_5I$$

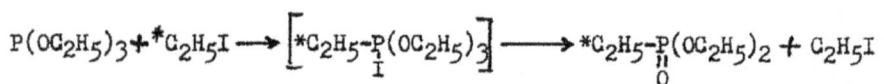

The existence of the proposed intermediate was tested by using CH_3I instead of C_2H_5I in the above reaction, and the pentavalent phosphorus ester was found to have accepted a methyl group. From such experiments, Arbuzov proposed that the action of PCl_3 on alcohols occurs via the formation of trivalent phosphorus esters, which, in turn, decompose into the pentavalent phosphorus esters under the influence of the HCl formed:

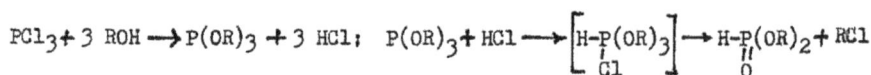

$$PCl_3 + 3\ ROH \longrightarrow P(OR)_3 + 3\ HCl;\quad P(OR)_3 + HCl \longrightarrow \left[H\overset{}{\underset{Cl}{P}}(OR)_3 \right] \longrightarrow H\overset{O}{\underset{\parallel}{P}}(OR)_2 + RCl$$

The trivalent phosphorus esters are unstable in water, being decomposed into the corresponding pentavalent phosphorus esters. For the same reason, $P(OH)_3$ or H_3PO_3 was unstable in water, being decomposed into its pentavalent isomer:

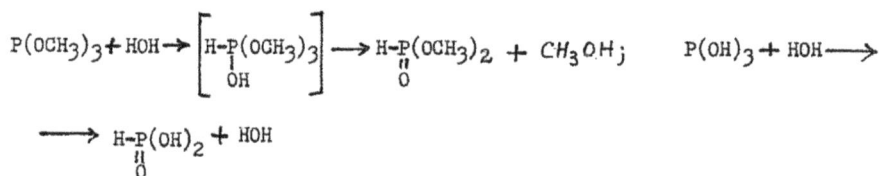

$$P(OCH_3)_3 + HOH \longrightarrow \left[H\overset{}{\underset{OH}{P}}(OCH_3)_3 \right] \longrightarrow H\overset{O}{\underset{\parallel}{P}}(OCH_3)_2 + CH_3OH;\quad P(OH)_3 + HOH \longrightarrow$$

$$\longrightarrow H\overset{O}{\underset{\parallel}{P}}(OH)_2 + HOH$$

The pinacol-pinacolone rearrangement has been a favorite subject of many chemists in several countries including Russia, where the most active group in this area was led by A. E. Favorsky (1860–1945). He was a student of Butlerov and a graduate of St. Petersburg University. His associate, Vladimir Ipatieff, was first to show the rearrangement of ethylene glycol into acetaldehyde under acidic conditions. Favorsky then investigated the mechanism of the reaction and proposed that dioxane may be an intermediate in the reaction. By warming ethylene glycol with 4% H_2SO_4, he was indeed able to isolate

1,4-dioxane as the principal product, which under more vigorous conditions was converted to acetaldehyde (82).

An interesting method for ring contraction was discovered by Favorsky in collaboration with Bozhovsky (83). This method was used to form five-membered rings from chlorocyclohexanone:

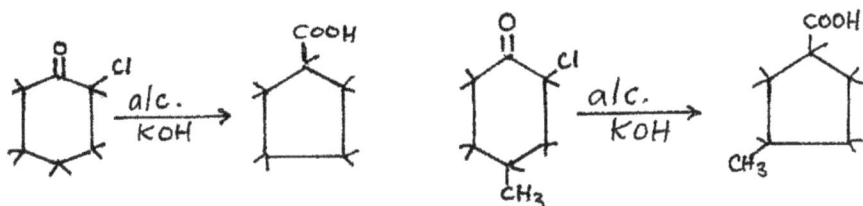

Favorsky was quite interested in the chemistry of acetylenes and the chemistry of various compounds containing triply-bonded carbon atoms. This interest undoubtedly originated because of Butlerov's work on this subject; his student, Myasnikov, had synthesized acetylene by treating sodium amyl alcoholate with vinyl bromide. At the same time, another Russian chemist by the name of Savich had synthesized the compound from ethylene bromide and alcoholic KOH (84). Favorsky made his acetylenes from ketones by treating them with PCl_5 in alcoholic KOH at 120–170°C. Thus, dimethyl acetylene was produced from ethylmethyl ketone by the action of PCl_5 in alcoholic KOH, whereas ethyl acetylene was produced from the same starting material at a lower temperature. In the same manner, Favorsky was able to prepare propyl acetylene and methylethyl acetylene from methylpropyl ketone by using temperatures of 120 and 170°C, respectively (85). This prompted him to investigate whether or not isomeric acetylenes were interconvertible, and indeed, such a rearrangement could be demonstrated. He thus produced dimethyl acetylene from ethyl acetylene by heating the former in alcoholic KOH at 170°C, and in a similar fashion, propyl acetylene was converted into methylethyl acetylene. The isomerization was believed to have occurred *via* the formation of vinyl ether intermediates:

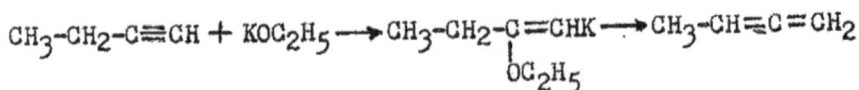

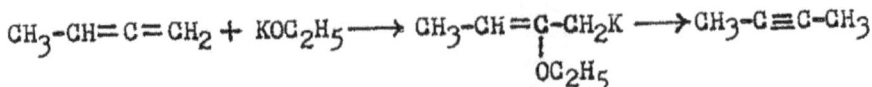

$$CH_3-CH_2-C\equiv CH + KOC_2H_5 \longrightarrow CH_3-CH_2-\underset{\underset{OC_2H_5}{|}}{C}=CHK \longrightarrow CH_3-CH\equiv C=CH_2$$

$$CH_3-CH=C=CH_2 + KOC_2H_5 \longrightarrow CH_3-CH=\underset{\underset{OC_2H_5}{|}}{C}-CH_2K \longrightarrow CH_3-C\equiv C-CH_3$$

Favorsky and his students later synthesized a number of vinyl ethers using olefins, KOH, and alcohols, and made an extensive study of the chemical properties of these types of compounds.

A convenient method for the synthesis of acetylenic alcohols was also discovered by Favorsky's group (86), which involved the reaction of a ketone and a substituted acetylene:

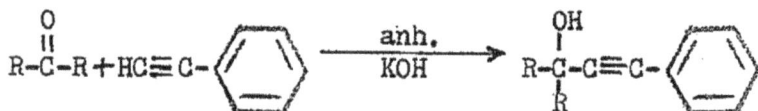

A general method for the synthesis of keto alcohols was also worked out by Favorsky (87). It involved the treatment of the ketone with PBr₅, followed by alkaline hydrolysis. He prepared a multitude of keto alcohols by this technique. Phosphorus pentachloride could not be used in the reaction, since it tended to attack the carbonyl group instead of vicinal carbon atom:

$$\text{Acetone} + PBr_5 \longrightarrow CH_3\overset{\overset{O}{\|}}{C}\text{-}CH_2Br, \text{ but Acetone} + PCl_5 \longrightarrow CH_3\overset{\overset{Cl}{|}}{\underset{|}{C}}\text{-}CH_3$$

$$\begin{array}{ccc}
\overset{CH_3}{\diagdown}\overset{/}{\underset{C\text{-}H}{C\text{-}H}}\overset{CH_3}{} & & \\
\underset{|}{C\text{-}H} + PBr_5 \longrightarrow & \overset{CH_3}{\diagdown}\underset{C\text{-}Br}{\overset{/}{C\text{-}Br}}\overset{CH_3}{} \xrightarrow[\text{heat}]{\text{K-formate}} & \overset{CH_3}{\diagdown}\underset{C\text{-}OH}{\overset{/}{C\text{-}OH}}\overset{CH_3}{} + POBr_3 \\
\underset{|}{C=O} & \underset{|}{C=O} & \underset{|}{C=O} \\
CH_3 & CH_3 & CH_3
\end{array}$$

There were several Russian investigators beside Favorsky who were interested in the synthesis and reactions of unsaturated as well as saturated hydrocarbons. One of the better-known procedures for the synthesis of olefins is the Chugayev reaction (88), described by Leo Chugayev (1873–1922), a graduate of the St. Petersburg University and a professor of the Moscow University. His procedure involved the formation of xanthate esters from alcohols, and was remarkable in that it did not bring about any rearrangements of the pinacol-pinacolone type. Chugayev represented this reaction as follows:

$$\underset{\text{Alcohol}}{C_nH_{2m-1}OH} + Na \longrightarrow C_nH_{2m-1}O\text{-}Na \xrightarrow{CS_2} C_nH_{2m-1}O\text{-}\overset{\overset{S}{\|}}{C}\text{-}S\text{-}Na \xrightarrow{CH_3I}$$

$$\longrightarrow \underset{\text{Xanthate ester}}{C_nH_{2m-1}O\text{-}\overset{\overset{S}{\|}}{C}\text{-}S\text{-}CH_3} \xrightarrow{\text{distill}} \underset{\text{Olefin}}{C_nH_{2m-2}} + CH_3SH + O=C=S$$

This method was useful in dehydrating menthol to menthene, which boiled at 167–168°C and had a specific rotation of 115–116 degrees, as well as to prepare limonene from carvone (89):

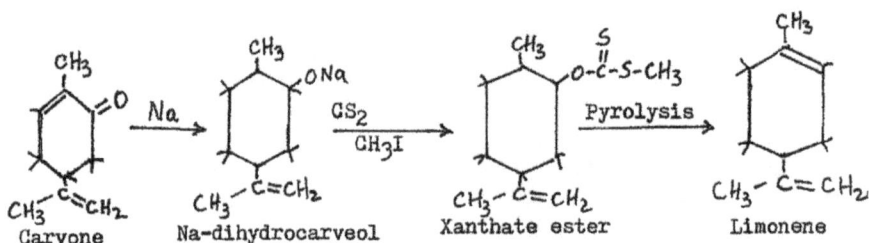

The utility of the Chugayev reaction was later proven by Fomin and Sokhansky from St. Petersburg Technological Institute, who were able to convert pinacolyl alcohol into tetramethylethylene without encountering the Wagner-Meerwein rearrangement observed by Zelinsky and Zelikov (73). The structure of the product was proven by oxidizing it with $KMnO_4$ to trimethylacetic acid, and by reducing it to trimethyethyl methane (90).

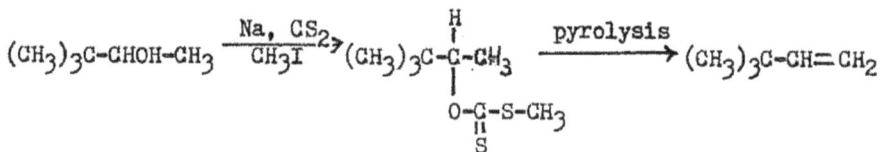

Chugaev was responsible for the method now used and termed the Zerevitinov procedure for the quantitation of alcoholic groups by the Grignard reagent (91). He determined that alcohols cause the evolution of methane when treated with CH_3MgI, and it was his student, T. Zerevitinov, who eventually derived the procedure to take advantage of this phenomenon in quantitating alcohols (92). The reaction was carried out in pyridine solution at room temperature as rapidly as possible, whereby the methane evolved was measured manometrically. The formula given for the calculation of the amount of –OH groups present was

$$\% \text{ OH} = 0.000719 \times V \times 17 \times 100/16 \text{ S} = 0.0764 \text{ V/S,}$$

where 0.000719 is the weight of 1 cm^3 of methane, 17 is the molar weight of the hydroxyl group, 16 is the molecular weight of methane, V is the volume of methane evolved, and S is the weight of the unknown substance. He determined the hydroxyl groups in a number of compounds, including sugars such as arabinose, sucrose, mannose, and glucose.

Chugayev eventually left the field of organic chemistry and became interested in organic-inorganic complexes. He discovered the extremely useful reaction between nickel and dimethylglyoxime (93). This reaction is now used extensively for the quantitation of nickel, and Chugayev himself reported that he could detect 1 part of Ni in 400,000 parts of water, and 0.1 mg of Ni in 500 mg of cobalt. He also studied the reactions of other transition metals with various types of dioximes, and was able to show that only alpha-dioximes and not beta-dioximes were able to form colored metal complexes. He proposed that this property be used to determine the configuration of unknown dioximes (94). During the Soviet era, Chugayev did important work on platinum complexes.

The synthesis of unsaturated hydrocarbons was also performed in connection with rubber research. The structure of isoprene was determined and synthesized by Ipatieff in 1897 (95). He used hydrobromic acid to add across the double bonds, then regenerated isoprene by treating the dibromo derivative with alcoholic KOH:

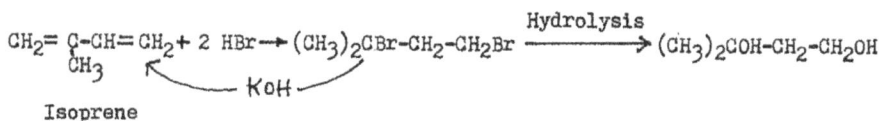

$$CH_2= \underset{\underset{\text{Isoprene}}{CH_3}}{C}-CH= CH_2 + 2\ HBr \longrightarrow (CH_3)_2CBr-CH_2-CH_2Br \xrightarrow{\text{Hydrolysis}} (CH_3)_2COH-CH_2-CH_2OH$$

Another distinguished rubber chemist in Russia was Ivan Ostromyslensky, who discovered a method for the synthesis of olefins from alcohols and aldehydes in the presence of dehydrating agents (96). It is commonly called the Ostromyslensky reaction.

Ostromyslensky, who was a professor at the Moscow Technological Institute, used this method for the synthesis of butadiene:

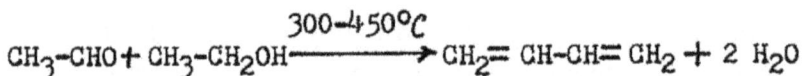

$$CH_3-CHO + CH_3-CH_2OH \xrightarrow{\text{300--450°C}} CH_2= CH-CH= CH_2 + 2\ H_2O$$

The conversion of carbonyl residues into saturated alkane groups was accomplished by N. Kizhner (1867–1935), a professor at Tomsk University (97). The method involves the formation of hydrazones, followed by hydrolysis or pyrolysis in an alkaline medium. Kizhner had observed this reaction as early as in 1899, though an extensive report on this did not appear until 1911. It is today known as the Wolff-Kishner reaction. In acid medium, hydrazones were converted to acidic substances, e. g., phenylhydrazine in the presence of HCl was converted to chlorobenzene. Kizhner gave several examples of the conversion of carbonyl compounds into hydrocarbons. One of these was the conversion of camphor into camphane:

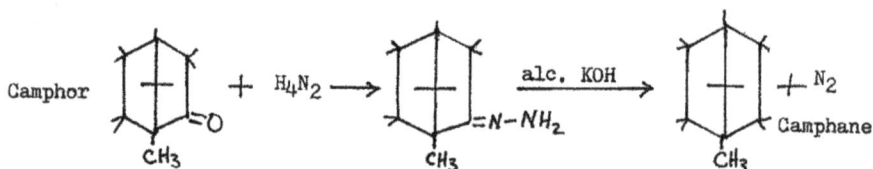

Another conversion from the terpene series that was accomplished was the synthesis of fenchane from fenchone, and a simpler reaction of converting cyclohexanone into cyclohexane.

The experience with hydrazine was used by Kizhner to devise a procedure for the synthesis of cyclopropane ring structures, commonly known as the Kizhner cyclopropane synthesis (98). Kizhner had first observed that he could convert pulegone into carane using the hydrazine method. He proposed the existence of a pyrazoline intermediate:

CH_3-C-CH_3

Pulegone $+ H_4N_2 \longrightarrow$ $(CH_3)_2-C$——$N-H$ $\longrightarrow$ $+ N_2$ Carane

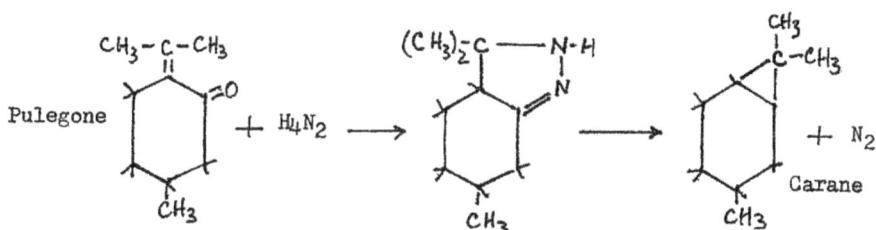

Kizhner then reasoned that the reaction might be common to all unsaturated ketones of the type

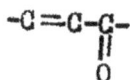

$$-C=C-\underset{\underset{O}{\|}}{C}-$$

He used mesityl oxide to test his hypothesis, and using the approach shown above with pulegone, was able to obtain 1,1,2-trimethylcyclopropane. When 3,5,5-trimethylpyrazole was treated with KOH, 1,1,2-trimethylcyclopropane was also obtained, indicating that the pyrazoline was an intermediate in the transition of mesityl oxide to the cyclopropane. It was also proposed that it might be possible to synthesize cyclopropanone from pyrazolones, but it is not known if Kizhner ever succeeded in executing this reaction.

Kizhner's life was pockmarked with unhappy events, much of it because of his own doings (99, 100). He was a student of Markovnikov in Moscow, then worked as an assistant at the university until he passed his doctorate exam. Thereupon he received a faculty appointment at Tomsk University, Tomsk being located in the Siberian area of Russia, and he relocated there in 1901. During the 1905–1906 "revolution" in Russia, Kizhner took the side of the rioters and participated in urging the students to strike and create havoc. Suntsov and Lewis characterized him as being "a political progressive with strong revolutionary leanings" (100). And so he was kicked out of his job and "exiled" to St. Petersburg, but allowed to return in 1907. Soon thereafter, in 1911 and 1912, he published his best-known research results, described above, in a Russian chemistry journal. By that time, he also lost portions of both of his legs to gangrene and apparently had to move with crutches with great difficulty. His presence in Tomsk was not exactly welcomed by the administration because of his behavior during the 1905–1906 disorders, and he was forced to resign his job due to his "disability" with a full pension. He left Tomsk for Moscow in 1914, where he taught in the Shanyavskii People's University, an unaccredited private college. During the Soviet era, he was put in charge of running Soviet Union's aniline dye industry. He died suddenly of a heart attack in 1935.

A method for the preparation of epoxides and glycols from olefins was proposed by N. Prilezhayev (1872–1944) of the Warsaw Polytechnicum (101). This is commonly known as the Prileschajew reaction, and involved the treatment of olefins with peroxides, especially benzoyl peroxide:

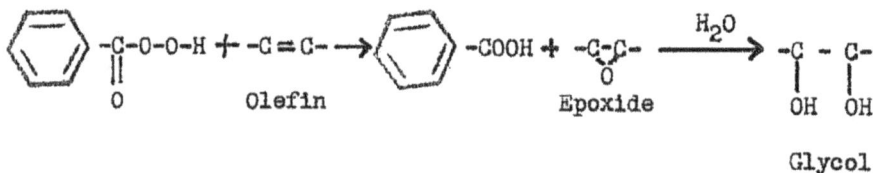

This reaction was carried out with limonene yielding the epoxide first, and later—the expected tetraol. Pinene, however, unexpectedly gave sobrerol.

A commercially important method for the synthesis of acetaldehyde was provided by M. Kucherov (1850–1911) of the St. Petersburg Agricultural Institute (102). Using dilute H_2SO_4, he was able to accomplish the hydration of acetylenes to either aldehydes or ketones. Thus, acetylene, when subjected to what we today know as rhe Kutcheroff reaction, yielded acetaldehyde, and ethymethyl ketone was formed by the same procedure from valerylene ($CH_3-CH_2-C=CH$).

A number of reactions involving olefins was studied by L. L. Kondakov (1857–1931), first of the Warsaw University, then of Dorpat University. Following the Bolshevik coup he emigrated to Czechoslovakia, where he taught at the Prague University. As early as in 1885, Kondakov described the halogenation of olefins, and was very upset when Hell and Wildermann published a similar paper without achnowledging Kondakov's contribution to the field (103). After studying the chlorination and bromination of a number of amylene and other unsaturated compound isomers, Kondakov proposed what is today known as the Kondakov rule: those olefins that react easily with mineral acids give unsaturated monohalides upon halogenation, whereas those that react with difficulty give bi-halides:

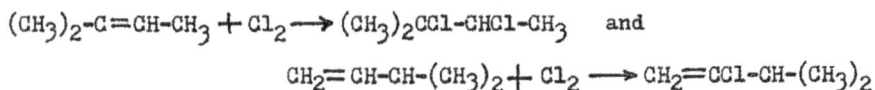

$$(CH_3)_2\text{-}C\!=\!CH\text{-}CH_3 + Cl_2 \longrightarrow (CH_3)_2CCl\text{-}CHCl\text{-}CH_3 \quad \text{and}$$

$$CH_2\!=\!CH\text{-}CH\text{-}(CH_3)_2 + Cl_2 \longrightarrow CH_2\!=\!CCl\text{-}CH\text{-}(CH_3)_2$$

In case of the latter reaction, Kondakov proposed that the chlorine adds across the double bond, and is followed by the abstraction of an HCl molecule. The halogenation of conjugated dienes was also studied by Kondakov, who discovered the now well-known "1,4 addition' (104):

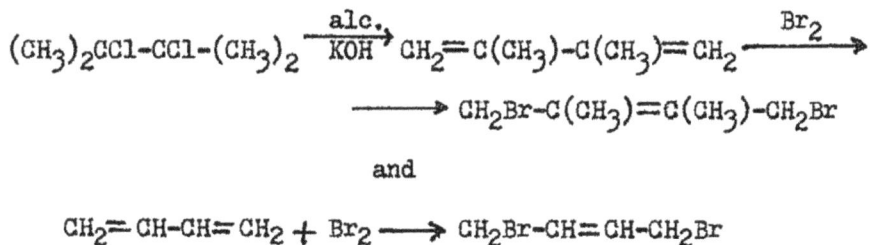

$$(CH_3)_2CCl\text{-}CCl\text{-}(CH_3)_2 \xrightarrow{\text{alc., KOH}} CH_2\!=\!C(CH_3)\text{-}C(CH_3)\!=\!CH_2 \xrightarrow{Br_2}$$

$$\longrightarrow CH_2Br\text{-}C(CH_3)\!=\!C(CH_3)\text{-}CH_2Br$$

and

$$CH_2\!=\!CH\text{-}CH\!=\!CH_2 + Br_2 \longrightarrow CH_2Br\text{-}CH\!=\!CH\text{-}CH_2Br$$

The function of $ZnCl_2$ in the catalysis of the reaction transforming olefins into alcohols and esters was discovered by Kondakov in 1893 (105). For instance, 1-methyl-1-ethylethylene, when treated with $ZnCl_2$, gave dimethyethylcarbinol; and when trimethylethylene was treated with $ZnCl_2$ in the presence of acetic acid, tert-amyl acetate was formed in a 20% yield. In a similar manner, tert-amyl esters of formic acid and propionic acid were obtained.

Kondakov's interests also included the chemistry of terpenes. He discovered the rearrangement of dihydrocarvone in the presence of HBr (106), and the rule whereby cyclic saturated compounds having a –CHR-C(OH)- grouping reacted with acid halides or phosphorus halides to form both secondary and tertiary halides. With fenchyl alcohol, the reaction proceeded as follows (107):

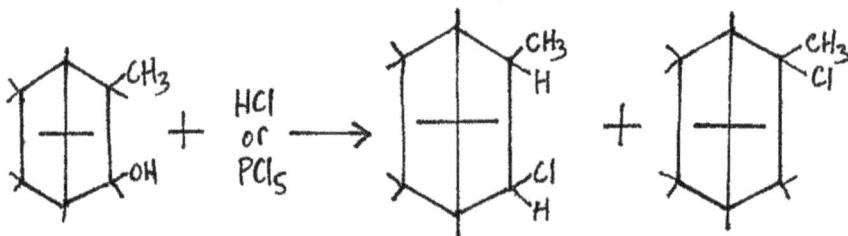

The nitration of hydrocarbons was investigated in some detail by Konovalov (1856–1929), a professor of St. Petersburg University. He found that nitrating aliphatic hydrocarbons with HNO_3 at elevated temperatures, the tertiary carbon atoms became nitrated first (108):

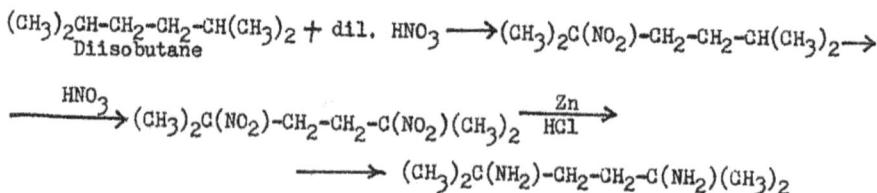

$$(CH_3)_2CH-CH_2-CH_2-CH(CH_3)_2 + \text{dil. } HNO_3 \longrightarrow (CH_3)_2C(NO_2)-CH_2-CH_2-CH(CH_3)_2 \longrightarrow$$

Diisobutane

$$\xrightarrow{HNO_3} (CH_3)_2C(NO_2)-CH_2-CH_2-C(NO_2)(CH_3)_2 \xrightarrow{Zn} _{HCl}$$

$$\longrightarrow (CH_3)_2C(NH_2)-CH_2-CH_2-C(NH_2)(CH_3)_2$$

Nitration of aromatic hydrocarbons with aliphatic side-chains resulted in the uptake of the nitro group by the side-chain, especially if the latter contained a tertiary carbon atom. Thus, 2-phenylpropane gave 2-phenyl-2-nitropropane. With longer side-chains, such as in phenylbutane, the carbon next to the phenyl ring was nitrated preferentially. The aliphatic methyl groups were very difficult to nitrate, though toluene could be nitrated with concentrated HNO_3 to give phenylnitromethane, which could be reduced to phenylmethylamine. Mesitylene was also susceptible to nitration with concentrated HNO_3 to give several products, and the conditions could be manipulated to get one product in preference to another (109). He was thus able to prepare nitromesitylene (m. p. 44°C), xylyl nitromethane (m. p. 46-47°C), dinitromesitylene (m. p. 86-87°C), and nitroxylyl nitromethane (m. p. 86°C). Extensive further nitration of xylyl nitromethane yielded two isomers of dinitroxylyl nitromethane. In the course of his studies on nitrated hydrocarbons, Konovalov discovered the fact that iron gives a red color with a variety of nitro compounds (110).

The synthesis of substituted carboxylic acids was accomplished by Reformatsky (beta-substitution) and by Zelinsky (alpha-substitution). The so-called Reformatsky reaction involves the treatment of an alpha-halo acid with zinc and a carbonyl compound (111). Reformatsky first observed this reaction when chloroacetic acid condensed with itself in the presence of zinc, and he followed up this by treating iodoacetic acid ethyl ester with zinc and acetone to form beta-hydroxyisovaleric acid ester:

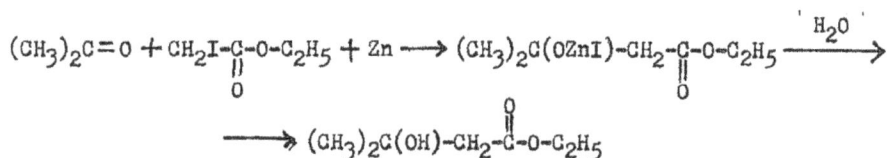

$$(CH_3)_2C=O + CH_2I\text{-}\underset{\underset{O}{\|}}{C}\text{-}O\text{-}C_2H_5 + Zn \longrightarrow (CH_3)_2C(OZnI)\text{-}CH_2\text{-}\underset{\underset{O}{\|}}{C}\text{-}O\text{-}C_2H_5 \xrightarrow{H_2O}$$

$$\longrightarrow (CH_3)_2C(OH)\text{-}CH_2\text{-}\underset{\underset{O}{\|}}{C}\text{-}O\text{-}C_2H_5$$

Reformatsky (1860–1934) was Zaitsev's student at Kazan University, and after studying abroad with Myer and Ostwald, settled at Kiev University where he remained to his death (112).

N. Zelinsky was responsible for developing improved methods for the general synthesis of alpha-substituted carboxylic acids, including amino acids. He was born in Odessa in 1861, where he received his education and had his first faculty appointment. He eventually went to Moscow University. In addition to his work on synthetic organic chemistry, Zelinsky is also known for his construction of a charcoal gas mask that was used by the Russian troops in the First World War. Zelinsky's method for the preparation of alpha-bromopropionic acid consisted of adding bromine dropwise at room temperature to a mixture of propionic acid and phosphorus, the product being propionyl bromide (113). The second step involved the dropwise addition of bromine to the propionyl bromide while the latter was refluxing. The final product was then alpha-bromopropionyl bromide. This general method for the preparation of alpha-halo acids was also described by others at the same time, and it is today known as the Hell-Volhard-Zelinsky method. Zelinsky also modified the classical Strecker amino acid synthesis procedure, and it is today known as the Zelinsky-Stadnikoff modification of the Strecker synthesis (114). This procedure involved the simultaneous mixing of KCN, NH_4Cl, and an appropriate aldehyde or ketone. The resulting aminonitrile was then extracted with ether and hydrolyzed with HCl. He prepared phenylglycine from benzaldehyde, dimethylglycine from acetone, cyclohexylglycine from hexahydrobenzaldehyde, and a number of other cyclic amino acids.

An interesting modification of the Cannizzaro reaction was reported by V. Tishchenko (1861–1941) of the St. Petersburg University, where instead of the acid and alcohol, the formation of an ester was observed in the presence of aluminum alcoholate (115). This reaction carries Tishchenko's name in organic chemistry texts. He was able to prepare benzyl benzoate in 80% yield from benzaldehyde, ethyl acetate in 60% yield from acetaldehyde, propyl propionate from propionaldehyde, and isobutyl isobutyrate in 85% yield from isobutylaldehyde. Ketones were inactive in this reaction.

The use of metalloorganic complexes in organic synthesis reactions has already been mentioned in connection with Zaitsev's method for alcohol synthesis. Another pioneer in this field was Paul Shorigin (1881–1939) from Moscow Technological Institute. He employed sodium to displace halogens from alkyl halides, then used the sodium complexes for a variety of synthetic reactions. These are commonly known as Shorigin reactions (116):

$$Rx + 2\,Na \rightarrow R\text{-}Na + NaX$$

The R-Na complexes could then react, like Grignard reagents, with ketones, aldehydes, esters, and CO_2 to form various alcohols and carboxylic acids, respectively. Another method for the preparation of sodium complexes was to use mercuric alkylates:

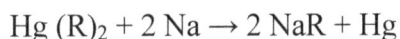

$$Hg\,(R)_2 + 2\,Na \rightarrow 2\,NaR + Hg$$

Shorgin also noted that various hydrocarbons were able to exchange sodium (117):

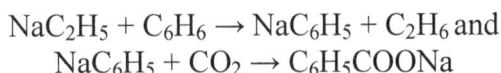

$$NaC_2H_5 + C_6H_6 \rightarrow NaC_6H_5 + C_2H_6 \text{ and}$$
$$NaC_6H_5 + CO_2 \rightarrow C_6H_5COONa$$

Using this method, he was able to prepare phenylacetic acid from toluene, m-methylphenylacetic acid from m-xyline, and 2-phenylpropionic acid from ethylbenzene. It is now known that sodium forms metallates with the hydrocarbons in the order of their increasing acidities:

$$C_2H_6 < benzene < toluene < diphenymethyline < triphenylmethane$$

The use of zinc complexes to synthesize branched hydrocarbons from alkyl halides was pioneered by N. Kursanov, who was working in Markovnikov's laboratory in Moscow. When cyclohexyl chloride was treated with dimethyl zinc, methylcyclohexane was obtained in a 25% yield. In the same fashion, using diethyl zinc, ethylcyclohexane was obtained in a 30% yield (118). A reaction of historical interest is the so-called Nencki reaction that utilizes chlorides for the synthesis of ketones (119). Nencki reacted benzoyl chloride with benzene in the presence of iron chloride, obtaining an intermediate with an empirical formula of $(C_6H_5COC_6H_5 . FeCl_3)_2 . C_6H_5COCl . FeCl_3$. It was then hydrolyzed to benzophenone.

There were a number of Russian investigators who were interested in the chemistry of nitrogenous organic compounds. One of the earlier workers in this field was J. Ponomarev, who in 1872 published a method for the synthesis of parabamic acid, which analyzed for $C_3H_2N_2O_3 . H_2O$, from urea, oxalic acid, and PCl_3 (120). This work was thus a forerunner of a more general set of reactions involving the condensation of urea with various dicarboxylic acids to give hydantoins and barbiturates.

An early investigator of the structure of alkaloids was Alexey Vishnegradsky (1851–1880), a graduate of the St. Petersburg University and an associate of Butlerov. One of his interests was the structure of cinchonine, an alkaloid isolated from cinchona bark (121). Alkali was able to split the compound into what was believed to be methylquinoline, and an unknown volatile base with an empirical formula of C_7H_9N and a boiling point of $166°C$. It was oxidized by chromic acid to $C_6H_5O_2N$, and was thus identified as ethylpyridine. Vishnegradsky concluded that cinchonine consisted of methylquinoline and ethylpyridine joined by an "acidic radical."

$C_2H_5C_5H_5N\text{-}C_2H_4CO\text{-}NC_9H_7CH_3$

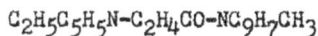

Ethyl- acid methyl-
pyridine radical quinoline

Vishnegradsky's cinchonine

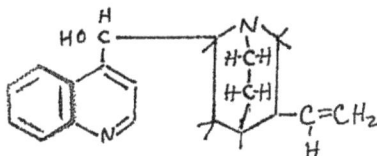

Modern structure of cinchonine

Another alkaloid studied by Vishnegradsky was quinine (122). Alkaline hydrolysis of the alkaloid yielded a derivative of quinoline containing oxygen. The new base boiled at 280°C and had an empirical formula of $C_{10}H_9NO$. He proposed that quinine differed from cinchonine only in the nature of its quinoline base. It is today known that indeed quinine differs from cinchonine only in that it has a methoxy group attached to the quinoline ring. Vishnegradsky was also interested in the chemistry of pyridine derivatives and showed that collidine was a trimethylpyridine, $C_5H_2N(CH_3)_3$, and that picoline was a methylpyridine, C_6H_7N (123). His analyses were, of course, correct, but he did not distinguish among the various isomers of these compounds. In the same manner, he attempted to structure-proof quinoline and pyridine by reducing both with tin in the presence of HCl and alkylating the nitrogen atom with C_2H_5I (124).

The most distinguished Russian researcher in the field of heterocyclic chemistry was Aleksey Chichibabin (1871–1945). He was a professor of chemistry at the Moscow Agricultural Academy, and after the Bolshevik coup, he emigrated to France. His best contribution is the so-called Chichibabin reaction for the synthesis of alpha-aminopyridine, which involved the treatment of pyridine with sodium azide (NaN_2) at 120°C in toluene as solvent (125). In a similar manner, aminopicoline and aminoquinoline were synthesized from picoline and quinoline, respectively. Chichibabin also found that the introduction of the amino group into pyridine made it much easier to nitrate the pyridine ring with HNO_3. The main product was p-nitro-alpha-aminopyridine. When HNO_2 was used instead of HNO_3, p-nitro-alpha-hyroxypyridine was obtained. Chichibabin also discovered a method for the synthesis of the pyridine ring structure from aliphatic aldehydes and ammonia (126):

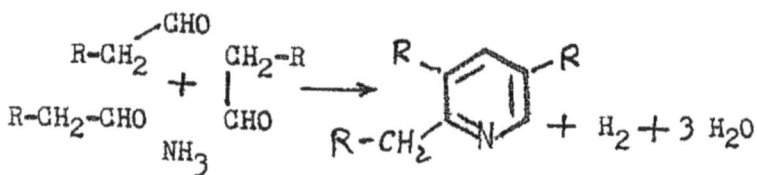

Thus, acetaldehyde and ammonia gave alpha-picolene, isovaleraldehyde gave alpha-isobutyl-beta, beta-diisopropylpyridine, and butylaldehyde gave alpha-propyl-beta, beta-diethylpyridine.

In addition to his work on heterocyclic compounds, Chichibabin also proposed a convenient general method for the synthesis of aldehydes using Grignard reagents and orthoformic acid esters (127). This procedure is commonly referred to as the Chichibabin-Bodroux synthesis:

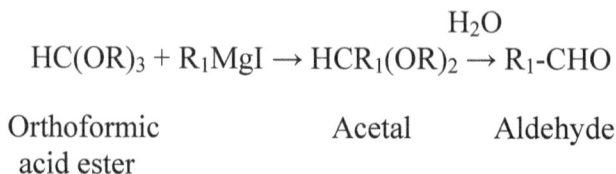

$$\overset{\text{H}_2\text{O}}{HC(OR)_3 + R_1MgI \rightarrow HCR_1(OR)_2 \rightarrow R_1\text{-}CHO}$$

Orthoformic acid ester	Acetal	Aldehyde

Beside preparing simple aliphatic aldehydes, the reaction was used to make cyclic carbonyl compounds (128):

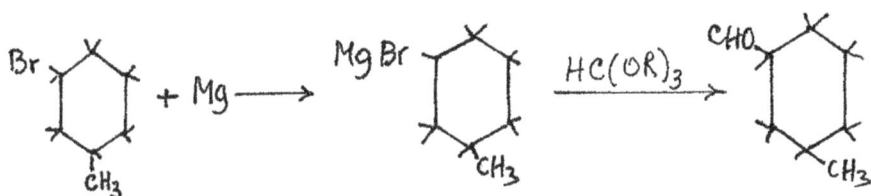

Piperidine synthesis was developed by Petrenko-Krichenko (1866–1944), a University of Odessa professor who used an ingenious method for condensation of acetone, benzaldehyde, and ammonia for this purpose (129). In addition, he was able to prepare pyridones from pyrans by exchanging the oxygen by ammonia:

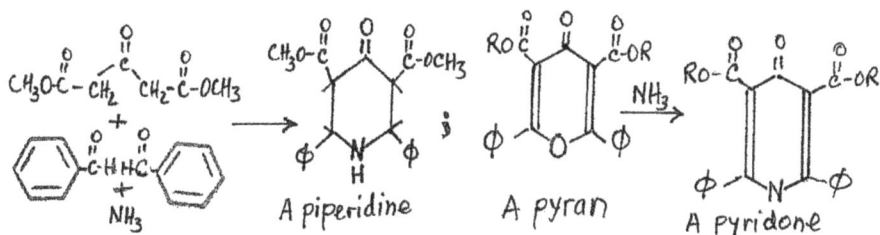

A piperidine A pyran A pyridone

This general procedure was then modified by substituting various amines for ammonia (130). The products were of course piperidines with substituted nitrogen atoms. Moreover, the piperidines were convertible into pyridones by chromic acid oxidation, whereby products identical to those obtained from pyrans were observed. In addition, the carboxyl groups could be removed by heating the pyridones above their melting points:

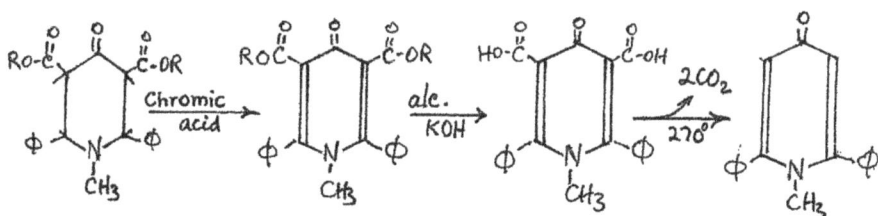

In addition to his work on heterocyclic compounds, Petrenko-Krichenko was interested in the effects of side-chains on the reactivity of carbonyl groups (131). He subjected various compounds to the action of bisulfite, phenylhydrazine, and hydroxylamine, and measured the extent to which the compound reacted after one hour. The reactivities, in decreasing order, were as follows:

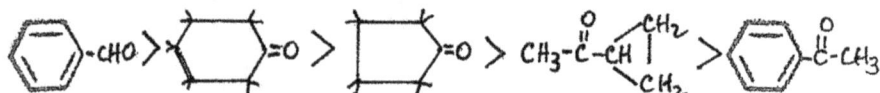

He proposed that the difference in reactivities of the various carbonyl compounds were due to the angle at which the carbonyl group was situated with respect to the neighboring atoms, and if the angle was not optimal, the side-chains would prevent the reagent from reaching the reactive group. Petrenko-Krichenko's data are now considered to be classical in the field of steric hindrance investigations.

Another Russian investigator who did much of the early work that was later used to formulate the concept of steric hindrance as well as the electronic theory of organic reaction mechanisms was Nicholas A. Menshutkin, who used mainly aliphatic amines for his investigations. Menshutkin (132) was the son of a merchant, born in 1842 in St. Petersburg, and a student of Voskresensky and Mendeleyev. Upon graduation from the St. Petersburg University in 1862, he went abroad to work with Strecker, Wurz, and Kolbe, and upon return home, became assistant to Mendeleyev. He remained in this position until 1869, when, upon passing his doctoral examinations, he became professor of analytical chemistry at St. Petersburg University, the other two chairs (general and organic chemistry) being occupied by Mendeleyev and Butlerov, respectively. In 1871, he published a book on analytical chemistry, which saw several editions and was translated into German and English. In 1885, Menshutkin took over the chair of organic chemistry as Butlerov left the University. In 1902, Menshutkin moved to the newly-established St. Petersburg Polytechnic Institute, in whose planning he participated along with Walden and Mendeleyev. In addition to his duties as professor of analytical chemistry, Menshutkin was also acting as the dean, whose duties included the planning of curriculum and hiring of faculty. Yet he found time to act as the editor of the *Journal of the Russian Chemical Society* (a job he held for thirty-two years), and to work as a deputy from the provincial county of the St. Petersburg Zemstvo. His son characterizes him as being politically a member of the "opposition," probably meaning opposition to the tight controls of educational institutions by the government. He publicly called for more academic freedom for university faculty, and after the 1905 disorders, actively participated in the election of the First Duma as a member of a "democratic reform party." Menchutkin died suddenly in 1907 of a stroke.

As indicated above, Menshutkin's main chemistry interest was the reactions of amines, as well as other nitrogenous compounds. Early in his career, he attempted to provide a classification of nitrogenous compounds (133), and had occasion to synthesize a number of compounds containing the elements of urea. He distinguished three groups of nitrogenous compounds: 1. Those that are hydrolyzed either to a monobasic acid and ammonia, or monobasic acid and urea. Examples given were acetamide, acetylurea, and other amides; 2. Those that are hydrolyzed into amino acids, e. g., aminobenzoic acid amide; and 3. "Dibasic" compounds, such as the barbiturates, alloxan, and oxalourea. This classification is of historical interest only. However, during his attempts at synthesis of the various nitrogenous substances in his quest for classification thereof, Menshutkin was able to develop a general method for the synthesis of quaternary amines from tertiary amines and alkyl halides. This method is commonly known as the Menshutkin reaction.

Menshutkin's main contributions to chemistry are of course his kinetic studies on the reactions between amines and alkyl halides. He determined that such reactions were of the second order type, obeying the equation

$$dx/dt = (A-x)(B-x)k$$

where A was the amine, B—the alkyl halide, t was time, x-- the amount of product formed, and k—the proportionality constant. Since the reactions were carried out with a twofold molar excess of the amine, so that A = 2 B, then

$$k = \log \frac{A - x/2}{A - x} \; 1/t$$

and all reactions were thus compared with respect to their k-values (134). All other factors being equal, the reaction proceeded fastest with methyl bromide, followed by ethyl bromide, and ending with propyl bromide. On the other hand, the speed with which ammonia was alkylated could be altered upwards or downwards by substituting the hydrogen atoms with alkyl residues. Thus, with respect to methyl bromide, dimethylamine reacted most rapidly, followed by trimethylamine, and finally by ammonia. On the other hand, ethylamine reacted faster than diethylamine, which was followed by ammonia, and finally by triethylamine. The same order of reactivities was seen with ethylamines with respect to ethyl bromide. In case of propyl bromide, propylamine was most reactive, followed by ammonia, dipropylamine, and tripropylamine. When allyl bromide was used (135), methylamine showed the highest reactivity with a k-value of 8302, whereas the reactivity of ethylamine was drastically lowered (k=3807). The k-value remained relatively constant for a series of other aliphatic amines, where heptylamine had a k-value of 3537. A methyl side-chain on the amines served to reduce the k-value: for propylamine and isopropylamine, the k-values were 3783 and 1257, respectively. The k-value remained about 1200 for 2-butylamine and 2-pentylamine. Amino groups attached to tertiary carbon atoms, such as in tert-butylamine, were least reactive, the latter compound showing a k-value of only 314. An amine with an ethyl side-chain generally showed a lower k-value than that with a methyl group, e. g., 2-butylamine and 3-pentylamine had k-values of 1240 and 672, respectively. However, if the side-chain was present on a carbon atom other than that attached to the amino group, the k-value was then much higher than that for the corresponding compound with a side-chain attached to the carbon atom holding the amino group. Thus, 2-butylamine, 1-amino-2-methylpropane, and 1-amino-3-methylbutane had k-values of 1200, 2759, and 2985, respectively. It may be noted that as the side-chain was moved away from the amino group, the k-value tended to approach that of the unbranched aliphatic amine.

A similar situation was observed with the rates of esterificastion of alcohols (135): methyl alcohol was esterified most rapidly with a k-value of 11,180. This was followed by ethyl alcohol, k=5420; propyl alcohol, k=4800, etc. Side-chains were inhibitory to the esterification process, and the slowest rates were observed with tertiary alcohols. As the side-chains were moved to carbon atoms other than those holding the hydroxyl groups, the rates of esterification tended to rise.

Menshutkin attempted to perform the same types of experiments with the aromatic compounds (136). Allyl bromide was permitted to react with the three toluidines, and the reaction rates were as follows: m-toluidine para-toluidine o-toluidine. On the other hand, when nitrobromobenzenes were permitted to react with dipropylamine, the opposite relationship was observed: o-nitrobromobenzene p-nitrobenzene m-nitrobenzene. Menshutkin investigated a number of such cases and came to the conclusion that in order to explain the reaction rates observed, it is necessary to consider the structural formulae of the benzene ring: the Kekule and the Claus versions. He of course did not have the benefit of the electronic theory to explain his results, though he did provide a wealth of data for the development of such a theory that was eventually formulated in the 1930s and 1940s.

Finally, Menshutkin provided an interesting correlation between the physical properties of compounds and their structures (137), one of the first such studies to be made. He measured the boiling points of various isomers of pentyl- or higher alcohols, and noted the difference between the boiling points of the highest and lowest boiling isomers (he called this the amplitude) was increased with increasing number of carbon atoms in the alcohol. For the butyl alcohols the amplitude was $34.6^{\circ}C$, and for pentyl, hexyl, and heptyl alcohols it was 35.4, 40.2, and 44.5. He also noted that the lowest boiling points were seen with the most branched alcohols, and the highest—with the most unbranched isomers.

As a final commentary, one can, arguably, state that organic chemistry research was the most successful basic science in prerevolutionary Russia, spanning some 100 years between its beginnings in the 1810s at Kazan University and ending with the disaster of 1917. A rather primitive way of identifying the number of pioneer organic chemists can be to look at the number of name reactions mentioned in this and other texts: only the number of German names is greater than that of the Russian ones. A number of organic chemists with international reputations left Russia after the revolution of 1917, including Ipatieff, Chichibabin, Arbusov, Kondakov, Walden, and others. And with the Soviet persecution of the resonance theory, even though it wasn't as severe as that on genetics, organic chemistry never resumed the stature it had in Imperial Russia before 1917.

Physical and Inorganic Chemistry

Physical chemistry had not yet come of age by the end of the nineteenth century. For instance, the ideas of Willard Gibbs and von Helmolz regarding the thermodynamics of chemical reactions were not formulated until the end of 1870s and beginning of 1880s. And the chemical nature of solvents and solutions was also a subject of incessant discussions. The ionization of solutes in solutions was not accepted by everyone. In Russia, especially St. Petersburg, where chemistry was under the heavy influence of Mendeleyev, ionization was not accepted as a fact, because Mendeleyev did not believe in it and insisted that the solutes somehow interact with water or other solvent molecules when the solutes are dissolved therein. On the other hand, Ostwald in Leipzig, Germany, believed that solutes become ionized in solution, and nothing happens beyond that. Both schools of thought had their acolytes. Enter into the fray two young Russian chemists, Ivan A. Koblukov (1857–1942) and Vladimir A. Kistiakovsky (1865–1952) (138), who both believed that a solute will, if possible, dissociate into its constituent ions and the ions will then interact with the solvent's molecules to form complexes. In other words, both processes proposed (ionization and hydration in case the solvent is water) will occur when a solution is formed. It was a novel concept, and both candidates wrote their doctoral dissertations promoting it. Koblukov passed his exam, which he took in Moscow, whereas Kistiakovsky, who took the exam in St. Petersburg close to Mendeleyev's presence, failed to pass. It turned out eventually that both Kistiakovsky and Kablukov interpreted their data correctly. Nevertheless, both did very well after the revolution in the USSR: both were admitted to the Soviet Academy of Sciences in Leningrad, and neither was arrested or sent to the gulag. Incidentally, Vladimir Kistiakovsky was an uncle of the American physical chemist George

A. Kistiakovsky (1900–1982), who fought against the Bolsheviks in the Russian civil war, and thereafter received his chemistry education in Berlin, Germany. He then emigrated to the US and became American citizen in 1933. He was a professor at the Harvard University and worked on explosives during World War II.

And thus, at the end of nineteenth century, physical chemistry was still much less popular in Russia than was organic chemistry. Kistiakovsky's examiners simply didn't know what to do with it when he was defending his doctoral dissertation. And to be safe, they did not award him the degree. There were some other Russian physical chemists who left Russia to pursue their chemistry careers, such as Gustav Tammann (1861–1938). He was born in Yamburg, near St. Petersburg, and was a graduate of the Dorpat University. He spent most of his professional life in Goettingen, Germany, where he developed the method for molecular weight determination by measurement of vapor pressures and other colligative properties. Then there was Robert Luther (1868–1945), who was born in Moscow and graduated also from the Dorpat University. He spent most of his life in Leipzig, where he became a distinguished electrochemist and photochemist. And perhaps the most famous Russian expatriate physical chemist was Wilhelm Ostwald (139, 140). He was born in Riga (Russia, now capital of Latvia) in 1853 in a tradesman's family, and graduated from the Dorpat University in 1875, where his mentor was principally Carl Schmidt (see chapter IV). In 1878, Ostwald obtained his doctorate in chemistry, and in 1882 he became a professor of chemistry at the Riga Polytechnicum. It was in Riga where he did most of the work that resulted in his winning the Nobel Prize in chemistry in 1909. He moved to Leipzig in 1887 to occupy the chair of physical chemistry and later, to direct a large institute concerned with physical-chemical problems. Ostwald's most distinguished student in Riga was Paul Walden, whereas in Leipzig he acted as an advisor to such luminaries as Beckman and Nernst. In 1904, Ostwald resigned his position in Leipzig to devote his time to philosophy, but following World War I he returned to science to work on the theory of colors. He even proposed himself as a candidate for another Nobel Prize in physics for the development of his color theory, but his recommendation was apparently not well received. Throughout his later life, Ostwald was interested in Monism, a metaphysically-oriented sect, and even established a camp/farm for his followers. It was a financial disaster, almost ruining the well-meaning professor. He died in 1932.

Ostwald's contributions were numerous, for in addition to his own work, he did much to promote and popularize the theories of other chemists, such as the electric conductance theory of Arrhenius, the thermodynamic concepts of Gibbs, and the energy relationships proposed by J. R. Mayer. While still in Riga, Ostwald, in collaboration with van't Hoff, established the first physical-chemical journal, the *Zeitschrift fuer physikalische Chemie.* In 1894, he established the German Electrochemical Society, later renamed the Bunsen Society. Ostwald is credited by many to be the "father of physical chemistry." He laid down the principles inherent in the practice of physical chemistry in his first book, the *Lehrbuch der allgemeiner Chemie,* which he started to write in Riga and finished in Leipzig. He popularized the term "physical chemistry," though it had been used earlier by Graham Otto. He did much of the original work in chemical catalysis, providing in 1884 the following definition: "Catalysis is the acceleration by a foreign substance of a chemical reaction, which is taking place slowly"; and also, "a substance—a catalyst—can affect a chemical reaction's speed, but is not included in its end-products." "This understanding of a catalyst shed a great light on chemical reactions occurring in both industrial processes and living organisms" (the *Internet* in its Ostwald's biography). It is more than surprising that until the 1890s, the world had no concept of what a catalyst was and why it was important, especially in the maintenance of living organisms and their enzymes. As stated above, Ostwald won the Nobel Prize in 1909 for his work on chemical catalysis. His other achievements include the development of the calomel reference electrode, and the formulation of the so-called "Ostwald dilution law," which provides a relationship between the concentration of an electrolyte, its dissociation constant, and its conductivity. The Ostwald dilution law can be stated by

$$K = L^2c / L_0 (L_0 - L),$$

where K is the dissociation constant, c is the electrolyte concentration, and L and L_0 are equivalent conductivity and the equivalent conductivity at infinite dilution, respectively. The law holds for salt solutions with concentrations of less than 0.1. The Ostwald viscometer is also well known to any student of physical chemistry.

There were, of course, a number of physical chemists who practiced their trade in Russia without going abroad. One here may mention A. L. Potilitsin, who studied heat relationships in chemical reactions; Vladimir Kistiakovsky of Kiev University, who was already mentioned above; and A. Titov from Moscow University, who made the first rigorous measurements of absorption of gasses by solids such as charcoal (141). His work undoubtedly contributed to the development of Russian charcoal gas masks during the First World War, which were developed by Zelinsky in 1915. It will also be remembered that Hess did his thermochemical work in the laboratories of St. Petersburg Academy of Sciences.

In America, the best known physical chemist of Russian origin is probably Vladimir Ipatieff, whose career spanned both the nineteenth and twentieth centuries and both the European and American continents (142). His name was already mentioned above in the section on organic reactions. Ipatieff was born in 1867 in Moscow, and received his education in the various military schools and academies of Moscow and St. Petersburg. He was eventually appointed to the faculty of the Michael Artillery Academy, where most of his research was conducted. Though much of Ipatieff's knowledge of chemistry was acquired on his own, he nevertheless sought the assistance of Favorsky in choosing a problem for his dissertation that he had to provide at the academy. He later studied with von Baeyer in Germany, and was awarded a doctorate from St. Petersburg University. Many honors were bestowed upon Ipatieff by his own country as well as foreign organizations. By 1914, he already had the rank of lieutenant general in the Russian army (through his association with the Artillery Academy). He was elected a member of the St. Petersburg Academy of Sciences in 1916, and was also elected to the French Legion of Honor. After the Soviet coup, he remained to rebuild Russia's chemical industry, and received much authority from Lenin to accomplish his task. Upon Lenin's death and ascendance of Stalin to the Soviet throne, suspicion fell upon Ipatieff as it did on numerous Russian intellectuals, and he was obliged to leave the country to avoid arrest. He settled in Chicago, where he was associated with the Universal Oil Company and Northwestern University. He died in Chicago in 1952.

Ipatieff is best known for his pioneering work on high pressure catalysis, though he had a number of important papers in the area of classical organic chemistry that generally reflected Favorsky's or von Baeyer's influences (e. g., 95). He constructed a portable high-pressure reaction vessel that is sometimes referred to as the "chemist's test tube," and using such an apparatus, he carried out numerous catalytic reactions involving liquids. For example, using nickel oxide as a catalyst, he was able to hydrogenate aniline to give cyclohexylamine; diphenylamine to give dicyclohexylamine; and quinoline to give either tetrahydroquinoline or dekahydroquinoline, depending on conditions. His conditions were a hydrogen gas pressure of 100–150 atm. at 220–250°C for 10–50 hours (143). In 1900, Ipatieff discovered the phenomenon of contact catalysis, and the methodology may be represented by his paper describing the hydrogenation of benzyl alcohol in an iron pipe at a hydrogen gas pressure of 96 atm. and 350–360°C (144). The products obtained with such a reaction were toluene, benzaldehyde, and dibenzyl. A copper pipe was much less effective. Ipatieff's work on the hydrogenation of olefins and fats eventually led to the development of the margarine industry. His research was, of course, of great interest to industry, a number of reactions being adapted for industrial production. As a result, Ipatieff was a consultant to numerous industrial organizations both in Russia and abroad, and this gave him an opportunity to travel widely and to meet many of his professional colleagues. These contacts enabled him to flee Stalin's persecution in the Soviet Union, for which he was expelled from the Academy of Sciences and made a nonperson in his native land. His existence was reacknowledged only after Stalin's death, and he was posthumously readmitted into the Academy of Sciences.

In the United States, Ipatieff worked principally on the catalytic reactions of saturated and unsaturated hydrocarbons and the application of these to petroleum industry. He was able to effect the dehydrogenation of saturated hydrocarbons by oxides of the transition elements at high temperatures to produce olefins, which, in turn, could be used to alkylate paraffins either using high temperatures only, or using various halides as catalysts. Thus, cumene was synthesized from benzene and propylene, and proved to be an important additive in aviation fuels. Ipatieff also developed the production of 2,2,4-trimethylpentene, an important component of high octane fuels, by dehydrogenating the appropriate alcohol, then catalytically hydrogenating the resulting olefin. His work contributed greatly to the

Allied efforts during World War II and to the development of high compression engines.

The man who did most to introduce physical chemistry as a separate subject into the Russian university chemistry curriculum was Nicholas S. Kurnakov (1860–1941), a graduate of and a professor at the St. Petersburg Mining Institute (145). His professional life spanned both the imperial and Soviet eras of Russian history, and he was a recipient of the highest awards from both regimes. Before the Bolshevik coup, Kurnakov was a member of numerous committees and commissions on the industrialization of Russia and utilization of its mineral and other natural resources. During the Soviet period, his activity in the public service sector was not diminished. Kurnakov's interests encompassed several areas of physical chemistry, as well as inorganic chemistry. He was first to distinguish between the cis- and trans-isomers of platinum coordination compounds by using thiourea as follows:

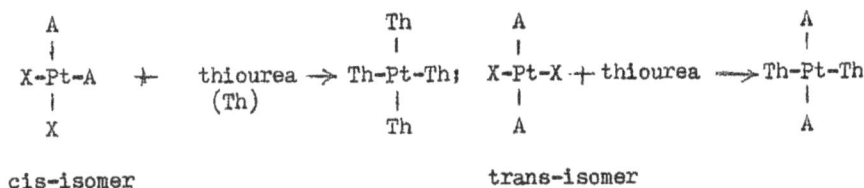

$$
\begin{array}{c}
\text{A} \\
| \\
\text{X-Pt-A} \\
| \\
\text{X}
\end{array}
+ \;\; \begin{array}{c} \text{thiourea} \\ \text{(Th)} \end{array} \rightarrow
\begin{array}{c}
\text{Th} \\
| \\
\text{Th-Pt-Th;} \\
| \\
\text{Th}
\end{array}
\quad
\begin{array}{c}
\text{A} \\
| \\
\text{X-Pt-X} \\
| \\
\text{A}
\end{array}
+ \text{thiourea} \longrightarrow
\begin{array}{c}
\text{A} \\
| \\
\text{Th-Pt-Th} \\
| \\
\text{A}
\end{array}
$$

cis-isomer   trans-isomer

He also worked on the significance of refractive indeces, color, and solubilities of inorganic compounds and complexes, and pioneered the field of halurgy, where he elucidated the equilibria observed among the various salts in salt lakes, e. g., $MgCl_2 + Na_2SO_4 \rightarrow MgSO_4 + 2\,NaCl$. His work in the field of metallurgy was also far-reaching. He described the construction of a pyrometer, which is still in use in many metallurgical plants and laboratories, he recognized the importance of the vapor phase in entectic fusion, and was first to correlate and systematize physical-chemical analyses of alloys and other complex mixtures with their physical properties such as electroconductivity, hardness, internal friction, pressure of flow, etc. He considered alloys to be simply compounds of variable composition, terming them berthollides, whereas compounds of constant composition were termed by him as daltonides.

The Russian chemist who is most popular among his American colleagues is without doubt Dimitry Mendeleyev. Though his periodic law is now some 150 years old, interest in its discoverer, both as a chemist and as a person, has not diminished, and we now have information as to the kind of chess player he was, what he thought about women in the professions, what his views were on secondary education, his opinions on America, and his writings on demographics. He is even credited with influencing Marie Curie with taking up chemistry as her profession (146). Most books on the history of chemistry or science devote a large chunk of space on Mendeleyev, so that it would appear that his life and work have been well researched in the West (147, 148, 149). Mendeleyev was born in the Siberian city of Tobolsk in 1834 of a Russian father and a Tatar mother. He was the seventeenth child in the family. He was educated in the Pedagogical Institute in St. Petersburg, where Voskressensky's lectures were apparently of most interest to him. After serving as a teacher in Odessa, he went abroad to work with Bunsen, and upon his return to Russia, he joined the St. Petersburg Technological Institute. He remained there until 1866, then moved to St. Petersburg University only to resign in 1890 in protest of a reprimand he received from the Minister of Education for his support of rebellious students. Yet soon thereafter, he was appointed head of the newly established Bureau of Weights and Measures, the equivalent of American Bureau of Standards. He remained there until his death in 1907.

Many honors were bestowed upon Mendeleyev both at home and abroad in recognition of his contributions to science. He was the recipient of the Davy and Copley medals from the Royal Society, the Faraday medal from the British Chemical Society, he was elected to membership of the Royal Society in 1890, and was member of the Prussian Academy of Sciences. Yet he never made it into the St. Petersburg Academy of Sciences. This unfortunate fact is interpreted by several historians as being a reflection of Mendeleyev's liberal views (149, 150). The story is told as follows: membership of the academy had separated into the so-called German and Russian factions. The former was supposedly conservative, even reactionary and chauvinistic. The Russian faction was liberal and progressive. Thus, when a liberal-minded scientist of Russian extraction was proposed for membership, the German faction would attempt to defeat the nomination, often with the collusion of government bureaucrats. Mendeleyev was classified as a liberal Russian, hence his nomination for membership in 1880 was defeated. Beilstein, being of German origin, was elected instead. It should be noted that only a limited number of new members from each discipline could be elected to the academy each year. A decisive role in the rejection of Mendeleyev was apparently played by the then Minister of Education, Count Dimitry Tolstoy. It is quite possible that Mendeleyev's liberalism contributed to his being rejected for the academy's membership. Yet it should also be pointed out that there were many liberal Russian members in the academy, and furthermore, Mendeleyev was far from being a revolutionary. He was a devoted monarchist, and was even received by the tsar. It is likely that a major contributing factor to his being rejected for the academy's membership was his unpleasant personality, i. e., his volatile temper and lack of prowess in the art of public relations. This created a number of enemies and non-well-wishers, especially in the person of Dimitry Tolstoy. Mendeleyev's difficult personality has been well documented (151). He was again proposed for membership in 1886, but refused to be considered.

Mendeleyev was often obliged to travel both domestically and abroad for the purpose of inspecting various industrial and mining operations, and to make recommendations for their improvement. The reason for such travels is often stated to be the fact that the government wanted him away from St. Petersburg whenever trouble was brewing among the students. Again, this may have been true, but it should also be pointed out that it was the practice of the Russian government to depend heavily on the opinions of its scientists for advice regarding development of industry and use of natural resources (perhaps it should have depended more on the views of its engineers and businessmen, but that is another matter). For this reason, the more accomplished scientists were often asked to undertake such travels and to make reports and recommendations on their findings. Thus, von Baer, the embryologist, was in his time asked to travel to the Caspian Sea to inspect Russian fishing industry, though it is not exactly clear as to what von Baer ever knew about commercial fishing. Beilstein was asked to inspect the oil wells of Baku, and again, it is unclear exactly what a chemical documentation specialist of Beilstein's repute should know about pumping oil from under the earth's crust. Nobody had endeavored to speculate whether or not von Baer's or Beilstein's trips were synchronized with student unrests, or whether or not their trips resulted in any improvements in the inspected industries.

One of Mendeleyev's assignments (in 1876) was to visit the United States and inspect the oil fields of Pennsylvania (152). Though he admired American technology in the oil exploration and exploitation fields, he decried the lack of planning, the disregard of environmental effects in the areas of oil fields, and the lack of scientific inquiry in the area of petroleum chemistry. The entire scientific atmosphere of the United States did not appear to him to be of high caliber. Though he liked Philadelphia well enough, New York, on the other hand, appeared to him to resemble a typical Russian provincial town in regard to drabness of its architecture, city planning, and the quality of shops. Yet Mendeleyev felt that the adoption of some American technological and commercial methods would be of great benefit to Russia, and, in fact, believed that America was the land of the future. He had no use for the socialistic form of economy, in spite of Soviet claims to the contrary (153).

Mendeleyev held definite views on a number of social issues. In the area of equality of the sexes, he was convinced that women deserved equal opportunities, yet he doubted that their temperament made them suitable for positions in science, technology, and commerce. On the other hand, he felt that women were eminently suited for the practice of fine arts (154). Nevertheless, Mendeleyev was one of the first chemistry professors who agreed to give lectures at the St. Petersburg Women's College. Concerning the demographics and population control, Mendeleyev was apparently anti-Malthusian, and felt that sciences and the people themselves would eventually bring the problem of overpopulation under control. He predicted that the population of Russia would be 280 million by 1950, though it reached barely the 200 million mark in that year (153). The reason: losses during World War I and the Russian civil war, followed by Soviet-generated famines, the gulags, executions, and losses during World War II when Russia lost over 20 million lives. Mendeleyev's ideas on education were characterized by their anticlassical and utilitarian principles. He felt that along with improving its educational system, Russia should develop its industry. Otherwise, trained technical and scientific personnel would occupy themselves with such useless activities as metaphysics and dialectics. All educational institutions, including secondary schools, were to be under the Ministry of Education, and the curricula of all secondary schools were to be standardized. However, he felt that the government should only pay the bills, not control the education, which should be administered by the faculty (155).

The formulation of the periodic law by Mendeleyev was one of the most important scientific discoveries of mankind. It undoubtedly required a genius of the kind that would have that rare and peculiar gift of being able to generalize from scattered and tentative pieces of information, often buried in obscure and inaccessible publications. Mendeleyev apparently possessed such a gift (156). In addition, the 1860s were a period when a sufficient number of elements had been discovered and their properties determined with a fair degree of accuracy. Mendeleyev had also attended the 1860 Karlsruhe chemical convention, where he heard Cannizarro's now classical paper correlating Avogadro's principle with the concept of atoms, molecules, and atomic weights. And so, in March of 1869, the periodic table was presented to the world at a meeting of the Russian Chemical Society. It was printed in several variations in the April issue of the *Journal of the Russian Chemical Society* (157) and is reproduced at the end of this chapter. The same paper contained eight conclusions reached by Mendeleyev regarding the properties of elements by studying their arrangement in this table: 1. Elements arranged in the order of increasing atomic weights represent a periodicity of properties; 2. Elements having similar atomic weights (e. g., Pt, Ir, Os), or those whose atomic weights show a regular increase (e. g., K, Rb, Cs) show similar chemical properties; 3. It is possible to predict the valence of elements from their positions in the table; 4. The most abundant elements have the lowest atomic weights, and the greatest differences are observed among such low-atomic weight elements; 5. The atomic weight of the element determines its properties; 6. It is expected that numerous other elements will be discovered in the future, e. g., elements similar to Al and Si with atomic weights of 65 to 75; 7. The atomic weights of some known elements should be corrected, e. g., that of Te should be 123–126 instead of 128; 8. Certain analogies among elements may be drawn from their atomic weights, e. g., uranium is an analogue of boron and aluminum.

In a later publication, Mendeleyev predicted the properties of three as yet undiscovered elements of that time: the eka-boron, eka-aluminum, and eka-silicon. His table attracted very little attention at first, and only after his predictions began to materialize did his colleagues begin to take him seriously. It may be noted that a German chemist, Lothar Meyer, published a periodic table of his own some seven months after Mendeleyev's, but Meyer's paper did not make the type of bold predictions that Mendeleyev's did. Lothar Meyer is therefore not generally honored for his discovery to the same extent that Mendeleyev is.

From time to time, Mendeleyev improved upon his original table, but it was not until 1913 when a major revision of the table appeared, when the elements became arranged on the basis of their atomic numbers instead of atomic weights. The greatest triumph of Mendeleyev's discovery came, of course, when the existence of eka-boron (atomic weight 45), eka-aluminum (atomic weight 68), and eka-silicon (atomic weight 75) were proven through the discovery of what we today call scandium (atomic weight 45) by Nilson in 1879, of gallium (atomic weight 69.8) in 1874 by de Boisbaudran, and of germanium (atomic weight 72.6) in 1885 by Winkler, respectively. The properties of germanium and eka-silicon were strikingly similar, as seen from the following table:

Properties of Mendeleyev's eka-silicon and Winkler's germanium

	Eka-silicon	Germanium
Atomic weight	70	72.6
Specific gravity	5.5	5.47
Atomic volume	13	13.22

Valence	4	4
Specific heat	0.073	0.076

The periodic table was not Mendeleyev's only contribution to his profession. He was, for instance, concerned with the structure of organic compounds (158) and the study of the properties of vapors and the expansion of liquids with increasing temperatures. He was first to recognize the existence of the critical point, i. e., temperature at which a vapor cannot be liquefied no matter what the pressure is. His term for the critical point was "absolute Siedentemperatur," and he defined the same as follows (159): *"Als absolute Siedentemperatur muessen wir den Punkt betrachten, bei welchem 1. die Cohaesion der Fluessigkeit 0° ist und $a^2 = 0$, bei welcher 2. die latente Verdampfungswaerme auch 0 ist und bei welcher sich 3. die Fluessigkeit in Dampf verwandelt, unabhaengig von Druck und Volum."* He determined the critical temperatures to be 190, 250, and 580°C for ether, alcohol, and water, respectively.

In closing, it may be stated that Russian chemists were a pragmatic lot who, contrary to the biologists, had no intention to manipulate human nature through conditioned reflexes or otherwise, or to replace the theocratic-autocratic nature of Russian society by an equally if not more dogmatic ethic based on Darwinism as perceived by them. For the most part, the chemists were loyal subjects of the tsar and sought only reforms that would contribute to the economic and intellectual development of their people. They were especially interested in contributing to the development of Russian industry, believing that the social and political problems would take care of themselves once the industrialization of Russia would be accomplished.

REFERENCES

1. N. A. Figurovski. *The history of chemistry in ancient Russia.* Chymia 11:45, 1966.
2. P. M. Luk'yanov. *The first chemical laboratories in Russia.* Chymia 9:59, 1966.
3. P. Walden. *The Gmelin chemical dynasty.* J. Chem. Ed. 31:534, 1954.
4. H. M. Leicester. *The history of chemistry in Russia prior to 1900.* J. Chem. Ed. 24:438, 1947.
5. B. N. Menshutkin. *Russia's Lomonosov.* Princeton Univ. Press, Princeton, 1952.
6. M. Leicester. *Mikhail Vasilyevich Lomonosov on the corpuscular theory.* Harvard Univ. Press, Cambridge, 1970. Quoted by R. Siegfried in Science 171: 1231, 1971.
7. H. M. Leicester. *The spread of the theory of Lavoisier in Russia.* Chymia 5:138, 1959.
8. H. M. Leicester. *Tobias Lowitz—discoverer of basic laboratory methods.* J. Chem. Ed. 22:149, 1945.
9. B. N. Menshutkin. *Discovery and early history of platinum in Russia.* J. Chem. Ed. 11:226, 1934.
10. H. M. Leicester. *Germain Henri Hess and the foundations of thermochemistry.* J. Chem. Ed. 28:581, 1951.
11. H. Hess. *Thermochemische Untersuchungen.* Poggendorf's Annalen d. Physik u. Chemie 50:385, 1840.
12. F. E. Sheibley. *Carl Julius Fritzsche and the discovery of anthranilic acid.* J. Chem. Ed. 20:115, 1943.
13. J. Fritzsche. *Ueber das Anilin, ein neues Zersetzungsproduct des Indigo.* Ann. Chem. Pharm. 36:84, 1840.
14. J. Fritzsche. *Ueber die Producte der Einwirkung von Kali auf Indigoblau.* Ann. Chem. Pharm. 39:76, 1841.
15. P. Friedlaender and E. Schwenk. *Ueber die Zersetzung von Indigoblau und Indigorot durch Alkalien.* Ber. 43:1971, 1910.
16. J. Fritzsche. *Ueber die Producte der Einwirkung der Salpetersaeure auf die Phensaeure.* Ann. Chem Pharm. 110:150, 1859.
17. J. Fritzsche. *Ueber Verbindungen von Kohlenwasserstoffen mit Pikrinsaeure.* Ann. Chem. Pharm. 109:247, 1859.
18. J. Fritzsche. *Ueber einen Kohlenwasserstoff aus Holztheer und seine Verbindung mit Pikrinsaeure.* Ann. Chem. Pharm. 109:250, 1859.
19. N. M. Brooks. *Nikolai Zinin and synthetic dyes: the road not taken.* Bull. Hist. Chem. 27:26, 2002.
20. D. E. Lewis. *Klaus at Kazan: the discovery of Ruthenium.* Bull. Hist. Chem. 41:3, 2016.

21. C. Claus. *Beitraege zur Chemie der Platinmetalle.* Ann. Chem. Pharm. 63: 337, 1847.

22. M. E. Weeks. *The discovery of elements.* VIII. The platinum metals. Ruthenium. J. Chem. Ed. 9:1028, 1932.

23. D. E. Lewis. *The beginnings of synthetic organic chemistry: Zinc alkyls* and the Kazan's school. Bull. Hist. Chem. 27:37, 2002.

24. M. Kolesnikov. *Lobachevskii.* Molodaia Gvardia, Publ., Moscow, 1965 (in Russian).

25. H. M. Leicester. *N. N. Zinin, an early Russian chemist.* J. Chem. Ed. 17: 303, 1940.

26. N. Zinin. *Organische Saltzbasen aus Nitronaphthalose und Nitrobenzid mittelst Schwefelwasserstoff entstehend.* Ann. Chem Pharm. 44:283, 1842.

27. N. Zinin. *Ueber das Benzil.* Ann. Chem. Pharm. 119:177, 1861.

28. N. Zinin. *Ueber die Einfuerhrung von Wasserstoff in organische Verbindungen.* Ann. Chem. Pharm. 119:179, 1861.

29. N. Zinin. *Ueber desoxydirtes Binzoin, ein Product der Einwirkung des Wasserstoffs auf Benzoin.* Ann. Chem. Pharm. 125:218, 1863.

30. A. Vucinich. *Science in Russian culture, 1861–1917.* Stanford Univ. Press, Stanford, 1970.

31. D. Joravsky. *Scientists and autocracy.* Science 172:550, 1971.

32. H. M. Leicester. *Alexander Mikhailovich Butlerow.* J. Chem. Ed.17:203, 1940.

33. A. Butlerow. *Einiges ueber die chemische Struktur der Koerper.* Zeitschr. f. Chem. U. Pharm. 4:549, 1861.

34. G. V. Bykov. *The origin of the theory of chemical structure.* J. Chem. Ed. 39:220, 1962.

35. A. Butlerow. *Bemerkungen ueber A. S. Couper's neue chemische Theorie.* Ann. Chem. Pharm. 110:51, 1859.

36. A. Butlerow. *Isomerie der gesaettigten Kohlenwasserstoffe C_4H_{10} und der Butylene C_4H_8. Isobutylalkohol (der primaere Pseudobutylalkohol oder Pseudopropylcarbinol).* Ann. Chem. Pharm. 144:1, 1867.

37. A. Butlerow. *Ueber einige Eigenschaften des Trimethylcarbinols.* Ann. Chem. Pharm. 162:228, 1872.

38. F. Walker. *Early history of acetaldehyde and formaldehyde.* J. Chem. Ed. 10:546, 1933.

39. A. Butlerow. *Ueber Isodibutylen.* Liebig's Ann. 189:44, 1877.

40. A. Butlerow and M. Ossokin. *Ueber eine synthetische Bildungsart der Alkohole und die chemische Struktur des Aethylens.* Ann. Chem. Pharm. 145:256, 1868.

41. A. Popoff. *Ueber die Isomerie der Ketone.* Ann. Chem. Pharm. 145:283, 1868.

42. A. Popoff. *Ueber das Aethyldimethylcarbinol.* Ann. Chem. Pharm. 145:292, 1868.

43. H. M. Leicester. *Vladimir V. Markovnikov.* J. Chem. Ed. 18:53, 1941.

44. W. Markownikoff. *Ueber die Acetonsaeure.* Ann. Chem. Pharm. 146:339, 1868.

45. W. Markownikoff. *Ueber die Abhaengigkeit der verschiedenen Vertretbarkeit des Radicalwasserstoffs in den Buttersaeuren.* Ann. Chem. Pharm. 153:228, 1870.

46. W. Markownikoff. *Einwirkung von Salpetersaeure und Nitroschwefelsaeure auf verschiedene Grenzkohlenwasserstoffe.* Ber. 32:1441, 1899.

47. A. Saytzeff. *Zur Kenntniss der Reihenfolge der Anlagerung und Ausscheidung der Jodwasserstoffelemente in organischen Verbindungen.* Liebig's Ann. 179:296, 1875.

48. G. Wagner and A. Saytzeff. *Ueber Amylenbromuer und Amyglycol aus Diaethylcarbinol.* Liebig's Ann. 179:302, 1875.

49. G. Wagner and A. Saytzeff. *Umwandlung des Diaethylcarbinols in Methylpropylcarbinol.* Liebig's Ann. 179:313, 1875.

50. G. Wagner and A. Saytzeff. *Synthese des Diaethylcarbinols, eines neuen Isomeren des Amylalkohols.* Liebog's Ann. 175:351, 1875.

51. J. Kanonnikoff and A. Saytzeff. *Neue Synthese des secundaeren Butylalkohols.* Liebig's Ann. 175:374, 1875.

52. (A.). Eltekoff. *Ueber die Abscheidung von Haliodwasserstoffelelementen aus den Haloidverbindungen der Kohlenwasserstoffe C_nH_{2n}.* Chem. Centralbl. 1878:85.

53. (A.) Eltekoff. *Ueber eine neue Synthese der Olefine.* Chem. Centralbl. 1878: 290.

54. A. Wischnegradsky. *Ueber die Isomerie der Amylene.* Chem. Centralbl. 1878: 403.

55. P. Walden. *Notes from the life of a chemist.* J. Chem. Ed. 28:160, 1951.

56. P. Walden. *Ueber Optisch active Halogenverbindungen.* Ber. 28:1287, 1895.

57. P. Walden. *Ueber die gegenseitige Umwandlung optischer Antipoden.* Ber. 29:133, 1896.

58. P. Walden. *Ueber die gegenseitige Umwandlung optischer Antipoden.* Ber. 30:3146, 1897.

59. P. Walden. *Ueber die Spaltung racemischer Verbindungen in ihre active Bestandtheile.* Ber. 32:2703, 1899.

60. P. Walden. *Ueber das optische Verhalten des Tannins.* Ber. 30:3151, 1897.

61. E. Hjelt. *Friedrich Konrad Beilstein.* Ber. 40:5041, 1907.

62. D. N, Kursanov, M. G. Gonikberg, B. M. Dubinin, M. I. Kabachnik, E. D. Kaverzneva, E. N. Prilezhaeva, N. D. Sokolov, and R. Kh. Freilina (transl. By I. S. Bengelsdorf). *The present state of the chemical structural theory.* J. Chem. Ed. 29:1, 1952.

63. M. Hunsberger. *Theoretical chemistry in Russia.* J. Chem. Ed. 31:504, 1954.

64. V. M. Tatevskii and M. I. *Shakhparanov.* (trans. by I. S. Bengelsdorf). *About A Machistic theory in chemistryand its propagandists.* J. Chem. Ed. 29:13, 1952.

65. H. B. Friedman. *Alexander Borodin—musician and chemist.* J. Chem. Ed. 18:521, 1941.

66. F. H. Getman. *Alexander Borodin—chemist and musician.* J. Chem. Ed. 8:1763, 1931.

67. A. Borodine. *Ueber die Constitution des Hydrobenzamids und des Amarins.* Ann. Chem. Pharm. 110:78, 1859.

68. A. Borodine. *Ueber die Einwirkung des Jodaethyls auf Benzoylanilid.* Ann. Chem. Pharm. 111:254, 1859.

69. A. Borodine. *Ueber Bromvaleriansaeure und Brombuttersaeure.* Ann. Chem. Pharm. 119:121, 1861.

70. A. Borodine. *Ueber die Einwirkung des Zinkaethyls auf das Chlorojodoform.* Ann. Chem. Pharm. 126:239, 1863.

71. A. Borodin. *Ueber einen neuen Abkoemmling des Valerals.* Ber. 5:480, 1872 and Ber. 6:892, 1873.

72. A. Borodine. *Zur Geschichte der Fluorverbindungen und ueber das Fluorbenzoyl.* Ann. Chem. Pharm. 126:58, 1863.

73. N. Zelinsky and J. Zelikow. *Ueber Umwandlung von Alkoholen in ungesaettigte Kohlenwasserstoffe unter Einwirkung der Oxalsaeure.* Ber. 34:3249, 1901.

74. G. Wagner. J. Russ. Phys. Chem. Soc. 31:680, 1899. Quoted by H. Meerweinin in *Ueber den Reaktionsmechanismus der Umwandlung von Borneol in Camphen* Liebig's Ann. 405:129, 1914.

75. G. Wagner and K. Slawinski. *Zur Constitution des Pinens.* Ber. 32:2064, 1899.

76. G. Wagner and W. Brickner. *Ueber die Beziehung der Pinenhaloidhydrate zu den Haloidanhydrided des Borneols.* Ber. 32:2302, 1899.

77. A. Sementsov. *Egor Egorovich Vagner and his role in terpene chemistry.* Chymia 11:151, 1966.

78. N. J. Demjanow. *Die Ringerweiterung bei den cyclischen Aminen mit der Seitenkette CH_2-NH_2.* Ber. 40:4392, 1907.

79. N. J. Demjanow. *Cyclobutylcarbinol und seine Isomerisation zu Pentamethylenederivaten.* Ber. 40:4959, 1907.

80. .N. J. Demjanow. *Die Umwandlung des Tetramethylenringes in den Trimethylenring.* Ber. 40:4961, 1907.

81. A. Arbuzov. *Ueber die Struktur der phosphorigen Saeure und ihrer Derivate.* Chem. Centralbl. 77 (II):748, 750, and 1639, 1906.

82. A. Faworski. *Ueber den Diaethylenaether, den einfachen Aether des Aethylenglycols. Zur Frage der Umwandlung des Aethylenglycols in Acetaldehyd.* Chem. Zentralbl. 78(I):15, 1907.

83. A. Faworski and W. Boschowski. *Uener isomere Umwandlungen der Cyclischen Alpha-Monochloroketonen.* Chem. Zentralbl. 86(I):894, 1915.

84. A. Butlerow. *Zur Geschichte der Darstellung des Acetylens.* Ann. Chem. Pharm. 136:354, 1865.

85. A. Favorskyj. *Isomerisation in der Kohlenwasserstoffreihe C_nH_{2n-2}.* Chem. Centralbl. 1887:1539.

86. A. Favorski. *Ueber die Einwirkung von Kaliumhydrat auf Gemische von Ketonen und Phenylacetylen.* Chem. Centralbl. 76 (II):1018, 1905.

87. A. Faworski. *Die Einwirkung von Halogenverbindungen des Phosphors auf Ketone, Bromketone, und Ketoalkohole.* Chem. Zentralbl. 84(I):1004, 1913.

88. A. L. Tschugaeff. *Ueber eine neue Methode zur Darstellung ungesaettigter Kohlewasserstoffe.* Ber. 32:3332, 1899.

89. L. Tschugaeff. *Ueber die Umwandlung von Carvon in Limone.* Ber. 33: 735, 1900.

90. W. Fomin and N. Sochanski. *Ueber die Wasserstoffabspaltung aus Pinakolin-alkohol und ueber Tertiarbutylaethylen.* Ber. 46:244, 1913.

91. L. Tschugaeff. *Magnesium-organische Verbindungen als Reagens auf die Hydroxylgruppe.* Ber. 35:3912, 1902.

92. Th. Zerewitinoff. Quantitatice *Bestimmung von Hydroxylgruppen mit Hilfe magnesiumorganischer Verbindungen.* Ber. 40:2023, 1907.

93. L. Tschugaeff. *Ueber ein neues, empfindliches Reagens auf Nickel.* Ber. 38:2520, 1905.

94. L. Tschugaeff. *Ueber eine Methode zur Konfigurationsbestimmung bei Dioximen.* Ber. 41:1678, 1908.

95. Wl. Ipatiew and N. Wittorf. *Zur Konstitution von Isopren.* Chem Centralbl. 68(I):457, 1897.

96. (I.) Ostromyslensky. Chem. Abstr. 10:3178, 1916.

97. N. Kizhner. *Catalytic decomposition of alkyl hydrazones as a method for the preparation of hydrocarbons* (in Russian). J. Russ. Phys. Chem. Soc. (Chem. Sec.) 43:582:1911.

98. N. Kizhner. *The decomposition of pyrazoline bases as a method for the preparation of cyclopropane derivatives* (in Russian). J. Russ. Phys. Chem. Soc. (Chem. Sec.) 44:165, 1912.

99. V. Suntsov and D. E. Lewis. *A century of base-promoted decompositions of hydrazones: the early career of Nikolai Matveevich Kizhner (1867-1935).* Bull. Hist. Chem. 39:43, 2014.

100. V. Suntsov and D. E. Lewis. *After the revolution: Nikolai Matveevich Kizhner (1867-1935) in Soviet Moscow.* Bull. Hist. Chem. 42:46, 2017.

101. N. Prileschajew. *Oxydation ungesaettigter Verbindungen mittels organischer Superoxyde.* Ber. 42:4811, 1909.

102. M. Kutscheroff. *Ueber eine neue Methode directer Addition von Wasser (Hydration) an die Kohlenwasserstoffe der Acetylenreihe.* Chem. Centralbl. 1881:610.

103. J. Kondakoff. *Bemerkungen zu Carl Hell und M. Wildermann's Abhandlung: "Ueber Halogenderivate des Amylens."* Ber. 24:929, 1891.

104. I. Kondakow. *Ueber das anormale Verhalten des Polyhalidverbindungen zu alkoholischer Kalilauge.* Chem Centralb. 71(II):1061, 1900.

105. J. Kondakow. *Ueber die Synthesen bei Einwirkung von Chlorzink.* Chem. Centralbl. 64(II):857 and 1082, 1893.

106. I. Kondakow and T. Gorbunow. *Ueber einen neuen Fall der Isomerisation des Dihydrocarvons zu Carvenon.* Chem. Centralbl. 69(I):105, 1898.

107. J. Kondakow and A. Lutschinin. *Zur Frage der Isomerisation in der Mentholreihe.* Chem. Centralbl. 71(II):675, 1900.

108. M. Konowalow. *Nitrirende Wirkung der Salpetersaeure auf den Charakter gesaettigter Verbindungen besitzende Kohlenwasserstoffe und deren Derivate.* Ber. 28:1852, 1895.

109. M. Konowalow. *Ueber die nitrirende Wirkung der Salpetersaeure auf \gesaettingten Charakter besitzende Kohlenwasserstoffe und deren Derivate.* Ber. 29:2199, 1896.

110. M. Konowalow. *Ueber eine empfindliche Reaction der Primaeren und secundaeren Nitroverbindungen.* Ber. 28:1850, 1895.

111. S. Reformatsky. *Neue Synthese zweiatomiger einbasischer Saeuren aus den Ketonen.* Ber. 20:1210, 1887.

112. A. Sementsov. *S. N. Reformatskii and his reaction.* J. Chem. Ed. 34:530, 1957.

113. N. Zelinsky. *Ueber eine bequeme Darstellungweise von alpha-Brompropion-saeureester.* Ber. 20:2026, 1887.

114. N. Zelinsky and G. Stadnikoff. *Ueber eine einfache allgemeine synthetische Darstellungsmethode fuer alpha-Aminosaeuren.* Ber. 39:1722, 1906.

115. W. Tischtschenko. *Ueber die Einwirkung von Aluminiumalkoholaten auf Aldehyde. Die Esterkondensation als neue Kondensationsform der Aldehyde.* Chem. Centralbl. 77(II):1309 and 1552, 1906.

116. P. Schorigin. *Synthesen durch Natrium und Ahalogenalkylen.* Ber. 41:2711 and 2717, 1908.

117. P. Schorigin. *Neue Synthese aromatischer Carbonsaeuren aus den Kohlenwasserstoffen.* Ber. 41:2723, 1908.

118. N. Kursanoff. *Ueber die Einwirkung von Zinkmethyl und Zinkaethyl auf Chlorhexnaphthen.* Ber. 32:2972, 1899.

119. M. Nencki. *Ueber organische Synthesen mittels Eisenchlorid.* Ber. 32:2412, 1899.

120. J. Ponomareff. *Synthese der Parabansaeure.* Chem. Centralbl. 1872:644.

121. A. Wischnegradsky. *Ueber einige Derivate des Cinchonins.* Chem Centralbl. 1879:614.

122. A Wischnegradsky and A. Butlerov. *Ueber eine neue Base aus Chinin.* Chem. Centralbl. 1879:820.

123. A. Wischnegradsky. *Ueber Aldehydcollidin.* Chem. Centralbl. 1879:646.

124. A. Wischnegradsky. *Ueber die Reduction von Chinolin und Aethylpyridin.* Chem. Centralbl. 1880:630.

125. A. Tschitschibabin and O. Seide. *Eine neue Reaktion der Verbindungen welche den Pyridinring enthalten.* Chem. Zentralbl. 86(I):1064, 1915.

126. A. Tschitschibabin. *Ueber die Synthese von Pyridinbasen aus den Aldehyden gesaettigten Characters und Ammoniak.* Chem. Centralbl. 77(I):1438, 1906.

127. A. E. Tschitschibabin. *Eine neue allgemeine Darstellungsmethode der Aldehyde.* Ber. 37:186, 1904.

128. A. Tschitschibabin. *Ueber den Hexahydro-m-toluylaldehyd.* Ber. 37: 850, 1904.

129. P. Petrenko-Kritschenko and N. Zoneff. *Ueber die Condensation von Aceton-dicarbonsaeureestern mit Benzaldehyd unter Anwendung von Ammoniak.* Ber. 39:1358, 1906.

130. P. Petrenko-Kritschenko. *Ueber die Kondensation der Acetondicarbonsaeure-ester mit Aldehyden vermittelst Ammoniak und Aminen.* Ber. 42:3683, 1909.

131. P. Petrenko-Kritschenko and W. Kantscheff. *Ueber die Reaktionsgeschwindig-keit bei der Bildung von Oximen.* Ber. 39:1452, 1906.

132. B. Menschutkin. Nikolai *Aleksandrowitsch Menschutkin.* Ber. 40:5087, 1907.

133. N. Menschutkin. *Zur Kenntniss der Harnstoffverbindung.* Ann. Chem. Pharm. 153:83, 1870.

134. N. Menschutkin. Zur Chemie des Stickstoffs: *ueber die Bildungsgeschwindig-keiten der Amine und der Alkylammoniumsaltze.* Ber. 28:1398, 1895.

135. N. Menschutkin. *Zur Kenntniss der aliphatischen Kohlenstoffketten.* Ber. 30: 2775, 1897.

136. N. Menschutkin. *Ueber den Einfluss der Seitenketten auf die Vertheilung der Umsetzungsgeschwindigkeit im Benzolring.* Ber. 30:2966, 1897.

137. N. Menschutkin. *Ueber die Regelmaessigkeit der Siedepunkte der isomeren aliphatischen Verbindungen.* Ber. 30:2784, 1897.

138. Richard E. Rice. *Hydrating ions in St. Petersburg and Moscow. Ignoring them in Leipzig and Baltimore.* Bull. Hist. Chem. 27:17, 2002.

139. W. D. Bankroft. *Wilhelm Ostwald, the great protagonist.* J. Chem Ed. 10: 539, 1933.

140. F. E. Wall. *Wilhelm Ostwald.* J. Chem. Ed. 25:2, 1948.
141. A. Titow. *Die Adsorption von Gasen durch Kohle.* Chem. Zentralbl. 82(I): 113, 1910.
142. H. Pines. *Ipatieff: man and scientist.* Science 157:166, 1967.
143. W. Ipatiew. *Katalitische Reaktionen bei hohen Temperaturen und Drucken.* Ber. 41:991,996, and 1001, 1908.
144. W. Ipatiew. *Katalytische Reaktionen bei hohen Temperaturen und Drucken.* Ber. 41:993, 1908.
145. G. B. Kaufmann and A. Beck. *Nikolai Semenovich Kurnakov.* J. Chem. Ed. 39:44, 1962.
146. B. Jaffe. *Crucibles.* Tudor Publ. Co., New York, 1934, p. 244.
147. H. M. *Leicester. The historical background of chemistry.* Wiley, NewYork, 1956, p. 194.
148. R. Partington. *A history of chemistry, vol. 4.* Macmillan, London, 1964, p. 891.
149. B. Jaffe. *Crucibles.* Tudor Publ. Co., New York, 1934, p. 199.
150. H. M. Leicester. *Mendeleev and the Russian Academy of Sciences.* J. Chem. Ed. 25:439, 1948.
151. W. R. Winicov. *Some of Mendeleeff's personal characteristics.* J. Chem. Ed., 14:372, 1937.
152. H. M. Leicester. *Mendeleev's visit to America.* J. Chem. Ed. 34:331, 1957.
153. G. Siemeincow. *Mendeleev: demographics in addition to chemistry.* C & En, Sept. 28, 1970, p. 43.
154. L. White. *Mendeleeff and women.* C & EN, April 12, 1971, p. 8.
155. G. Siemeincow. *Mendeleev's views on education.* Abst. 1, Sect. Hist. Chem., 162-nd Am. Chem. Soc. Meeting, Sept. 12–17, 1971, Washinton, D.C.
156. H, M. Leicester. *Factors which led Mendeleev to the periodic law.* Chymia 1:67, 1948.
157. D. Mendeleev. *The relationship between the properties and atomic weights of elements.* J. Russ. Chem. Soc. 1:60, 1869 (in Russian).
158. D. Mendeljeff. *Versuch einer Theorie Ueber die Grenzen der organischen Verbindungen.* Zeitschr. f. Chem. 4:560, 1861.
159. D. Mendeljeff. *Ueber die Ausdehnung der Fluessigkeiten beim Erwaermen ueber ihren Siedepunkt.* Ann. Chem. Pharm. 119:1, 1861.

General References

I. V. Kuznetsov. *Lyudi Russkoy nauki.* Government Publ. House, Moscow, 1961.

A. A. Zvorykin, ed. *Biografitsheskiy slovar' deyateley estestvoznaniya i tekhniki.* Government Publ. House, Moscow, 1959.

Illustrations

Dimitry Mendeleyev (1834-1907), discoverer of the periodic table; in his senior years (left), and younger years (above) from a Soviet postage stamp.

Figure 1. *Karl Karlovich Klaus (Клаус Карл Карлович, Carl Ernst Claus) in Dorpat (daguerreotype, ca. 1852).*

Karl K. Klaus (1796-1864), discoverer of the element of ruthenium. From *Bull. Hist. Chem.* 41 (1-2):3, 2016.

Alexander M. Butlerov (1828-1886), organic chemist and originator of the theory of structures of organic compounds.

Alexander M. Zaitsev (1841–1910). Author of the *Saytzeff's Rule* for the elimination reactions. From the *Bull. Hist. Chem.* 17/18, 21, 1995.

Friedrich K. Beilstein (1838–1906), originator of the encyclopedia of organic compounds. From ACS' *Chemical Heritage*, vol. 21, No. 4, Winter 2003–2004.

N. N. Zinin (1812–1880), organic chemist,
discoverer of aniline synthesis,
which started the dye industry in Western Europe.

A. A. Voskresensky (1809-1880), discoverer of theobromine
and "grand-father of Russian chemistry."

Vladimir Ipatieff. [Photograph by Universal Oil Products Company]

Vladimir Ipatieff (1867-1952)
Russian/American chemist,
developer of industrial
catalytic processes.

The late Vladimir Ipatieff had no plans to return permanently to the U.S.S.R.

Vladimir Ipatieff in America.
From ACS's *Chemical and
Engineering News.*

Mikhail V. Lomonosov (1711-1765).
Father of Russian sciences and chemist
par excellence. From a Rumanian
postage stamp.

Kazan University in the
mid-19th century, "power
house of organic chemistry"
in Russia.

191   THE HISTORICAL BACKGROUND OF CHEMISTRY

Ueber die Beziehungen der Eigenschaften zu den Atomgewichter. der Elemente. Von D. Mendelejeff. — Ordnet man Elemente nach zunehmenden Atomgewichten in verticale Reihen so, dass die Horizontalreihen analoge Elemente enthalten, wieder nach zunehmendem Atomgewicht geordnet, so erhält man folgende Zusammenstellung, aus der sich einige allgemeinere Folgerungen ableiten lassen.

			Ti = 50	Zr = 90	? = 180
			V = 51	Nb = 94	Ta = 182
			Cr = 52	Mo = 96	W = 186
			Mn = 55	Rh = 104,4	Pt = 197,4
			Fe = 56	Ru = 104,4	Ir = 198
		Ni = Co = 59		Pd = 106,6	Os = 199
H = 1			Cu = 63,4	Ag = 108	Hg = 200
	Be = 9,4	Mg = 24	Zn = 65,2	Cd = 112	
	B = 11	Al = 27,4	? = 68	Ur = 116	Au = 197?
	C = 12	Si = 28	? = 70	Sn = 118	
	N = 14	P = 31	As = 75	Sb = 122	Bi = 210?
	O = 16	S = 32	Se = 79,4	Te = 128?	
	F = 19	Cl = 35,5	Br = 80	J = 127	
Li = 7	Na = 23	K = 39	Rb = 85,4	Cs = 133	Tl = 204
		Ca = 40	Sr = 87,6	Ba = 137	Pb = 207
		? = 45	Ce = 92		
		?Er = 56	La = 94		
		?Yt = 60	Di = 95		
		?In = 75,6	Th = 118?		

1. Die nach der Grösse des Atomgewichts geordneten Elemente zeigen eine stufenweise Abänderung in den Eigenschaften.
2. Chemisch-analoge Elemente haben entweder übereinstimmende Atomgewichte (Pt, Ir, Os), oder letztere nehmen gleichviel zu (K, Rb, Cs).
3. Das Anordnen nach den Atomgewichten entspricht der *Werthigkeit* der Elemente und bis zu einem gewissen Grade der Verschiedenheit im chemischen Verhalten, z. B. Li, Be, B, C, N, O, F.
4. Die in der Natur verbreitetsten Elemente haben *kleine* Atomgewichte

Fig. 15. First form of Mendeleev's periodic table. (From *Zeitschrift für Chemie*, 12, 405 (1869).)

Mendeleyev's periodic table published in a German chemical journal in 1869.

по въ ней, мнѣ кажется, уже ясно выражается примѣнимость вы-
ставляемаго мною начала ко всей совокупности элементовъ, паи
которыхъ извѣстенъ съ достовѣрностію. На этотъ разъ я желалъ
преимущественно найдти общую систему элементовъ. Вотъ этотъ
опытъ:

			Ti=50	Zr=90	?=180.
			V=51	Nb=94	Ta=182.
			Cr=52	Mo=96	W=186.
			Mn=55	Rh=104,4	Pt=197,4
			Fe=56	Ru=104,4	Ir=198.
			Ni=Co=59	Pl=106,6	Os=199.
H=1			Cu=63,4	Ag=108	Hg=200.
	Be=9,4	Mg=24	Zn=65,2	Cd=112	
	B=11	Al=27,4	?=68	Ur=116	Au=197?
	C=12	Si=28	?=70	Sn=118	
	N=14	P=31	As=75	Sb=122	Bi=210
	O=16	S=32	Se=79,4	Te=128?	
	F=19	Cl=35,5	Br=80	I=127	
Li=7	Na=23	K=39	Rb=85,4	Cs=133	Tl=204
		Ca=40	Sr=87,6	Ba=137	Pb=207.
		?=45	Ce=92		
		?Er=56	La=94		
		?Yt=60	Di=95		
		?In=75,6	Th=118?		

а потому приходится въ разныхъ рядахъ имѣть различное измѣненіе разностей,
чего нѣтъ въ главныхъ числахъ предлагаемой таблицы. Или же придется пред-
полагать при составленіи системы очень много недостающихъ членовъ. То и
другое мало выгодно. Мнѣ кажется притомъ, наиболѣе естественнымъ составить
кубическую систему (предлагаемая есть плоскостная), но и попытки для ея образо-
ванія не повели къ надлежащимъ результатамъ. Слѣдующія двѣ попытки могутъ по-
казать то разнообразіе сопоставленій, какое возможно при допущеніи основнаго
начала, высказаннаго въ этой статьѣ:

Li	Na	K	Cu	Rb	Ag	Cs	—	Tl
7	23	89	63,4	85,4	108	133		204
Be	Mg	Ca	Zn	Sr	Cd	Ba	—	Pb
B	Al			—	Ur	—	—	Di?
C	Si	Ti		Zr	Sn	—	—	—
N	P	V	As	Nb	Sb	—	Ta	—
O	S	—	Se	—	Te	—	W	—
F	Cl	—	Br	—	J	—	—	—
19	95,5	58	80	190	127	166	190	220

Mendeleyev's periodic table published in the Russian language in the Russian journal *Journal of the Russian Chemical Society* in 1869.

EPILOGUE

T his volume has briefly covered the development of certain scientific disciplines in Imperial Russia. It should be clear that Russian accomplishments in the medical, biological, and chemical branches of sciences were impressive indeed, and it is only regrettable that a similar account of the fields of engineering, physics, the earth sciences, mathematics, and astronomy could not be presented. The high quality of the Russian university and technical school faculty is evident from the fact that scientists trained in Russia were sought after to occupy positions of authority and responsibility abroad, e. g., Ostwald, Schmiedeberg, von Bergmann, Mechnikoff and others. Similarly, scientists who left Russia to escape Bolshevik atrocities were welcomed to foreign laboratories and educational institutions. These included men like Chichibabin, Ipatieff, Babkin, Walden, Maximow, Kondakov, and many others.

There seems to be little evidence to indicate that the tsarist regime was responsible for stifling or discouraging scientific inquiry or science education. Scientists were decorated, ennobled, and otherwise honored for their accomplishments just like members of other professions were. In addition, scientists were in great demand by the government to serve on committees and commissions for the planning and execution of educational programs, industrial development, and the utilization of natural resources. Neither were there attempts to corrupt scientific inquiry to fit political or social points of view of the tsarist regimes, as was later the case with genetics and structural organic chemistry during the Soviet era (see above). There were occasional exceptions, as in a case reported by Leicester (1), where a scientist got a promotion because he delivered a lecture on the reconciliation of scientific principles with Christian religious beliefs (Orthodoxy was Russia's state religion). When one looks back, that scientist may have been ahead of his time. But the fact is that the science historians were apparently unable to unearth anything more serious than the above trifling incident in regard to any bureaucratic interference in scientific inquiry. There was also no attempt, after the death of Nicholas I, to hinder the free exchange of ideas and visits among scientists of Western Europe and Russia. And indeed, it was a rare scientist who did not undertake at least one journey abroad during his professional life, to work, sometimes for a period of years, in a foreign laboratory. In fact, the government usually funded such visits abroad. The above does not mean that scientists were not "punished" for certain forms of misbehaviors. An example is Kizhner's leadership of student revolt at Tomsk University during the 1905–1906 revolutionary disorder period in Russia. He was removed from his job and "expelled" to St. Petersburg. When things quieted down a year later, he was able to return to the university in Tomsk.

Let us now drive our time-machine to twenty to thirty years later, after the tsar and his family had been murdered by the Bolsheviks, and the Comintern with Joseph Stalin at its head was in charge of Russia. A hunt on wayward or otherwise undesirable scientists was carried out first in 1925–1929 period, and then more seriously between 1934 and the 1950s (2, p. 138). Nicholas Vavilov was arrested in 1942 (see chapter III), along with numerous of his coworkers (2, p. 77). He was supposed to be executed, but that verdict was changed to twenty years in the Gulag. He died shortly after his verdicts were announced. Among other well-known scientists who were arrested and died shortly thereafter was Jacob O. Parnas (1884–1949) (2, p. 233). He was well known for his work on glycolysis and glycogen synthesis and metabolism. He was considered for sharing the Nobel Prize eventually received by the Coris, but he was already deceased when the prize was awarded. Such was the fate of hundreds of Russian scientists in Russia during the Stalinist era. Not much has been written about their fates in the West by science historians. But then we have the report on that scientist (during the tsarist era, see above), who, in a seminar at a university, attempted to compromise religious beliefs with scientific principles, and apparently received a promotion thereafter. And our historians are still upset about such an event!

On the other hand, the Russian government maintained pretty strong administrative controls over Russian educational institutions. After all, such institutions were all financed by the government. After the assassination of Alexander II university statutes were changed, giving the government tighter administrative controls, so that most appointments could not be made without governmental approval. In this way, it was hoped to keep revolutionary agitators from the campuses. This policy, of course, elicited a chronic high degree of unrest among both the students and faculty. If a professor was too vocal in his support of student rebels, he was in some instances dismissed, or he resigned in protest after being reprimanded. Yet few of those who were dismissed or resigned remained without a job for long, and fewer yet emigrated for this reason, though no barriers to emigration were then in existence.

Many of Russia's problems can be traced to its underdeveloped industry, and hence its antiquated agricultural methods. It is safe to say that the development of science as well as technology on the pilot plant level were far ahead of Russia's general industrial development or its public works program. The know-how was certainly there, as for instance in case of Ipatieff, who had developed numerous industrial processes; or the highly imaginative concept of Zemstvo medicine, which was in its infancy when the Bolsheviks took over. Even going back to nineteenth century, there were men like Fritsche and Zinin who were interested in aniline and dye chemistry. Yet Russia never developed a dye industry. Then there were metallurgists like Chernov who, in 1869, showed that steel could be hardened only if it was heated to a certain temperature (critical temperature). He also studied the dendritic structure of steel ingots. Yet Russia's steel industry was not well developed as that of other European countries. The vast oil fields of Baku were poorly utilized even though Shukhov had patented a process for the cracking of oil before Burton did, and Letny had developed a method for pyrolysis of oil in 1877 (3,4). What Russia lacked were human and monetary resources. It will be recalled that in 1908, there was only one working individual for every three nonworking persons in Russia. And to raise revenue, private industry was subjected to intolerable taxation. The lack of money also hindered electrification. Ipatieff thus reports (4) that there were Russian inventors who constructed electric lighting devices well before Edison did in 1878 (Ladygin in 1874 and Yablochkov in 1876). However, other than lighting a bridge over the Neva River in St. Petersburg, this technology did not spread very far. Similarly, A. S. Popov, a professor at the Naval Engineering Academy, demonstrated the method of radio communications in 1895, well before Marconi's announcement in 1897. Popov asked for a grant of 35,000 rubles to develop a prototype communications system linking St. Petersburg with Kronstadt, but received only 5,000 rubles. The failure to develop wireless communications in Russian navy cost the Russians dearly during the Russo-Japanese War, when due to a primitive naval communications system, the entire Russian Baltic fleet was destroyed by the Japanese navy in the Straights of Tsushima. In desperation, Russia turned abroad for loans, and France became one of the biggest investors in Russian economy. This fact, of course, entangled Russia into a military alliance with France and ultimately England against Germany, which led to the disasters of World War I and the revolutions that followed.

In spite of all the shortcomings, industrialization of Russia was proceeding at a rapid pace on the eve of World War I (5). Given this momentum, and given the technological and scientific base developed during the tsarist era, it is not surprising that on the eve of World War II, the Soviet Union was perhaps the world's third greatest industrial power. Another factor accounting for the Soviet success in World War II was its use of large armies of forced laborers (6), a phenomenon unknown in Tsarist Russia and the rest of civilized world.

There are a few conclusions that can be reached regarding the social aspects of life under the tsars by studying Russia's scientific and medical establishment. The various minority groups were represented among the more prominent Russian scientists, hence it would appear that opportunities for education and employment did exist for nearly every one who was qualified. Among scientists or doctors of Polish background one may mention the Nencki brothers, Talko, and Szokalski. Jews were represented by men like London, de Cyon, Mandelstamm, and Haffkine. A disproportionate number was of German origin, and it seems that it would be more logical to restrict the number of German students in the universities rather than those of Jewish nationality. However, regardless of the above arguments and statistics, it was a fact that admission to Russia's educational institutions, secondary and university-level, was highly competitive no matter what the nationality of the applicant. Such a limited educational opportunity was again due to chronic fiscal difficulties. Illiteracy in Russia in 1897 was high, some 76 percent overall and 45 percent in the cities. Progress in the area of medical care was also lagging, so that the frequent cholera, typhoid, and other epidemics acted to debilitate and demoralize the population. To solve this problem, a rudimentary system of medical districts (Zemstvo districts) was established both in the cities and the countryside, with a network of hospitals, dispensaries, and drugstores. This was indeed what we today would call socialized medicine. It was a task of enormous proportion, as medical care had to be brought to a population of 180 million souls who populated a land stretching from the Baltic Sea to the Pacific Ocean and from beyond the Arctic circle to the borders of India. It should be noted that throughout their history, the Russian people, in contrast to their Western cousins, depended on the government—local or national—for various services and improvements in their way of life. This included medical care, communications, transportation, and many industries. At this point it becomes easier to understand Mendeleyev's respect for American economic and technological systems, which differed from that of Russia by its minimal governmental control and reliance on private initiative. Mendeleyev thought that given more freedom and tax relief, private enterprise could transform Russia from a primarily agricultural society to an industrial one. With respect to its economics and social services, the tsarist system thus differed from the Soviet system only in terms of degrees.

REFERENCES

1. H. M. Leicester. *The history of chemistry in Russia prior to 1900.* J. Chem. Ed. 24:438, 1947.
2. S. E. Shnol'. *Heroes and villains of the Russian Science.* Cron-Press, Moscow, 1997 (in Russian).
3. H. K. Work. *Metallurgy in the nineteenth century.* J. Chem. Ed. 28:364, 1947.
4. V. N. Ipatieff. *Modern science in Russia.* J. Chem. Ed. 20:159, 1943.
5. A. de Goulevich. *Czarism and revolution.* Omni Press, Hawthorne, 1962.
6. D. J. Dallin *Forced labor in Soviet Russia.* Hollis & Carter, London, 1947.

Illustration

Fig. 1—A. S. Popov (1859–1905).

A. S. Popov, professor at the Naval
Engineering Academy in Kronstadt,
demonstrated radio communications
in 1895, some 2 years before Marconi
did. His request for funds to continue
research was not funded appropriately.

INDEX OF SCIENTISTS AND PHYSICIANS MENTIONED IN THIS VOLUME WHO HAVE BEEN ASSOCIATED WITH IMPERIAL RUSSIAN INSTITUTIONS*

* il. indicates illustration.